DATE DUE

DEMCO 38-296

Microphone Engineering Handbook

Microphone Engineering Handbook

Edited by

Michael Gayford

ann Ltd

OX2 8DP

rier group

OXFORD   LONDON   BOSTON
MUNICH   NEW DELHI   SINGAPORE   SYDNEY
TOKYO   TORONTO   WELLINGTON

First published 1994

British Library Cataloguing in Publication Data
Microphone Engineering Handbook
  I. Gayford, Michael
  621.38284

ISBN 0 7506 1199 5

Library of Congress Cataloguing in Publication Data
Microphone engineering handbook/edited by Michael Gayford.
  p.   cm.
Includes bibliographical references and index.
ISBN 0 7506 1199 5
1. Microphone – Handbooks, manuals, etc. I. Gayford, M. L.
TK6478.M53   93–33681
621.382'84–dc20   CIP

Cover photograph kindly supplied by Bruel & Kjaer

Composition by Genesis Typesetting, Laser Quay, Rochester, Kent
Printed in Great Britain by Redwood Books, Trowbridge, Wiltshire

Contents

Preface   viii
Acknowledgements   ix
List of Contributors   x

1 Microphone techniques   *Michael Gayford*   1
    1.1   Sound waves and microphones   1
    1.2   Microphone developments   12
    1.3   Microphone design techniques   13
    1.4   Directional effects   13
    1.5   Limitations on directional characteristics   16
    1.6   Types of directional microphones   17
    1.7   Microphone powering and connections   18
    1.8   Microphone windshields and anti-noise precautions   22
    1.9   Antivibration, shock and handling noise protection   23
    1.10 Microphone evolution and design: the use of analogues   25
    1.11 Acoustic boundary microphones   52
    1.12 Lavalier microphones   53
    1.13 Highly directional microphones   54
    1.14 Integrated circuit microphones   56
    1.15 Specialist microphone types   56
    1.16 Electrical and radio interference   57
    1.17 Microphone appearance and finish   59
    1.18 Microphone responses and linearity   59
          References   59

2 Precision microphones for measurements and sound reproduction
    Torben G. Nielsen   62
    2.1 Introduction   62
    2.2 Condenser microphone theory   69
    2.3 Cartridge sizes and applications   73
    2.4 Factors affecting cartridge response   77
    2.5 Cartridge and preamplifier response   80
    2.6 Field calibration equipment and accessories   100
    2.7 Piezoelectric microphones and hydrophones   125
    2.8 Precision condenser microphones and hydrophones   130
          References   139

3 Optical microphones   *D.A. Keating*   140
    3.1 Introduction   140
    3.2 Optical components   140
    3.3 Optical measurement techniques   142
    3.4 Force feedback   155
          References   157

4 High-quality RF condenser microphones *Manfred Hibbing*   158
   4.1 Introduction   158
   4.2 Historical review   159
   4.3 General design concepts   160
   4.4 Phase-modulation design   166
   4.5 AM push–pull bridge design   175
   4.6 Summary   185
      References   185

5 Radio microphones and infra-red systems *Erhard Werner*   187
   5.1 Introduction   187
   5.2 Wireless links on radio frequencies   189
   5.3 Wireless links on infra-red   224
      References   234

6 Microphone testing *Harald Sander-Röttcher* and *Kersten Tams*   236
   6.1 Introduction   236
   6.2 Measurement of acoustic parameters   236
   6.3 Measurement of non-acoustic parameters   256
   References   265

7 Ribbon Microphones *G. Rosen*   267
   7.1 History of ribbon microphones   267
   7.2 Function   268
   7.3 Directivity pattern at high frequencies   271
   7.4 Pulse characteristics   271
   7.5 Transformers   275
   7.6 Stereo recording using ribbon microphones   276

8 Microphone amplifiers and transformers *Peter J. Baxandall*   277
   8.1 Introduction   277
   8.2 History   277
   8.3 Some facts about random noise   290
   8.4 Electronically balanced input circuits and noise   305
   8.5 Transformers   312
      References   347

9 Microphones in stereophonic applications *Francis Rumsey*   349
   9.1 Introduction   349
   9.2 Directional perception   349
   9.3 Principles of directional sound reproduction   359
   9.4 Stereo signals   372
   9.5 Stereo microphone configurations   377
   9.6 Ambisonics   406
   9.7 Conclusion   418
      References   419

10 International, regional and national standards *J.M. Woodgate*   421
   10.1 Introduction   421
   10.2 Origins of standardization   421
   10.3 Safety standards   425
   10.4 Electromagnetic compatibility (EMC)   426
   10.5 Requirements for radio microphones   428
   10.6 Sources of published standards   429

11 Glossary *J.M. Woodgate*   430
    11.1 Introduction   430
    11.2 General terms in acoustics and electroacoustics   430
    11.3 Electroacoustic systems and their elements   431
    11.4 Transduction principles of microphones   434
    11.5 Acoustic principles of microphones   435
    11.6 Classification by application   436
    11.7 Microphone accessories   436
    11.8 Electroacoustic characteristics   437
    References   438

Index   439

Preface

Today more than ever before the microphone represents a vital interface between the sound field and the multitude of electrical circuits now in use.

A number of excellent books have appeared which give a good grounding in the general theory and applications of microphones. However, it is apparent that there is an advantage in producing an up-to-date work in which the various specialized aspects of microphone design and use are covered in detail by acknowledged experts involved today. Special emphasis is placed on inclusion of comprehensive references to other work. The aim is to give clear and detailed treatment of microphone design and performance in fields such as communications, sound measurements, sound reinforcement and public address, broadcasting and sound recording etc., including the latest stereophonic techniques. It is hoped that all concerned with the engineering design and applications of microphones in the diverse fields of use today, and many students, will find interest and enlightenment from this book.

Michael Gayford
Harlow Essex

viii

Acknowledgements

In compiling a technical handbook of this nature a considerable number of manufacturers are involved world-wide in providing technical information both directly and through reviews, catalogues and technical articles. The following are thanked in particular:

AKG Acoustics, AMS (Calrec), Bruel & Kjaer, beyerdynamic, Cannon, Coles Electroacoustics, ACO Pacific Inc, Crown International, Gotham Cables, Neumann, F.O Bauch Ltd, Magnetic Developments Ltd, Neutric, Jensen Transformers Inc, Sanken Microphones, Shure Inc, Sowter Audio Transformers, Stirling Sound Distributors, Switchcraft, Sennheiser, Electrovoice Inc.

Illustrations provided are acknowledged on the page. These are all gratefully acknowledged.

Contributors

Peter Baxandall BSc (Eng), CEng, FIEE, FAES
Electro-acoustic consultant, UK

Michael Gayford BSc, MIEE, MAES
Electro-acoustic/electronic consultant, UK

Manfred Hibbing
Sennheiser Electronic, Germany

David Keating BSc, PhD, MAES
Department of Cybernetics, University of Reading, UK

Torben Nielsen MSc
Bruel & Kjaer, Denmark

Günter Rosen Dipl Ing
beyerdynamic GmbH, Germany

Francis Rumsey BMus, PhD
Music Department, University of Surrey, UK

Harald Sander-Röttcher Dipl Ing
Sennheiser Electronic, Germany

Kersten Tams Dipl Ing
Sennheiser Electronic, Germany

Erhard Werner DrIng, MAES, MASA
Sennheiser Electronic, Germany

John Woodgate BSc, CEng, MIEE, MAES, FInstSCE
J. M. Woodgate and Associates, UK

1 Microphone Techniques

Michael Gayford

1.1 Sound waves and microphones

Sound is primarily a small pressure wave disturbance in the air with a frequency range from a few herz up to ultrasonic frequencies. The human audible frequency range is often taken to cover some 20 Hz to 20 kHz. Sub-audible and supersonic sounds can produce physical effects and sensations other than what is accepted as normal hearing, the sensations being 'felt' rather than heard. Some tests show that subsonic and supersonic sound waves may produce extra-auditory clues due to intermodulation processes inherent in the hearing process. Comparisons between Hi-Fi systems with extended response below some 20 Hz and above 20 kHz, and systems covering the more usual 20 Hz to 20 kHz would appear to suggest that extra realism can be obtained by such extensions of the frequency range. It is accepted that the basic hearing process is due to motion of the ear drum or tympanic membrane, induced by sound pressure. It is not surprising that since early days most microphones have been based on exploitation of the motion of a membrane. Any membrane has mechanical mass and elasticity and some degree of non-linearity, and can exhibit many resonances in addition to the fundamental response. The human aural system shows resonances and non-linearities, but the brain learns to accept the total aural response as a normal hearing sensation. Good microphone design traditionally involves controlling or damping the microphone diaphragm resonances whilst retaining the desired frequency and transient response. It is notable that the highest-fidelity microphones use very light thin diaphragms whose resonances may be controlled by the coupled air load, similar to the manner in which the resonances of high-quality electrostatic loudspeaker membranes are damped out by the coupled air radiation resistance load.

It is accepted that different applications require particular types of microphone. Cheaper and more robust microphones suitable for mass production are needed for general purposes, telephony and communications. There is necessarily some trade-off in absolute fidelity, but the responses may well be specially tailored to suit the systems and conditions of use, e.g. restricted bandwidth. Present microphones must use a suitable mechano-electrical transduction system in conjunction with a suitably large diaphragm to give an electrical output large enough to give clearance over the inevitable circuit noise in the analogue first amplifier stages associated with the microphone. Diaphragms of present sizes are usually comparable to the wavelength of the sound at the highest frequencies, and this can cause errors in the directional response. Some omnidirectional condenser microphones, with sacrifice in sensitivity, can be made small enough to be omnidirectional up to supersonic frequencies.

Microphones are increasingly used for acoustic measurements in air and in water (hydrophones). Section 2 deals largely with precision microphones for measurements in air and water.

1.1.1 Sound waves and their effects on microphones

Sound waves are initiated by a vibrating surface or air front, e.g. the output of a human voice or an organ pipe. Figure 1.1 shows the physical relationships between sound pressure, particle displacement and velocity in a basic plane sound wave. Figure 1.2 shows how the components of a sound wave, which in practice might be a pulse or a complicated waveform such as the note from a musical instrument, are propagated. The pressure excursions are seen to be a series of successive positive and negative waveforms accompanied by movement of the air particles from points of higher pressure to lower pressure, i.e. sound particle velocity. These quantities are the physical actuators which are exploited by microphones in reproducing sounds. The most basic sound source would be a

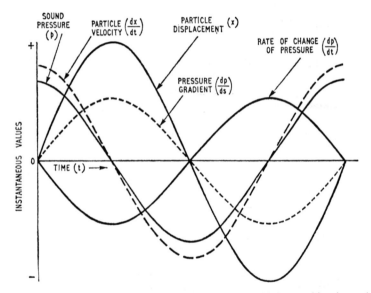

Figure 1.1   *Phase relationship between sinusoidal quantities in a plane wave.*

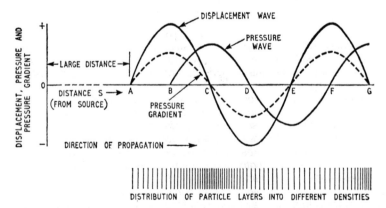

Figure 1.2   *Displacement, pressure and pressure-gradient waves.*

pulsating sphere which launches a spherically expanding wave. A small section of this front at a considerable distance would be substantially flat and could constitute a plane wave front. A flat piston filling a smooth pipe could also launch a plane wave down the pipe. Plane-wave concepts are useful for studying wave propagation over relatively large distances, e.g. in large auditoria, or alternatively in enclosed pipes and passages inside microphone structures etc. Spherical waves often better represent the conditions around a microphone relatively close to a small sound source such as a voice. We see that there can be two kinds of microphone, i.e. those responding to the sound pressure and those actuated by the sound pressure gradient between two points in the line of wave propagation, which gives rise to particle velocity in the wave. Hence so-called pressure and velocity microphones. Tables 1.1, 1.2 and 1.3 summarize some of these relationships and the relevant equations of motion. Theoretically, a plane wave is

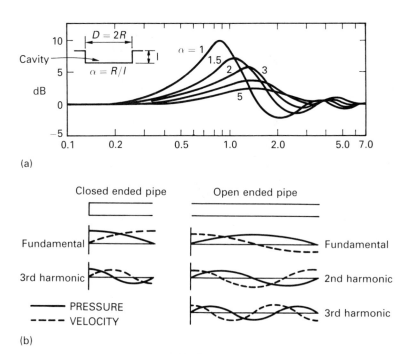

(a)

(b)

Figure 1.3 *Acoustic cavity and pipe resonances. (a) Pressure gain due to resonant effects in a shallow cavity before a cylindrical diaphragm. The ratio of the pressure p at the bottom of the cavity to the free field pressure p_o is plotted against the normalizing parameter $kR = 2\pi R/\lambda = \pi D/\lambda$.*
(b) The air column resonance conditions for a pipe closed at one end. Resonances occur at the open end whenever a velocity maximum (antinode) is presented, the corresponding pressure being a minimum (node). This occurs at pipe lengths equal to odd numbers of quarter-wavelengths and gives a low impedance at the open end. Anti-resonances occur when the length is equal to even numbers of quarter-wavelengths. Here the pressure is a maximum and the velocity a minimum, representing a high impedance at the open end.

The resonant conditions for a pipe open at each end. Resonances occur for lengths corresponding to multiples of half-wavelengths. Nodes occur inside the pipe at lengths corresponding to odd numbers of quarter wave-lengths.

Table 1.1 Transducer types and operating principles

Microphone transduction principle	Method of Operation	Sensitivity (Efficiency)	Effective frequency range	Application	Typical source input impedance
Resistive					
(a) Contact-resistance variation (A)	(i) Carbon granules compressed	High output 0.1–5 V	50 Hz–5 kHz	Telephone or communication pressure or gradient microphones	Low impedance (100–300 Ω)
	(ii) Carbon discs pile under pressure	High output	0–50 Hz	V.L.F. microphones and pressure measuring transducers	
(b) Strain gauge (A)	(i) Carbon strips or graphite paint (stretched) (ii) Bonded or unbonded silicon or metal wires (stretched) (iii) Silicon junction	Low output	0–3 kHz	Not normally used for microphones owing to coupling difficulties; used for pressure and displacement transducers	Low impedance (0.1–10 Ω) (500–1000 Ω)
Capacitive					
Capacitor microphones	(i) Pressure moves diaphragm relative to backplate (ii) Pressure gradient moves diaphragm via phase-shifting rear opening	Low output	0–1 MHz	Pressure-measuring standard or studio microphones; ultrasonic microphones and hydrophones; directional studio microphones	Small capacitive source impedance (5 pF–100 pF) FET impedance converter
Variable reluctance					
(a) Moving armature with	(i) Single-sided longitudinal modulation	Low output as a microphone	50 Hz–10 kHz	Telephone or communications	Low or medium impedance

Type	Description	Output	Frequency response	Application	Impedance
fixed permanent magnet polarization (P)	air-gap reluctance (ii) Flux relay parallel action system (iii) Flux relay rocking-armature system (iv) Double-sided symmetrical balanced-armature system	Moderate output as a receiver		microphone or receivers	(50–2000 Ω)
(b) Moving permanent magnet near soft magnetic coil yoke (P)	(i) Flux relay structure with parallel action (ii) Flux relay with longitudinal balanced action	Low output	50 Hz–10 kHz	Telephone or communications microphone or receivers	Low or medium impedance (50–2000 Ω)
(i) The piezo junction microphone (A) (ii) Movable grid in triode thermionic valve (A)	(i) 12 μm radius probe applies pressure directly to a pn junction (ii) Flexible glass/metal grid seal required	Medium to high output; used on experimental communication microphones	0–10 kHz theoretically possible response	Experimental application to date for telephone or communications microphones	
The piezo-electric transducer (P)	(i) Direct action on a longitudinal expander crystal or ceramic bar by end loading or hydrostatic action (ii) Diaphragm coupled to torque or bending mode bimorph or single crystal or ceramic laminae or multimorph (iii) Hydrostatic action on a hollow bimorph 'sound-cell'	Low to medium output microphones and receivers of high acoustical and electrical impedance	0–2 MHz	Telephone, communications and high-quality microphones, receivers, hydrophones and hydrosounders; calibrated pressure-measuring microphones	

Table 1.1 *Continued*

Microphone transduction principle	Method of Operation	Sensitivity (Efficiency)	Effective frequency range	Application
The electro dynamic transducer (P)	(i) Copper or aluminium wire or tape multi-turn speech coil fixed to conical or domed diaphragm with a flexible surround (moving-coil units)	Low output microphone Medium output receiver	1–50 kHz	Telephone, communications and high-quality microphones and receivers
	(ii) Moving-ribbon dynamic microphone or receivers	High-quality microphones		
Magneto-strictive transducers (P)	Coil windings closely coupled to rods or laminations of suitable soft or ceramic magnetic material	Low-output sound generator or microphone	1–100 kHz	Underwater sound source or microphone
Laser scan systems	Doppler processing of the velocity signal of a small light-reflecting diaphragm or part of a system.			Sound and vibration measurements on microphones and parts of machines

(A) Active transducers
(P) Passive transducers

Table 1.2

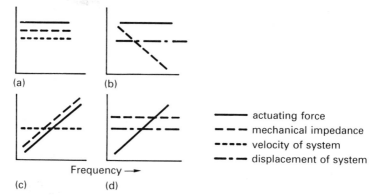

(a)

(b)

—— actuating force
– – – mechanical impedance
····· velocity of system
—·— displacement of system

Frequency ——

(c)

(d)

(a) pressure-operated microphone   with resistance control
(b) pressure-operated microphone   with stiffness control
(c) gradient-operated microphone   with mass control, e.g. ribbon microphone
(d) gradient-operated microphone   with resistance control, e.g. condenser
     microphone

Table 1.3   *Summary of analogue quantities, units, symbols and relationships*

Mechanical	Acoustical	Electrical
Mass M (kg)	Mass M_a, inertance	Inductance H (henries)
Force F (newtons)	Sound pressure, p	Volts, V
Displacement x (m)		Charge q (coulombs)
Velocity $\dot{x}$ (m/s)	Volume velocity	Current, q/s (amperes)
Energy, work (joules)	joules	joules
Power (watts, joules/s)	watts	watts
Stiffness F/x		
Compliance x/F	Compliance C_a	Capacitance C
Frequency Hz f (1/s)		
Resistance R_m	Acoustic resistance, R_a	Resistance, R
Temperature (K)		Impedance, Z
(SI/MKS units used)		

Impedance relationships:
  mechanical impedance $= F/\dot{x}$
  $Z_m = PxA/\dot{x}$
  where A = area of element
  Specific acoustic impedance, $Z_s = p/\dot{x}$
  Analogous acoustic impedance, $Z_a = p/\dot{x}A$

Sound pressure level SPL (pascals, newtons/m², Pa) referred to threshold of
hearing $= 2 \times 10^{-5}$ Pa (0 dB)

Admittance $= 1/z$;   Transformations: $F = Bli$; $V = Bl\dot{x} = $ e.m.f.

Newton's laws $F = M\ddot{x}$ Kinetic energy $= \frac{1}{2}M\dot{x}^2$
                                  Potential energy $= \frac{1}{2}sx^2$

propagated without loss unless it encounters resistance or energy absorption in the medium, but in a spherical wave the pressure is reduced in a more complex fashion as it expands, as shown in Figures 1.4 and 1.5. For higher frequencies and/or relatively large distances from the source, the air particle velocity is proportional to and in phase with the sound pressure, whilst for low frequencies and smaller distances, the particle velocity is proportional to $1/\omega$ times the pressure and is not in phase with it. This gives rise to the spherical wave proximity effect for microphones responsive to pressure gradient.

Figure 1.5 shows the effect for fully gradient-operated microphones, i.e. bidirectional ribbon microphones, etc. Partially gradient-operated microphones, such as cardioids, experience a reduced proximity effect, giving less low-frequency rise in response at close distances.[1] The proximity effect can be exploited to give an enhanced bass response or to reduce lower-frequency noise and reverberation by using an over-damped gradient microphone, such as a ribbon microphone with increased acoustic resistance gauzes closely coupled to the ribbon. This causes a considerable reduction in the middle- and low-frequency response to substantially plane waves, i.e. distant sounds, these frequencies being substantially brought up in level by the proximity effect for close speech.

Unfortunately, the proximity effect operating down to very low frequencies gives rise to severe 'pop' effects on speech plosives such as 'B' and 'T' and other low-frequency surges, unless effective wind/pop shields are fitted to the microphones.

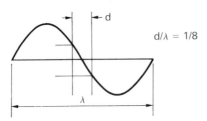
$d/\lambda = 1/8$

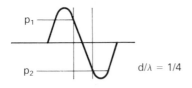
$d/\lambda = 1/4$

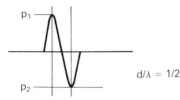
$d/\lambda = 1/2$

(a)

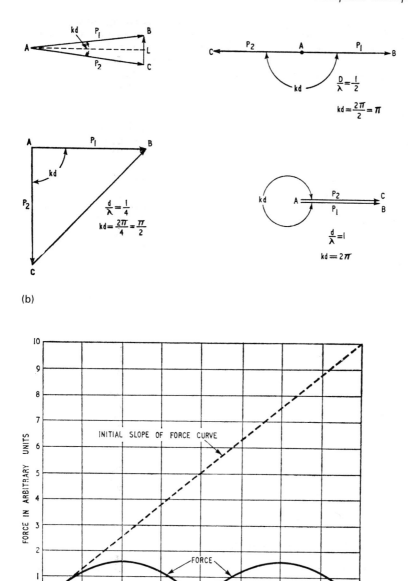

(b)

(c)

Figure 1.4 *Sound pressure gradient relationships. (a) Portion of sound wave intercepted by pressure gradient taken over the distance d; (b) vector diagrams of the force developed by the pressure gradient principle of operations; (c) pressure gradient operation with free progressive plane wave.*

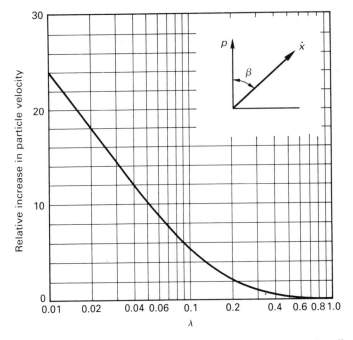

Figure 1.5   *Spherical wave proximity effects. The proximity effect gives a low frequency increase for gradient microphones at close distances. The effect for a cardioid is approximately 6 dB less over the straight part of the curve.*

Free field conditions are not strictly met in most acoustic environments. Sound waves encounter physical changes, obstacles or discontinuities, such as sharp edges in their path. In all these cases, a secondary reflected or diffracted wave is produced which interacts with the original incident wave. In the case of a large object, standing wave maxima and minima occur. For smaller objects, or sharp edges of a size still comparable to the wavelength, a diffracted interfering wave is generated, which again interferes with the incident wave to give rise to irregularities in the resultant sound pressure around and to the rear of the obstacle. This is particularly marked around the body of a microphone at higher frequencies, producing changes in response, particularly for sounds incident from the rear. Figure 1.6 shows the effects for a sphere and a cylinder; also, for sounds incident around 90° to a cylindrical front face incorporating a diaphragm, a portion of the wave is intercepted, causing a 'phase loss' effect at higher frequencies (Figure 1.7). These effects on sound pressure have a direct effect on pressure-responsive microphones. In addition, one must also consider the effects of the sound pressure gradient on microphones responsive to the gradient or the difference in pressure at different points on the casing, which are displaced in the line of the sound incidence. Figure 1.4 shows how these pressure gradient forces are related to this distance. It rises to a maximum over a distance of a half-wavelength and then varies erratically for higher frequencies. This particularly applies to bidirectional ribbon microphones and to unidirectional or cardioid microphones with additional rear or side sound entries.

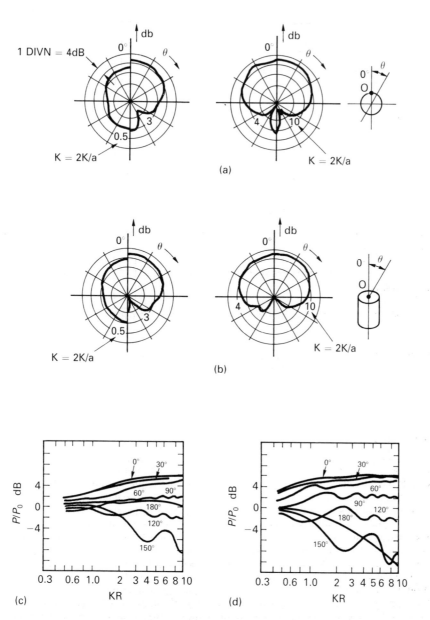

Figure 1.6   *The effect of diffraction on the sound pressure at a point on the surface of a solid body. (a) the pressure at a point O on a sphere for different angles of plane wave incidence for the indicated values of KR (K = 2π/λ = ω/c). (b) The pressure at the centre of the end face of a cylinder. One division = 4 db.*
P/P$_o$ plotted in rectangular co-ordinates against KR for different values of θ. (c) for a sphere; (d) for a cylinder. P/P$_o$ is the ratio of the pressure at the point O on the solid to the free field pressure P$_o$ which would exist if the solid body were absent.

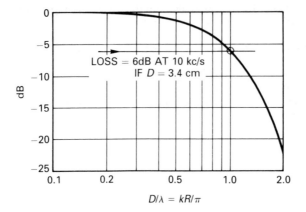

Figure 1.7 *Normalized pressure phase loss effects over the centre of a cylindrical end face for a plane wave incident at 90° to the normal.*

p/p$_o$ is the ratio of the mean effective pressure over the cylindrical face of diameter D to the free field pressure which would exist at O in the absence of the diaphragm.

In order to calculate the acoustic force on a diaphragm in the body of a microphone we must consider both the diffraction and the phase-loss effects for any given frequency and angle of incidence. Hence the acoustic pressure reduction is obtained due to phase loss. Diffraction effects may be estimated from Figure 1.6.

1.2 Microphone developments

Most sound reproducers use diaphragms as a driving element. Other physical effects have been exploited. Examples are the hot-wire microphone, where a low-frequency sound wave causes resistance variations, and the Rayleigh disc, where a small rigid disc is deflected torsionally to measure the sound wave particle velocity. Table 1.1 summarizes microphone principles. The types of transducer seen as reversible can also be used as sound sources. They all use diaphragms or some solid element which is moved by the action of sound wave pressure or pressure gradient, though the principle of mechano-electrical transduction can vary widely. Other systems come to mind, such as smoke particle movement monitoring, and movement of charged particles, analogous to the ionic loudspeaker. More interesting is recent laser-based development such as the laser velocity transducer, where a vibrating reflective surface is scanned by a low-power laser beam, using the Doppler velocity shift principle to give a measure of the sound wave. Developments in electronics and digital techniques may have a future here.

Historically, it is possible to identify major concepts which have constituted breakthroughs which amounted to a revolution in the art. Usually the result of a large amount of scattered scientific work has suddenly produced a key invention. An obvious case is the carbon granule self-amplifying microphone, which has formed the basis of the world's telephones virtually to recent times. Similarly, the moving-coil loudspeaker opened the way to home high-quality sound reproduction. Broadcasting depended mainly on versions of the carbon microphone, in spite of its limitations of noise and instability, until in the early 1920s, precision stretched diaphragm condenser microphones and high-impedance head amplifiers were made. These microphones were basically omnidirectional, until developments in Germany produced a dual diaphragm condenser microphone capable of a directional response, a desirable feature in order to reduce the troublesome pick-up of excessive noise and reverberation. A great many workers then devised moving-coil, moving-iron and ribbon microphones of many types, leading to the products widely used today. Similarly, piezoelectric transducers were developed following the original work by the Curies in the last century. Natural piezoelectric crystals were used, such as quartz and Rochelle salt, followed by specialized ceramics. Recently, piezoelectric foils have been made to form a combined diaphragm and transducer, with certain limitations in use for microphones and sound sources.

1.3 Microphone design techniques

Earlier microphones were undoubtedly evolved mainly by trial and error following many experiments, and the study of earlier results, possibly guided by such calculations that could be made on mechanical and electrical parts, as well as on certain idealized acoustical layouts. Basic mechanical theory analyses moving elements in terms of masses, compliances, frictional resistances, and the kinetic energy, potential energy and dissipation involved. It was observed some years ago that the mechanical equations of motion could also apply to equivalent electrical circuits and that the study of these circuits was greatly facilitated because of the advanced study and comprehension of electrical circuit performance. In particular, work at the USA Bell laboratories in the early 1920s established and applied the use of equivalent electrical circuits for electro-mechanical devices, such as the electromagnetic telephone receivers which complemented the carbon microphone as the basis of the rapidly expanding telephone networks. It was vital to improve the efficiency of the receiver so as to allow the use of longer telephone lines without the use of repeaters or amplifiers, which would have added considerably to the cost and complication of the urban lines which then formed the vast majority of the telephone systems. Further work at Bell laboratories by Maxfield and Harrison for Western Electric exemplified the use of electrical analogous circuits as a basis for the design of an electro-mechanical system to produce a greatly improved disc recording cutter head, needed to improve the rapidly growing disc record industry and to provide sound for the coming talking picture industry, before the advent of sound tracks on the edge of films. The use of these analogies is now widely adopted for the analysis and design of many electromechanical devices, including microphones. The procedure is described in more detail later.

1.4 Directional effects

We have noted that omnidirectional microphones were back-enclosed devices with a small rear leak for barometric pressure equalisation. The presence of a rear entry was later refined to provide extra low frequency response for omnidirectional moving coil microphones, and subsequently to give directional microphones by exploiting the rear entry path as a phase-shifting system, now a normal

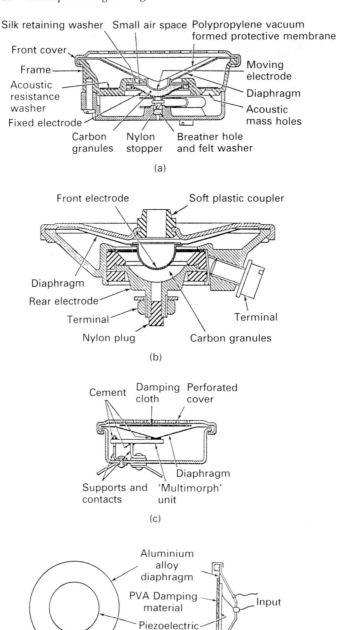

Figure 1.8   *Types of microphone. (a) A modern carbon telephone microphone; (b) a miniature carbon microphone for use with a tubular horn; (c) a typical piezoelectric microphone; (d) a piezoelectric earphone;*

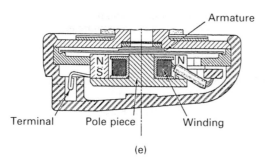

(e)

1. Front case.
2. Spacer.
3. Protective membrane.
4. Rear case.
5. Terminal.
6. Transducer unit.
7. Connector lead.

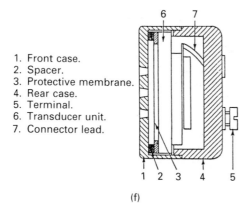

(f)

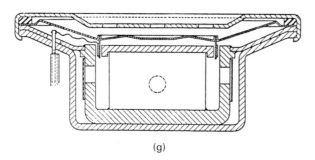

(g)

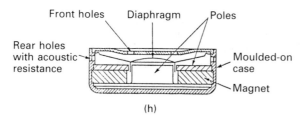

(h)

(e), (f) moving iron microphones; (g) a modern general-purpose earphone or microphone moving-coil insert; (h) a moving coil unit using a flat ceramic magnet.

design feature for many types of microphones. The original bidirectional ribbon microphone with symmetrical access to the front and back of the ribbon was modified by RCA's introduction of a ribbon system with the upper part of the ribbon open to sound and the lower part terminated in an acoustic phase-shifting network.

This amounted to the combination of a gradient unit with a pressure omnidirectional unit, a technique which was applied in a modified form in the Western Electric combined ribbon and moving coil microphone and the STC ribbon and moving coil microphone, which mounted the two transducer centres as close together as possible, the outputs being electrically combined in series. This provided a rugged type of directional microphone which found considerable use as a TV boom microphone. The concept then followed of obtaining a basically cardioid directional response by using a single unit with a rear phase-shifting network, as exploited in the RCA ribbon microphone with the rear of the ribbon coupled to a resistive back entry and a damped terminating tube. This was followed by single unit cardioid microphones using moving coil, ribbon, condenser units with rear phase shifting networks, produced by WE, Shure, Electro-Voice and others in the USA, as well as Neumann in Germany, AKG in Austria, STC in England and others, leading to what amounts to normal practice for obtaining unidirectional response.

More detailed design procedures are given later.

1.5 Limitations on directional characteristics

A fundamental disadvantage of all these microphones was that they could not focus on a distant sound source in the same way that the eye or a camera can focus on a distant object, to the exclusion of unwanted surrounding sources. Highly directional microphones were sought, and attempts were made to evolve these. One solution was to place an omnidirectional microphone at the centre of a large baffle or paraboloid reflector. These could be effective but became unwieldy when made suitably large. In the late 1930s RCA and WE in the US developed tubular microphones of high directivity by feeding sound to the front of a ribbon or diaphragm via a bundle of tubes graded in length. Sounds arriving on the axis of the tubes are combined in phase at the diaphragm whilst sounds at an angle suffer a large degree of cancellation. These were the fore-runners of the present directional line or 'shot gun' microphones in which the response has been smoothed by the use of a single front tube perforated and acoustically treated along its length, and a cardioid or directional transducer at the end. There are still limitations at low frequencies where high directivity demands inconveniently large lengths of tube, etc. Tubular microphones of various lengths are used. Shorter ones can give useful directivity at middle and high frequencies for audience participation, shows, etc., a low frequency cut-off being incorporated to reject low-frequency noise and reverberation. In fact, the question of the inevitable variation in response, particularly at higher frequencies for virtually all but the smallest condenser microphones, does represent a serious limitation to obtaining a perfect off-axis frequency response. Serious attempts have been made to solve this problem by novel or revolutionary ideas in microphone design. One proposal was to scan a line array of small microphones and to apply Doppler processing and advanced electronics to achieve constant polar characteristics at all frequencies.[2]

It may be that new techniques involving laser technology, microelectronics and digitization may ultimately bear fruit.

1.6 Types of directional microphones

We have noted that the pressure gradient force available in a microphone with side or rear openings rises to a maximum over a distance of a half-wavelength in an approximately linear relationship but becomes irregular at higher frequencies. Small condenser microphones are able to meet this half-wavelength distance requirement up to fairly high frequencies and ribbon microphones can operate with fairly small distances around their pole pieces, but difficulties arose with cardioid moving-coil microphones whose magnet and pole systems and control-of-response requirements complicated the phase-shift requirements. A comparatively long path distance was required at low frequencies and a smaller path distance at higher frequencies. This was met by adopting the so-called variable-distance principle, where paths of suitable acoustic impedance characteristics were arranged within the rear structure of the microphone so as to give, in effect, a variable entry point (Figure 1.9).

The detailed design procedures for various representative microphones are outlined in Section 1.10.

The previous description and Figure 1.4 deal with microphones operated by one gradient stage of sound. It is possible to construct second-order gradient microphones by connecting in phase opposition two first-order microphones which are physically displaced apart in the line of sound incidence by an amount which is small compared to the wavelength. The result is a narrower polar

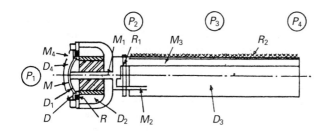

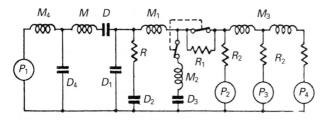

Figure 1.9 *The construction and analogous circuit of a variable phase shift microphone.*
The distributed mass M_3 and resistance R_2 of the rear-entry tube have impedance characteristics such that higher-frequency sound-pressure components entering at the rear (P_4) are subjected to a greater attenuation than are lower-frequency components. The switches represent a mechanical shutter on the microphone which can be arranged to cut the bass response for speech purposes.

response and increased discrimination against ambient noise. In the 1950s RCA produced a second-order gradient studio microphone consisting of two phase-shift cardioid ribbon microphones assembled in a case giving about 13 mm linear spacing between the units. It gave a narrow cardioid frontal lobe up to about 1500 Hz. Above this frequency electrical filtering was needed to reduce the output to one of the units only, the narrow cardioid being maintained by diffraction effects. Second-order noise reducing microphone assemblies have been made in more compact form for communication purposes by using more closely spaced gradient units. In theory such a microphone could give nearly 20 dB discrimination against random noise.

An interesting modern variant on gradient microphone techniques is the zoom microphone, which was evolved to work with TV cameras. Two cardioid microphone units facing forwards and spaced apart are connected to give second-order gradient characteristics, together with a third cardioid unit pointed to the rear, whose output could be added in or out of phase to the forward units. Variable levels and equalization characteristics can be introduced, to suit the various directional responses required. The overall directivity from omnidirectional to a fairly high value of directivity is available.[1,3] The controls may be interconnected to the camera zoom mechanism with the object of giving a compatible acoustic perspective in relation to the picture.

1.7 Microphone powering and connections, cables. Preamplifiers

Certain classes of microphone, carbon microphones and condenser microphones require DC power supplies. Carbon microphones have long been confined to telephones, where they get a nominal 50 V supply from the line interface circuits. Small electret condenser microphones replace the carbon microphones in more modern sets; an integrated circuit with an FET input amplifier and interface is used (see Figure 1.10). The original studio condenser microphones with valve impedance converters in the head amplifier need heater supplies as well as a higher voltage for polarization and valve operation. Modern condenser microphones normally operate from standard phantom supply circuits (P 48 V etc., DIN 45 596, IEC 268–15). Electret condenser microphones will operate from a 12 V phantom supply, whilst other conventional condenser microphones may use internal solid state generators for polarization. Many modern condenser microphones are designed to work from any phantom supplies from 12 V to 52 V from studio mixers etc., having the advantage that moving-coil and other microphones are not affected by the phantom DC supply voltages. The older phantom A–B supply circuits (DIN 45 595) are less suitable.[1] Cheaper electret microphones may use an FET and a small internal battery.

1.7.1 Microphone connections, cables and input circuits

The connectors, cables and input circuits to the preamplifier line circuits are of paramount importance. All microphones must have a sensitivity to sound inputs (i.e. a conversion efficiency) in relation to a signal that is high enough compared to all forms of noise and interference that may be encountered.

Early stages and the initial amplifier stages all have to be analogue, even in digital sound systems. A reasonable voltage level will be required for the A/D conversion process. Plug and socket connectors for microphones should be robust and durable, using the best contact metals and springs. The original Cannon XLR designs and some DIN types have formed the basis for present good connector

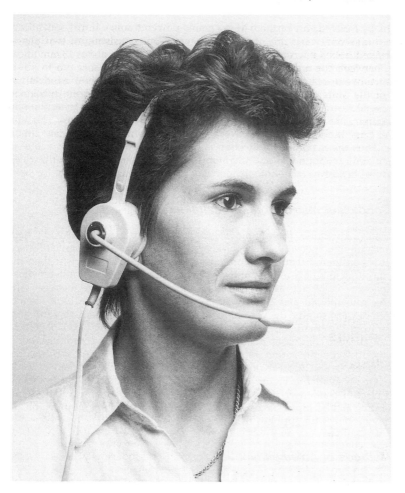

Figure 1.10   *A light-weight telephone operators' headset which uses a small electret condenser microphone capsule on an adjustable boom. An integrated circuit mounted behind the earphone includes an FET preamplifier/impedance matcher and the interface circuitry for standard telephone networks, providing a replacement for a carbon microphone, with improved performance. (Courtesy Coles Electroacoustics)*

practice. Similarly, microphone cables should be carefully designed and made for the purpose. The insulation and arrangement of the conductors should minimize electrostatic and electromagnetic interference pick-up and any microphonic effects due to movement or handling. It may be necessary to have graphite or other conductive treatment on the outside of each individual core lead insulation to remove any static generation due to movement of the conductors. Similarly the outer earthed shield of the cable should be properly designed for its purpose. Conventions apply to microphone pin numbers and functions. In the XLR 3-pin convention pin 2 is positive for an inward movement of the microphone diaphragm.[1] It is good safety practice for each microphone input circuit to be individually isolated and earthed. Many mains-powered musical instruments are used in studios today and it is necessary to ensure that simultaneous touching of an instrument and a microphone will not cause risk of shock or hum generation. Microphone isolation amplifiers represent good isolation amplifier practice. It is common now to include ELCBs (earth leakage contact breakers) in electrical mains supply systems, so as to avoid electric shock risks for users who may handle metal cases of instruments and appliances. This can apply in studio and stage equipment. In order to avoid widespread studio disruption, isolation amplifier feed blocks are supplied, each with a separate ELCB mains circuit breaker. Thus the maximum protection is offered to personnel whilst avoiding effects on other microphones and phantoms in the event of a wiring fault, which might cause ELCBs to come into action.

1.7.2 Condenser microphone preamplifiers

Traditionally, low-noise thermionic valves have been used to provide high impedance inputs. Specially selected valves with low gas content are still used where it is considered that valves offer a better overload characteristic than the FET semiconductors now almost universally used for these preamplifiers whose main function is to act as an impedance transforming stage between the condenser microphone and the low-impedance output line. Feedback circuits have been devised to give the very high impedances needed to obtain a full bass response from the small source capacitance of the microphone (see Chapter 2). The small cheaper condenser microphone cartridges used for communications, telephony and general sound reproduction incorporate a simple FET source follower stage within the cartridge, with a small local battery when external supplies are not available. Figure 1.10 shows a telephone operator's headset using a small electret condenser microphone on a light boom, an integrated circuit incorporating an FET input stage and telephone line circuit being mounted inside the earphone case with supplies from the telephone circuit.

1.7.3 Methods of noise elimination

Larger systems may also provide separate 48 V phantom power supplies for each condenser microphone input in order to take the maximum precautions in use of the microphones. Balanced input transformers are conveniently used to provide voltage step-up and impedance matching of microphones to lines and amplifier input circuits. Balancing of input lines to earth can cancel induced longitudinal electromagnetic interference pick-up. Even the best transformers may introduce some slight deterioration in the form of noise, magnetic non-linearity at very high or low signal levels, possible hum induction, etc. Electronic transformerless circuits are now being used for some of the highest quality circuits, but the question of isolation still has to be considered. Chapter 8 deals with preamplifier design and noise problems in detail, in an authorative contribution.

The use of a separate earthed conductor in addition to the signal conductors is especially helpful in achieving a secure and noise-free phantom earth return, as the conductor can be earthed to the outer shield at both ends of the cable.

In addition to the basic noise appearing in the microphone element due to the Brownian movement of air particles on the diaphragm and the thermal movement of the electrons in conductors, there are many possible sources of noise which can appear in its output. Among these environmental hazards, in addition to electrical and radio interference of all kinds, are mechanical vibration and shocks, wind, draughts and breath (pop) noise, handling noise, clothes rustle etc.

Microphone units have to be shielded and isolated in various ways against these hazards. These forms of shielding and protection have to be designed so that the microphone response is not adversely affected to any marked degree and are described in connection with shock mounts, wind shields etc.

Microphone cables have to meet a number of special requirements over and above those that normal audio cables encounter:

(1) They have to be able to survive possible rough usage on floors and on mobile booms etc., and to have sufficient strength to support microphones in overhead use. They must not generate electrical or mechanical noise when moved. The requirements are thus considerably more stringent than those for general purpose cables.
(2) Electrical characteristics should be suitable, e.g. low-capacitance, low-resistance and high-grade copper conductors, so as to avoid deterioration of the response on long cable runs.
(3) The size and flexibility of the cable should be suitable for specific usage, e.g. small diameter for microphones worn on the person, unobtrusive colour, low friction and high flexibility. Some cables are now made with a very smooth woven fabric outer cover (e.g. Neumann).

High-quality microphone cables are usually 3–5 mm diameter. Great attention is paid to rejection of ES and EM interference. Firmly twisted inner conductors and a continuous outer braided shield aid the rejection of all types of electrical interfering fields. The outer braiding should resist bending and twisting stresses of the kind that may be encountered on boom operation. A good counter-wound twist braiding can achieve 95% shielding coverage. A type of mesh screening (Reussen) has been evolved which resists repeated bending round boom pulleys, etc. Some types of cable have also used a conductive coating over the insulation of the inner conductors to discharge to the braiding any static frictional charge built up when conductors move in use.

It will be appreciated that in high-quality work it is well worth paying for specialized microphone cables.

Special cables have been evolved to combat the additional hazards that may arise in unbalanced lines, where there is often no possibility of the noise signals generated by cable movement etc. being cancelled, as occurs in balanced twisted lines. A conductive plastic layer over the single inner conductor will harmlessly earth statically induced noise due to cable being moved. Double Reussen outer shields tightly counter-wound give maximum RF and EM shielding.

The use of four conductor cables with double Reussen outer shielding, together with the use of the star quad connection arrangement of diametrically opposed conductors at each end of the cable, can give a considerable extra isolation in areas of very high electrical interference.

It may be borne in mind that some microphones include mechanisms that can generate interference. For instance, condenser microphones may include transistor oscillators and rectifiers to generate polarizing voltages, and RF condenser microphones include their own oscillators.

Microphones may also include switches for equalization, on – off and other functions; these switches should be unobtrusive, positive and silent in action. Microphone cases should be smooth, stream-lined and free from sharp edges, as far as possible, with no gaps. Hum-bucking coils, vibration insulation and pop shields may be built into microphones or may be in the form of external fittings. In both forms, they have to be tested for effectiveness and any tendency to adversely affect the microphone response (Figure 1.11). A number of standardized tests are applied to microphones to check these and other effects. Chapter 6,

Microphone Testing, describes some of these. Standards are being evolved in some cases. In addition, moisture and dust exclusion should be checked. The complex questions of noise and the signal handling capability of microphones, and the interaction between them and preamplifier stages, is covered in Chapters 2 and 8. The advent of digital sound has made these factors of considerably increased importance.

1.8 Microphone windshields and anti-noise precautions

In use microphones have to be mounted on stands, booms, rods, desks, floors, as well as being hand-held or mounted on the person, or even on musical instruments. Winds, draughts, electrical power, TV and radio interference may all be encountered, as well as extreme climatic conditions. Manufacturers have to allow for these factors, conduct tests and meet such standards as may apply or be imposed by particular users.

It is not always appreciated to what extent microphones may be affected by windshields, mountings and the proximity of other objects. Anti-vibration mountings, in particular, may have to be strong and fairly elaborate structures. Standards may require new tests and specifications on such factors as microphone 'pop' response. In many cases modern microphones have wind, pop and anti-vibration precautions built-in and the response curves include any effects (Figure 1.11).

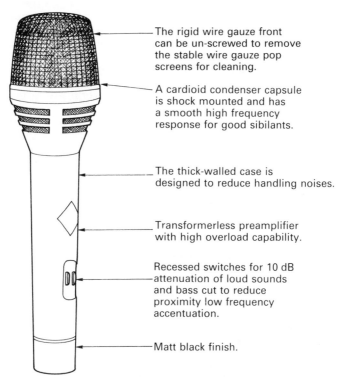

The rigid wire gauze front can be un-screwed to remove the stable wire gauze pop screens for cleaning.

A cardioid condenser capsule is shock mounted and has a smooth high frequency response for good sibilants.

The thick-walled case is designed to reduce handling noises.

Transformerless preamplifier with high overload capability.

Recessed switches for 10 dB attenuation of loud sounds and bass cut to reduce proximity low frequency accentuation.

Matt black finish.

Figure 1.11 *A high quality microphone for use by vocalists. (Courtesy of Neumann.)*

Windshields often have to be fitted to microphones for boom use and outdoors. The effect on frequency response is often not known, and can be severe even under good conditions and catastrophic in wet or with extremes of dust, etc. Typical older windshields often used on omnidirectional microphones can cause response irregularity of 6 dB or more at mid- or high frequencies. Gradient or part-gradient microphones may be even more affected by proximity effects reflections inside the shield.

The exact manner in which windshields function is not fully understood, though the effects are widely known through experience. The air velocity of even a moderate breeze is probably 1000 times the magnitude of the particle velocities involved in a sound. The air in the interstices of a windshield material exhibits nonlinearity and thus limits higher velocities to some extent. Adjacent air particles in the main air stream tend to slide over those in contact, but may form areas of turbulence which give rise to characteristic audible tones, as well as 'edge tones' around any sharp discontinuities. The practical designed shapes of microphones and windshields does not usually lend itself to a streamlined shape which could minimize turbulence. The exhalation of breath in plosives or pop producing speech is liable to cause air velocities several times those of a moderate wind. Microphones with delicate members such as ribbons should have some built-in guards against excessive wind and plosives.

The materials used for windshields necessarily depend on the microphones concerned and the uses to which they are put, indoor or outdoor conditions, close or distant speech etc. As we have noted, it is strictly necessary to check the effect on the response of given types of microphone when shielded, as well as the effect of moisture accumulations, where these are likely to occur. As well as the use of naturally moisture-repellent windshield materials or silicone treatment etc, means such as drainage holes may have to be provided in the casing if the microphone is classed as suitable for prolonged close-talking or outdoor use. Older windshields used for outdoor measurements used precision-woven fabrics such as flour sieving cloths, fine woven wire mesh etc. Fabrics must be well supported on frames or mesh. Any movement may cause non-linearity in the response. Other materials used include natural wools, lambswool etc., foam plastics, porous sintered metals. Foam plastics make convenient and economical windshields easily slipped on to many shapes of microphone, offering also a range of colours and a degree of mechanical protection. They may be made by freezing and machining flexible foam materials. Optimum porosity is achieved with an open cell construction with many inter-communicating cell paths. The material in bulk may have to be processed in various ways to ensure enough cell wall breakdown. Thus foam windshields from the main manufacturers are specialized products.

1.9 Anti-vibration, shock and handling noise protection

We see that microphones having very flexibly mounted diaphragms or ribbons may be the most susceptible to wind or pop noise; an effectively mass-controlled moving system may be seismically sensitive to transmitted mechanical vibrations from the casing or mountings. Basically, shock mountings consist of rubber or synthetic blocks or thong mountings supporting the microphone mass as a whole. Figure 1.13 shows typical mountings. Some general purpose microphones have wind and pop shields, shock and vibration isolation and anti-handling noise outer case treatment built-in. Boom mountings may have to be particularly specialized to combat motion through the air and mechanical noise generated by the boom and trolley mechanism.

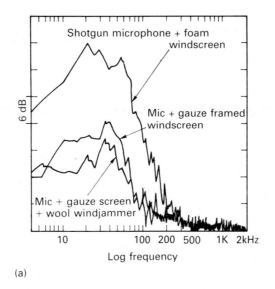

(a)

(b)

(c)

Figure 1.12 *Windscreens. (a) Relative wind noise for a shotgun microphone; (b) with various windshields; (c) the effect on frequency response of a microphone with various windshields. (Courtesy of Neumann.)*

Figure 1.13a shows the theoretical isolation response of a basic mass on a flexible mounting. This shows that the resonance of the system must be preferably below disturbing frequencies and that the mechanical resistance of the mountings must be within a certain range if magnification at resonance is to be avoided.

1.10 Microphone evolution and design; the use of analogues

Earlier microphones were undoubtedly evolved mainly by trial and error, aided by such calculations that could be made on the mechanical and electrical parts, as well as certain acoustic configurations. Basic mechanical vibration theory analyses moving elements in terms of masses, compliances and frictional resistances and the kinetic energy, potential energy and dissipation involved. We have observed that the equations of motion could also apply to equivalent electrical circuits and that these greatly facilitated analysis because of the advanced art of study and use of electrical circuit performance. In particular, work at the Bell Laboratories in the USA in the early 1920s established electrical equivalent circuits of electromechanical structures such as telephone instruments. This technique is thus widely adopted to understand and to design microphones. When the appropriate masses, stiffnesses and resistive elements have been identified in the planned microphone structure, the equivalent inductors, capacitors and resistors can be set up as the equivalent or analogous circuit. The response can then be calculated or measured, without the cost and delay of making complete microphone models. For example, the effect of changing the mass of a diaphragm or the volume of a cavity can be estimated and the effect of the various elements of the structure is apparent, leading to a much better understanding of the way in which the system operates. There is, however, a proviso. Lumped constants of the above type can only give an accurate analogy when the size of the element is a small fraction of the sound wavelength (1/10 to 1/4). Above this frequency range, particularly in cavities, standing wave effects can occur in three dimensions, which may preclude representation, even by an electrical transmission line, as this would represent only one dimension of sound propagation. These effects have to be estimated by experience and intuition in the absence of an actual model. The transducer output is a measure of the diaphragm motion; in the case of a moving-coil microphone this is represented by the current through the coil.

A distinction is drawn between the direct (impedance) analogue and the inverse (mobility or admittance) analogue. Where connected elements are acted on by a force or pressure and all move with a common velocity or volume velocity, it is apparent that they have a direct impedance analogue represented by a voltage and a common series current. However, when elements are acted on by the same force but acquire different velocities, the electrical analogue must show parallel connected circuit elements and the common potential replaces velocity; the admittance of each branch replaces impedance, the degree of opposition to motion then being described as mobility. A working knowledge of both types of analogue is desirable in order to allow one to interchange constants from one analogue to the other when required to do so. For instance, in the study of loudspeakers, if one starts from the electrical drive of the amplifier which is translated into the velocity of the diaphragm, an ideal transformer of ratio Bl, the force factor F has to be introduced; $F = Bli$ driving the diaphragm. Furthermore, the back-emf $e = Bl\dot{x}$ induced in the coil is a maximum at the resonance of the loudspeaker, implying a parallel resonant structure in the equivalent circuit.

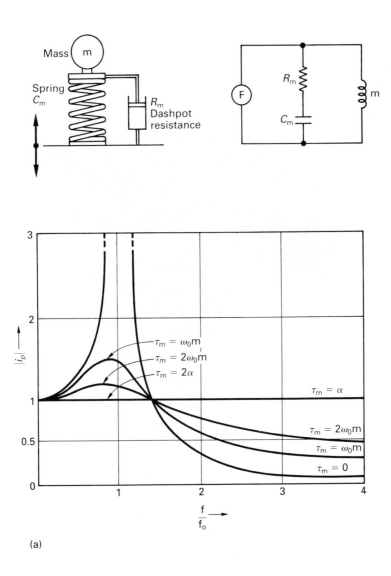

(a)

Figure 1.13   *(a) A basic system for vibration isolation. The larger the damping the greater the reduction in the resonance peak but the less effective the mounting as an isolating device; by replacing the spring by a resilient material which is not loss-free. The most commonly used materials of this type are rubber and certain plastics. Studio boom microphones are often suspended on soft rubber thong mountings, whilst stand units are mounted on concentric bonded-rubber shear-type mountings. (b) A microphone on an elastic suspension. (Courtesy of Neumann.)*

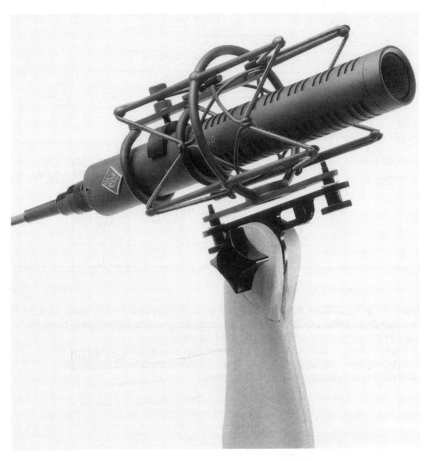

(b)

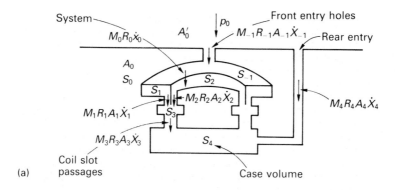

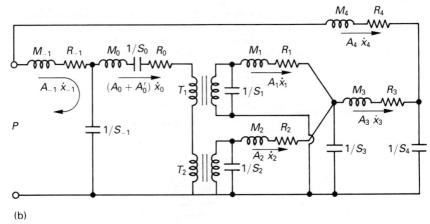

(b)

Figure 1.14   *(a) The functional diagram of a moving-coil microphone. Various mechanical and acoustical masses and compliances are identified in order to facilitate the derivation of the analogous electrical circuit.*
(b) The analogous circuit for the acoustical and mechanical elements of microphone (a). The current through the diaphragm element $M_0 1/S_0 R_0$ represents the response of the microphone.

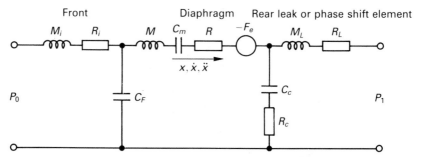

Figure 1.15   *The analogous circuit of a small cardioid condenser microphone as in Figure 1.20. The analogy gives a guide to the working of the microphone but cannot be very exact due to the distributed nature of some of the elements.*

Hence the *Bl* element represents a gyrator or ideal transformer with the property of inverting input current into force or voltage in the analogue, and the diaphragm resonance into a parallel shunt resonant circuit. In other words, the impedance analogue of the primary circuit is converted into an admittance (mobility) analogue in the secondary circuit. Further inverting ideal transformers may be needed in the secondary circuit to match the equivalent area of the diaphragm.[4] This makes for complicated circuit structures, not always easy to comprehend. A conversion from the mobility to the direct impedance expressions then gives series connections, which are easier to handle. A similar difficulty arises in the analysis of a moving coil microphone, where the inner and outer parts of the diaphragm on each side of the coil feed into different acoustic impedance terminations. Here a conversion from a mobility to a direct impedance analogue is achieved by the introduction of some ideal transformers (see Section 1.10.2 on moving-coil microphone design).

We note that the generalized construction and layout shown in Figure 1.16 can apply to both omnidirectional pressure-operated moving-coil microphones and to pressure-gradient operated microphones. For omnidirectional microphones, the rear leak path $M_4 R_4$ is usually made in the form of a long narrow tube with a high

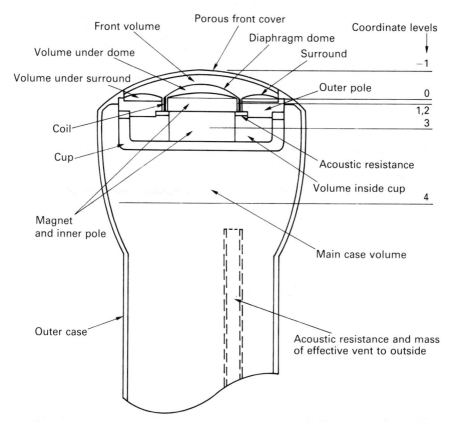

Figure 1.16 *Sectional view of a moving coil microphone identifying the mechano-acoustic constants for analysis.*

acoustic mass and resistance which effectively blocks sound entry except at very low frequencies, where its mass resonates with the main case volume and gives a useful boost to the low bass response, in a manner similar to the action of the vent in a reflex loudspeaker cabinet. (Some leak has always to be provided to equalise barometric pressure).

For directional microphones that are pressure-gradient operated, the rear entry path is given quite different constants and may even be of a distributed nature, as it functions as a phase-shifting network designed to give the proper delay to the sound pressure p_2 in the manner shown in Figure 1.17. The system thus becomes pressure-gradient operated and the other microphone constants such as the diaphragm stiffness and the controlling resistances are arranged to make the overall system substantially mass-controlled instead of resistance-controlled, as for the omnidirectional pressure-operated microphone.

We also note that, for small narrow passages such as where the edge of the diaphragm dome comes close to the inner pole piece, it may be difficult to decide whether this area should be assigned as an additional mass added to the inner magnet gap or should simply form part of the small volume of air under the diaphragm dome. Two ideal transformers of ratios A_1/A_2 have to be introduced to couple the inner and outer diaphragm areas into the appropriate sides of the coil slot.

The basic microphone transducer relationships, the limaçon curves of Figure 1.18 and the relevant vectors for the phase shift microphone systems of Figure 1.17 illustrate the fundamental concepts involved. The directional effects depend on the transit time of a wave front as it traverses the distance round the microphone to the rear opening and the time delay in the rear phase shift network. The polar response R is given by

$$R = A + B \cos \theta$$

If $A = B$ a cardioid results;
If $A = 0$ a bidirectional curve;
If $B = 0$ an omni curve;
If $B = 2A$ a supercardioid curve;
If $B = 3A$ a hypercardioid curve;
If $B < A$ a so-called wide cardioid.

These various polar curves are shown in Figure 1.18 and can be obtained from a dual unit switched condenser microphone, thus enabling various degrees of sharpness of the polar curve and nulls to be realized at certain other angles, to suit various set-ups.

The equivalent circuits for phase-shift microphones identify the elements acting as masses, compliances and resistances. Moving-coil systems cannot easily be made with a basic resonance below about 200 Hz, but it is possible to couple a low-frequency resonator to the rear of the diaphragm, the acoustic mass of its input tube effectively adding to the diaphragm system mass, thus making the diaphragm system mass controlled. The various acoustic resistances combine with the acoustic masses to give the desired phase shift. Unlike the more compact condenser microphone systems, some moving coil units resort to multiple entries spread along the rear tube so as to give appropriate distances over particular frequency ranges, the various acoustical masses, compliances and resistances being suitably proportioned (Figure 1.19).

This illustrates the basic microphone structures and elements leading to equivalent analogous circuit elements. The basis of the analogy is that voltage

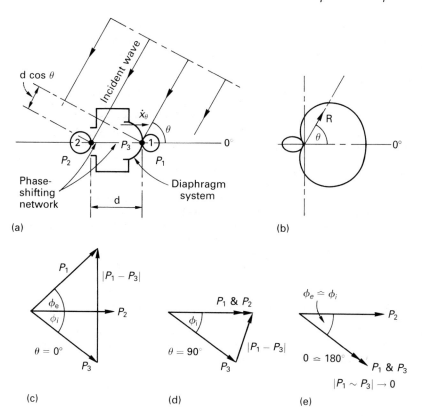

The generalized polar equation for the cardioid family of curves:-

R = A + B cos θ

B = 0	B < A	A = 0	A = B	B = 2A	B = 3A
Omni	wide cardioid	Bi-directional	cardioid	Super cardioid small lobe	Hyper cardioid wider lobe

The properties offered by the various curves above are summarized:-
Sound acceptance angle. The frontal angle over which the output is substantially constant.
Output at specific angles. e.g. 90° 180° 150°
Angle giving best sound rejection. e.g. 90° 150° 180°
Random energy efficiency (REE). Distance Factor, random/direct sound

Figure 1.17 *Cardioid response characteristics.*

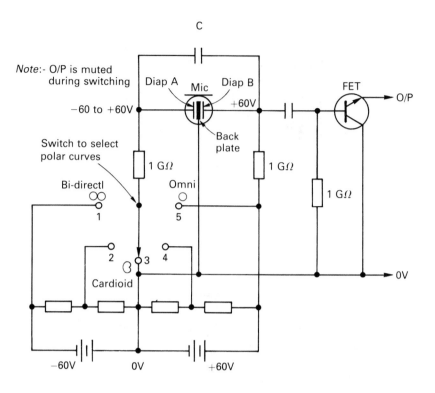

Polarizing voltage

Polar curves

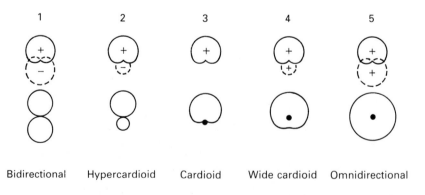

1	2	3	4	5
Bidirectional	Hypercardioid	Cardioid	Wide cardioid	Omnidirectional

Figure 1.18 *Condenser microphone with variable polar response.*

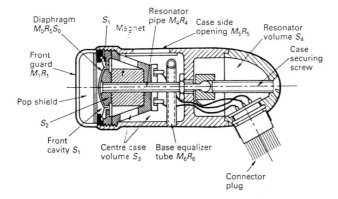

(a)

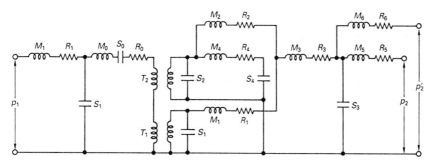

p_1 = Acoustic pressure on front openings.
p_2 = Acoustic pressure on case side openings.
p_3 = Acoustic pressure on bass equalizer pipe opening.
T_1 and T_2 are ideal transformers

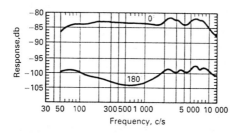

(c)

Figure 1.19 *(a) A unidirectional moving coil microphone with a fixed rear entry distance; (b) analogous electrical circuit; (c) free field response at 0° and 180°.*

represents sound pressure, current represents the volume velocity of the alternating air flow into an element, the values being in 'acoustic ohms':

$$Z_a = \frac{p}{\dot{x}A} \qquad \text{analogous impedance}$$

$$Z_s = \frac{p}{\dot{x}} \qquad \text{specific impedance}$$

$$Z_m = \frac{pA}{\dot{x}} \qquad \text{mechanical impedance}$$

where p = sound pressure
$\dot{x}A$ = sound volume velocity
$\dot{x}$ = sound particle velocity
A = area of element exposed to air flow.

The concept of acoustic impedance is thus dependent on three definitions, according to the mode of operation being considered. For sound propagation in free space or large enclosures, the specific acoustic impedance $Z_s = p/\dot{x}$ is used for wave calculations. In a structure such as a microphone, where the sound encounters elements of differing areas such as diaphragms, holes, slots, partially or completely enclosed air cavities etc., the volume and the velocity of this volume passing through a complex structure is a convenient measure to use. If particle velocity of the wave is used, a series of ideal transformers has to be introduced according to the different areas encountered.

For mechanical moving devices where air coupling is negligible, the mechanical force pA is used and $Z_m = pA/\dot{x}$ is the mechanical impedance.

1.10.1 Omnidirectional condenser microphones

Omnidirectional condenser microphones are single-sided, with diaphragms (metallized plastic or metal film) typically $5\,\mu m$ thick, spaced $25\,\mu m$ from an insulated back plate, grooved, drilled or fissured to add acoustic resistance. Electret charged material on diaphragm or back plate gives polarization of 40–100 V. The diaphragm is permanently secured to a rigid outer case by clamping, special adhesive or plating *in situ*, in order to maintain tension. Diaphragm resonance is 5–20 kHz. Capacitance is typically 5–50 pF. Moisture-sealed connections go to high input impedance low-noise valve or FET impedance transforming stages with output impedance 200 to 1000 ohms. Measuring standards specify 1 in (25 mm). ½ in (12 mm), ¼ in (6 mm), ⅛ in, (3 mm) dia. Omnidirectional response tends to unidirectional at high frequencies. Theory shows that constant amplitude response is required, i.e. stiffness control of the diaphragm, the resonance frequency being placed at the top of the frequency range.

Figures 2a–d of Chapter 2 (Precision Microphones) show the internal construction of typical omnidirectional condenser microphones. Design details, equivalent circuits and specifications are also given. The constants of the rear vent differ as between omni and cardioid units. Figure 1.20 shows the construction of a studio condenser microphone. The equations of motion are given by

$$F_A + F_E = M\ddot{x} + R\dot{x} + Sx,$$

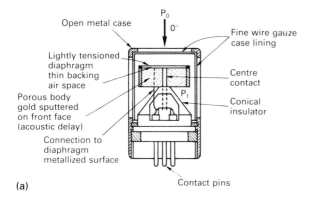

Open metal case
P₀
0°
Fine wire gauze
case lining
Lightly tensioned
diaphragm
thin backing
air space
Centre
contact
Porous body
gold sputtered
on front face
(acoustic delay)
P₁
Conical
insulator
Connection to
diaphragm
metallized surface
Contact pins

(a)

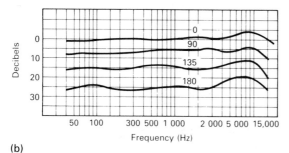

(b)

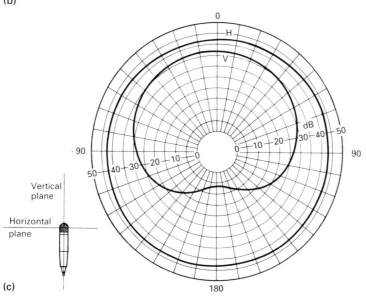

(c)

Figure 1.20 *(a) A small unidirectional condenser microphone. The required rear phase shift is provided by an acoustical labyrinth (porous body) which forms the insulated backing plate behind the diaphragm, the back of the labyrinth providing the rear entry point for sound. (b) The response of microphone (a); (c) polar-response curves of microphone (a).*

where A = diaphragm area
          Q = CE = charge
          F_A = acoustic force
          F_E = negative electrical force = $-Q^2/2A$
   M,R,S = mechanical constants of the microphone

The presence of Q^2 in the equation indicates non-linearity. If the polarizing voltage is made large and constant by providing high insulation and a high input impedance for the preamplifier, the F_E term will be a constant, representing a static deflection. The response of the microphone will then be substantially linear for normal working levels. Some non-linearity will occur for very large signal levels or uncorrected barometric surges, the limit being reached when the diaphragm collapses on to the back plate. (Figure 1.21). The diaphragm tension is made much higher for stiffness-controlled pressure-operated omnidirectional microphones, as compared to resistance-controlled gradient unidirectional studio microphones. Careful design can make both these substantially linear up to the highest sound levels normally encountered. Double-sided units using RF modulation are described in Chapter 4.

1.10.2 Unidirectional condenser microphones

For condenser microphones a flat frequency response requires constant amplitude, which implies a resistance control/pressure gradient actuation. Slits and fissures in the close-coupled back plate add acoustic resistance, whilst through holes in the back plate constitute a mass-resistance phase-shift network. Some specialized high-quality condenser microphones have used porous ceramic or plastic back plates to produce the same desired effect. Electret material applied to the back-plate surface adjacent to the diaphragm can give constant polarization, consisting possibly of a fluorethylene film about 25 μm thick, comparable to the back-plate-to-diaphragm spacing. With reasonable care in use the electret polarization can be constant at about 100 V.

The presence of small amounts of dirt or moisture in the interspace will not cause electrical breakdown, but will cause variations in the response of the capsule, particularly in the low-stiffness membranes used in gradient units such as the one described in Figure 1.20.

The porous member is chosen to have a high insulation resistance and moisture repellence and is fitted firmly inside the outer metal ring, its front face being accurately flat and placed precisely at the correct spacing distance from the outside peripheral clamping surface. Any swarf or particles must be removed from the porosities, and the active face is then rendered conducting by gold sputtering, a polarizing connection being taken out at the back.

The microphone capsule thus incorporates a porous back-plate construction which has been found to constitute a particularly effective acoustic phase-shift network. As we have seen in the previous discussions of the theory of phase-shift cardioid microphones, it is required to provide a phase delay which matches that of the external sound wave as it is propagated round the case of the capsule. A porous material of the ceramic or bonded-particle type can be made to operate so as to provide the correct degree of resistance control and to give the desired phase-shift characteristic.

It is also possible to achieve these properties by means of simple structures composed of discrete holes and slits etc.

The membrane or diaphragm acting in conjunction with the porous backing member essentially constitutes a system with distributed properties, and hence it

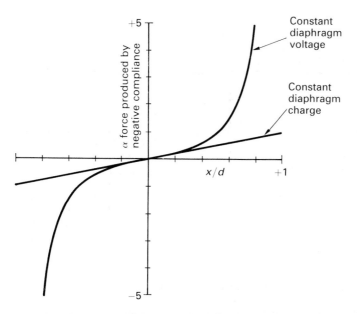

Figure 1.21 *In a condenser microphone a potential between the diaphragm and back plate produces a force on the diaphragm which acts against the restoring stiffness due to the tension of the stretched diaphragm. This force is equivalent to a negative compliance and is nonlinear unless the polarizing voltage is able to maintain a constant charge due to high insulation. This figure shows that substantially constant charge operation linearizes the force/deflection characteristic of the microphone.*

does not lend itself easily to representation by a lumped-constant analogous circuit. Figure 1.20 shows response curves.

The required resistance control of the diaphragm has to be achieved by the resistive components of the air in the slit between the diaphragm and back plate and the air displaced into and through holes and fissures in and around the back plate, leading to the rear entry holes in the microphone casing. The tension of the diaphragm has to be reduced as compared to that of the omnidirectional condenser microphone, whilst still retaining stability. Various pre-forming and ageing techniques have been used for this purpose. Figure 1.20 shows typical construction of back plate and case elements to produce the desired phase-shift elements leading to the rear entry holes. With such assemblies of small slits, holes and cavities, it is not easy to say how far mass, stiffness or resistance impedance modes predominate in a small distributed system of this kind. In particular, lumped constants cannot be accurately assigned to an acoustic system whose dimensions exceed $\frac{1}{20}$ of a wavelength. In some designs the porous labyrinth is a key element, the final design being arrived at experimentally. The system can work by acoustic phase shift up to a frequency where the total phase delay (internal and external) is equivalent to $\frac{1}{2}$ wavelength. It is usually impractical to reduce the effective distance concerned below about $\frac{1}{2}$ in (12 mm). Above about 10 kHz directionality will be due mainly to diffraction. At the lowest frequencies the stiffness of the diaphragm may have an influence, reducing the back-to-front ratio and tending towards a 'super cardioid' type of polar curve. Cardioid condenser microphones offer robustness and a good response in a small and unobtrusive microphone and are widely used for studio work, both in the form of phantom powered and electret-polarized units. Some models also offer internal shock mounts and windshielding within slightly larger casing. Some designs are described in detail, e.g. Figure 1.11. Applications to highly directional line microphones are also described in the literature. [1,5]

The curves of Figure 1.20 show that the response for different angles remains sensibly flat, thus ensuring that off-axis and reverberant sounds are reproduced with the correct frequency balance. The accuracy of the cardioid limaçon curve ensures uniform pick-up over a wide frontal angle and good unidirectional discrimination against sounds from the rear. The fact that the shape of the polar curves is maintained at high frequencies helps to maintain a good stereophonic image, free from shift with frequency when crossed or spaced microphone arrays are used for stereophony.

1.10.3 Dual unit microphones

We have seen that one of the shortcomings of some studio directional microphones is a tendency for the off-axis frequency response to become more irregular than the on-axis response due to the physical size of the unit becoming comparable to the wavelength at high frequencies. A solution is sought by using two capsules of different sizes with an electrical cross-over. Recently Sanken of Japan have produced a studio condenser microphone with two capsules mounted closely one above the other. 1 μm titanium diaphragms are used, the response being claimed as flat to 20 kHz with a good cardioid polar curve up to substantially the same frequencies (Figure 1.22).

The two capsules are designed to have optimum response over the low-to-middle frequencies and the middle-to-high frequencies respectively, their outputs being combined in an electronic cross-over network.

1.10.4 Moving-coil microphones

Omnidirectional microphones

For Moving-coil dynamic types with velocity transduction,

$$e = Blv$$

where e = e.m.f.
B = magnetic flux density in gap
V = velocity of coil
l = length of conductor.

Units normally have near-cylindrical cases with diaphragm close to the front face behind protective grilles or openings, with bass equalizing tube/barometric leak and connector at the far end. Some earlier units used a spherical case to minimize diffraction errors.

The diaphragm system is typically formed from plastic or thin metal foil with a domed centre and a fluted or roll surround, the outside diameter usually 17–30 mm, Coil diameter 12–15 mm.

The sensitivity is basically limited by the diaphragm size and the available centre-pole magnetic flux density. The coil is wound from flat or round section aluminium or copper wire, self-supporting and cemented to the base of the dome. The resonance of the moving system in free air is 200–500 Hz, but the overall

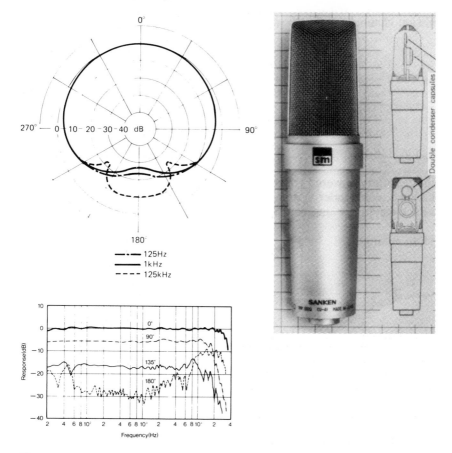

Figure 1.22 *Double condenser capsules. (a) Polar pattern; (b) frequency response.*

response is made substantially flat by adding acoustic resistance below the coil slot and elsewhere in the system, the whole being proportioned to function to this objective, by means of equivalent circuit analysis etc.

Analysis of a typical moving-coil microphone

Figure 1.16 shows the detailed construction of omnidirectional microphones with the various parts identified. The basic mechanical and acoustical system can then be approximated in the form of a 'flow diagram' from which the equations of motion can be derived.

Moving-coil microphones have strong cases and can be extremely robust to resist considerable abuse and adverse conditions of use. The coils are wound to an impedance of 30–200 ohms. The sensitivity is −60 to −70 dB relative to $1 V/Pa$.

The systematic detailed derivation of the analogous circuit of a moving-coil microphone affords an example of the analysis of a fairly complex mechano-acoustic system. A detailed sectional drawing of such a microphone is shown. From inspection of the mechanical layout, key mechanical and acoustic masses are identified which can be suitable for lumped constant representation. Each mass will oscillate about a mean position and will represent a resonant mode or 'degree of freedom' in the overall motion of the structure. If the section of the microphone is laid out across the page, then lines through the masses represent the degrees of freedom. It is convenient to take the '0' degree of freedom to represent the motion of the diaphragm and coil mass, the velocity of which is proportional to microphone output, given by $BL\dot{x}$ (B = radial flux density in the coil gap, l = length of coil conductor, $\dot{x}$ = velocity of coil perpendicular to the flux). The air in the holes of the front protective grille act as masses in the ' –1' degree of freedom, and form a resonator with the air volume before the diaphragm. Air masses are also identified in the coil slots and in other paths in the microphone case and rear access tube or apertures, giving other degrees of freedom (1, 2, etc.). The application of the well-known Lagrange's equation enables the equations of motion of the various masses to be derived as follows.

A functional diagram is drawn from a knowledge of the mechanical construction of the microphone. The various parts acting as masses are identified by reason of their ability to store kinetic energy, and their vibrational mode displacements are chosen as independent coordinates. The various constants in the form of lumped masses, resistances, and stiffnesses are identified, and the appropriate areas and linear velocities are designated. From Lagrange's equation

$$\frac{\mathrm{d}}{\mathrm{d}i}\left(\frac{\partial T}{\partial \dot{x}_n}\right) + \left(\frac{\partial F}{\partial \dot{x}_n}\right) + \left(\frac{\partial V}{\partial x_n}\right) = e_n(l)$$

where $e_n(l)$ = applied force for each coordinate;
$\dot{x}_n$ = linear coordinate velocities;
x_n = linear coordinate displacements;
n = –1, 0, 1, . . . ;
T = kinetic-energy function;
F = dissipational-energy function;
V = potential-energy function.

The kinetic-energy function is derived by inspection:

$$T = \tfrac{1}{2}M_{-1}A_{-1}^2\dot{x}_{-1}^2 + \tfrac{1}{2}M_0(A_0 + A_0')^2\dot{x}_0^2 + \tfrac{1}{2}M_1A_1^2\dot{x}_1^2$$
$$+ \tfrac{1}{2}M_2A_2^2\dot{x}_2^2 + \tfrac{1}{2}M_3A_3^2\dot{x}_3^2 + \tfrac{1}{2}M_4A_4^2\dot{x}_4^2$$

Similarly, for the dissipational-energy function:

$$F = \tfrac{1}{2}R_{-1}A_{-1}^2\dot{x}_{-1}^2 + \tfrac{1}{2}R_0(A_0 + A_0')^2\dot{x}_0^2 + \tfrac{1}{2}R_1A_1^2\dot{x}_1^2$$
$$+ \tfrac{1}{2}R_2A_2^2\dot{x}_2^2 + \tfrac{1}{2}R_3A_3^2\dot{x}_3^2 + \tfrac{1}{2}R_4A_4^2\dot{x}_4^2$$

and for the potential-energy function:

$$V = \tfrac{1}{2}S_{-1}[A_{-1}\dot{x}_{-1} - (A_0 + A_0')\dot{x}_0]^2 + \tfrac{1}{2}S_0(A_0 + A_0')^2\dot{x}_0^2$$
$$+ \tfrac{1}{2}S_1(A_0x_0 - A_1x_1)^2 + \tfrac{1}{2}S_2(A_0'x_0 - A_2x_2)^2$$
$$+ \tfrac{1}{2}S_3(A_1x_1 + A_2x_2 - A_3x_3)^2 + \tfrac{1}{2}S_4(A_3x_3 + A_2x_4)^2$$

The constants $M_{-1}, M_0, M_1, \ldots, R_{-1}, R_0, R_1, \ldots$ and $S_{-1}, S_0, S_1, \ldots$ are assumed to be in analogous acoustic impedance units and thus are multiplied by the square of the area concerned in order to give the equivalent mechanical constants.

Taking the x_{-1} co-ordinate, differentiating the equations with respect to $\dot{x}_{-1}$, x_{-1} and l as indicated and substituting in the appropriate equation, we obtain

$$j\omega M_{-1}A_{-1}\dot{x}_{-1} + R_{-1}A_{-1}\dot{x}_{-1} + (S_{-1}/j\omega)[A_{-1}\dot{x}_{-1} - (A_0 + A_0')\dot{x}_0] = p$$

Similarly, taking the x_0 coordinate, we obtain

$$\begin{aligned}
&j\omega M_0(A_0 + A_0')\dot{x}_0 + R_0(A_0 + A_0')\dot{x}_0 \\
&\quad + (S_{-1}/j\omega)[(A_0 + A_0')\dot{x}_0 - A_{-1}\dot{x}_{-1}] + (S_0/j\omega)(A_0 + A_0')\dot{x}_0 \\
&\quad + (S_1/j\omega)A_0/A_0 + A_0'(A_0\dot{x}_0 - A_1\dot{x}_1) \\
&\quad + (S_2/j\omega)/A_0'/A_0 + A_0'(A_0'\dot{x}_0 A_2\dot{x}_2) = 0
\end{aligned}$$

Taking the x_1 coordinate

$$\begin{aligned}
&(S_1/j\omega)(A_1\dot{x}_1 - A_0\dot{x}_0) + j\omega M_1 A_1 \dot{x}_1 + R_1 A_1 \dot{x}_1 \\
&\quad + (S_3/j\omega)(A_1\dot{x}_1 + A_2\dot{x}_2 - A_3\dot{x}_3) = 0
\end{aligned}$$

Taking the x_2 coordinate

$$\begin{aligned}
&(S_2/j\omega)(A_2\dot{x}_2 - A_0'\dot{x}_0) + j\omega M_2 A_2 \dot{x}_2 + R_2 A_2 \dot{x}_2 \\
&\quad + (S_3/j\omega)(A_2\dot{x}_2 + A_1\dot{x}_1 - A_3\dot{x}_3) = 0
\end{aligned}$$

Taking the x_3 coordinate:

$$\begin{aligned}
&j\omega M_3 A_3 \dot{x} + R_3 A_3 \dot{x}_3 + (S_3/j\omega)(A_3\dot{x}_3 - A_1\dot{x}_1 - A_2\dot{x}_2) \\
&\quad + (S_4/j\omega)(A_3\dot{x}_3 + A_4\dot{x}_4) = 0
\end{aligned}$$

Taking the x_4 coordinate:

$$j\omega M_4 A_4 \dot{x}_4 + R_4 A_4 \dot{x}_4 + (S_4/j\omega)(A_4\dot{x}_4 + A_3\dot{x}_3) = p$$

The equations are seen to be the mesh equations of the analogous electrical circuit. They show that the two separately acoustically loaded areas, represented by the rear surface of the diaphragm dome and the rear surface of the diaphragm surround, lead to the two ideal transformers, T_1 and T_2, which have turns ratios of $A_0/(A_0 + A_0')$ and $A_0'/(A_0 + A_0')$, respectively.

The input voltage applied to the network represents the sound pressure applied to the microphone front openings, whilst the current through M_0, the inductance representing the diaphragm and speech coil moving mass, is proportional to the e.m.f. generated or the response of the microphone. In this analysis, wave effects are assumed to be absent, and thus it is legitimate to assume that the same sound pressure is applied to the microphone front openings M_{-1} and to the rear tube M_4 openings, without any phase displacement, and also that the microphone main case volume can be represented by a single stiffness S_4. In practice, the fact that the leak tube is operative at frequencies below 100 Hz and the employment of distributed damping in the microphone case volume, mean that the approximation is valid.

It will be noted that the analogous circuit normally gives the pressure response rather than the free-field response of the microphone, as the effects of diffraction around the case cause a modification of the constant input voltage condition for the circuit.

1.10.5 The ribbon microphone

A flat moving conductor suspended in a transverse magnetic field can be regarded as the simplest type of electro-acoustic transducer. The combination of conductor and diaphragm in one element is obviously of basic value. Ribbon loudspeakers and earphones have long been used, but ribbon microphones are a more recent invention. The ribbon can be made from very thin light metal foil and meets one of the basic criteria for a microphone, which is that the moving system should be of low mass so as to allow a good transient response, and to be as near as possible a match to the air load. The most successful high-quality microphones have moving elements or diaphragms that are as light as possible. However, some important aspects of ribbon design must be borne in mind:

(1) A strong unidirectional magnetic field has to be maintained. This may involve massive permanent magnets applied to slender polepieces on each side of the ribbon. Such magnets may represent acoustic obstacles adversely affecting the response of the microphone. However, recently developed magnet alloys of much greater efficiency can effect a dramatic reduction in magnet size.
(2) A correct acoustic drive has to be applied to the ribbon to give pressure gradient or pressure drive, as required to give the desired response and efficiency.
(3) Ribbons, in common with any other mechanical elements exhibit a basic resonance as well as harmonics or higher modes of vibration, all of which have to be effectively controlled or damped. Ideally, if the ribbon is light enough and correctly proportioned, the resonances may be controlled by the radiation air load, augmented by suitably coupled acoustic resistance materials.

A microphone ribbon has a very low electrical resistance and, as a half-turn conductor, it generates a low voltage which requires amplification before being applied to an output line. There are designs that use plastic ribbons with multi-turn conductive tracks on the plastic base, similar to PCB techniques.[5] At present, high-quality ribbon microphones use fine beaten aluminium foil ribbons and a step-up transformer in the microphone. The voltage produced by even a closely addressed noise reducing type of ribbon microphone is still so low that some improvement in transistor noise technology would appear to be needed if transformerless ribbon microphones are to be realized. The frontal aspect ratio of the ribbon dimensions has been optimized to give shorter and wider ribbons compared to the long narrow ribbons of the earlier studio ribbon microphones, whose long 2 in (50 mm) ribbons caused considerable narrowing of the polar response in the vertical plane.

The more detailed design of ribbon microphones is given in subsequent sections.

Bidirectional microphones

Bidirectional microphones are pressure-gradient microphones responsive to the pressure difference between the two sides of a ribbon or diaphragm which is on the central axis of a symmetrical pole piece or back plate structure arranged so that maximum response occurs for 0° and 180° sound incidence, with a null or zero response at right angles on the axis of symmetry where the pressure resultant is zero (cosine polar curve). The pressure difference increases with frequency up to a maximum where the distance around the poles or structure is $\lambda/2$, with a falling succession of maxima and minima at higher frequencies. This type of microphone is subject to the maximum lower frequency proximity effect, due to the fact that a close source of sound approximates to a spherical rather than a plane wave generator. The basic physics shows that such a wave front contains an inverse square component dependent on distance and inverse of frequency (Figure 1.4). The consequent rise in bass response must be allowed for in use of the microphone or compensated in the design by extra damping etc. The dead 90° axis is a useful means of eliminating unwanted sound sources and the 'figure of 8' solid polar curve reduces random reverberation sound energy pick-up by over 60%. The mass control required of a ribbon system is liable to give a high response to vibrations, and the proximity effect gives enhanced wind and pop sensitivity. The ribbon material must be very thin (0.6 μm approximately) or it will show a sequence of resonant modes (Figure 1.23). Ribbons are now 1 in (25 mm) long and ⅛ to ¼ in (3 to 6 mm) wide. They must be laterally stiffened along the whole length to avoid edges curling up. The ends are clamped by the terminal blocks, the ribbon being lightly tensioned, stretching the corrugations slightly. All the resonances are damped out by applying woven metal or fabric gauzes to the whole length of the pole pieces, back and front. Increasing the value of these acoustic resistances overdamps the ribbon and reduces the bass response to any desired extent so as to compensate the proximity effect at any particular talking distance. This is done for very close talking distances in so-called lip microphones, to give a considerable reduction of low- and middle-frequency background noise. The side slits representing the clearances between the ribbon and the poles are usually about 0.007 in (200 μm). The air leakage through these slits causes an absolute loss of sensitivity.

The lightness and the low acoustic impedance of the ribbon mean that it is basically most suitable for an indoor studio microphone and precautions have to be taken to avoid air movements such as severe draughts, or mechanical vibration of the microphone mountings. Both can give rise to heavy low-frequency rumbling noises in a studio ribbon microphone.

The basic theory of ribbon microphones analyses a transducer of the type in which a light rectangular ribbon is suspended between two rod-like pole pieces energized by a permanent magnet across their ends, so that a transverse magnetic field is maintained across the ribbon. This has free access to the sound field in

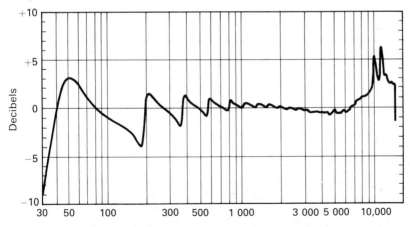

Figure 1.23 *Motional impedance curve of a thicker foil ribbon showing resonances insufficiently damped by the air load*

both forward and backward directions. As a sound wave strikes the poles, it flows around them. The fact that there is a mean effective distance d between the front and back surfaces of the ribbon ensures that there is, at any instant, a phase displacement between the acoustic pressures p_1 and p_2 at the front and rear of the ribbon. The resultant pressure difference $p_1 - p_2$ is available to move the ribbon. Such a pressure difference related to a distance is a pressure-gradient vector.

In mathematical terms, for sine waves,

$$p_1 = p_0 e^{-j\frac{\omega}{c}x}$$

$$p_2 = p_1 \left(1 - j \frac{\omega}{c} d \cos \theta \right)$$

therefore

$$p_1 - p_2 = j \frac{\omega}{c} p_1 d \cos \theta$$

where $\omega = 2\pi \times$ frequency
         d = acoustic distance around poles
         c = velocity of sound
         λ = wavelength of sound
         θ = angle of sound incidence

d is assumed to be small compared to λ. Thus, to a first approximation, the driving pressure on the ribbon is proportional to frequency. A flat frequency response requires the ribbon to be driven at constant r.m.s. velocity at all frequencies in accordance with the usual expressions:

$$e = BLv \times 10^{-8} \text{ volts}$$

$$v = \frac{P_1 - P_2}{z_m A}$$

where B = magnetic flux density (gauss) between pole pieces
         L = length of ribbon, cm
         v = velocity of ribbon, cm/s
         A = area of ribbon
         z_m = acoustic impedance of the ribbon, acoustic ohms.

If z_m is substantially a mass reactance $j\omega M_0$, then from the equations,

$$v = \frac{P_1 d}{c M_0 A} \cos \theta$$

where M_0 = mass of ribbon. In this case, the ribbon velocity is constant at all frequencies and the polar response is a cosine function of θ.

The general layout is illustrated in Figure 1.24. The basic design requirements for a ribbon microphone now become clearer. The ribbon must be very flexible and must have its fundamental resonance at the lowest bass frequency, so that it approximates to a mass-controlled element over substantially the entire audio-frequency range. The magnitude of the driving pressure $p_1 - p_2$ actually follows a sine function as the frequency increases, reaching a maximum at a frequency such that the distance d represents a half-wavelength. $p_1 - p_2$ becomes zero when

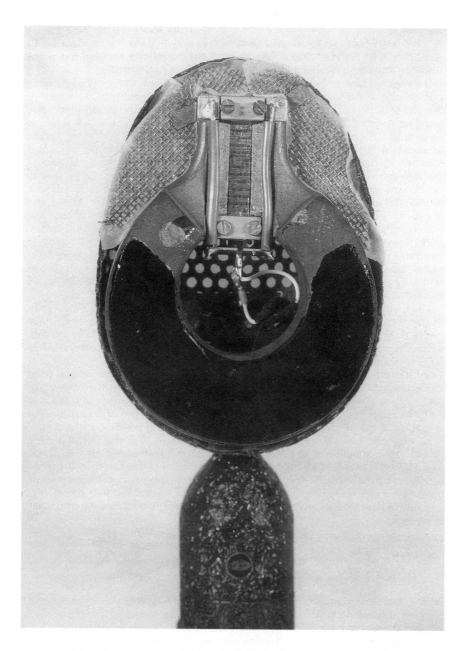

Figure 1.24 *Modern studio ribbon microphone with thin beaten foil ribbon. The acoustic resistance baffles around the poles enhance the low frequency response. The damping gauzes applied each side of the ribbon are omitted. The symmetrical hum-bucking connections to the top ribbon connections are shown. (Courtesy Coles Electroacoustics)*

a frequency is reached such that d is a complete wavelength. We can summarize by saying that the ribbon must be made to behave as a simple mass; all its resonances must be completely controlled, and the shape of the pole pieces, magnet, outer case and so on must be designed so as to make the incident sound wave produce a constant ribbon velocity at all frequencies; i.e. the tendency of the response must be compensated by various constructional artifices.

It is well known that the polar response exhibited by a simple ribbon microphone, as the source of sound is moved round it, is a bidirectional cosine or 'figure eight' curve in both horizontal and vertical planes, the solid polar figure being represented by two spheres touching at a point representing the origin of the graphs. This is a very useful sound pickup pattern, in that performers can be arranged within an angle of ± 60° on either side of the microphone, unwanted sounds sometimes being largely excluded by arranging that the 'dead' 90° axis of the microphone is in the plane in which the unwanted noises (often due to ventilators, air-conditioning units, and the like) may lie. Also random reverberant sounds and general background noises, which almost invariably are excessive in normal rooms and auditoria, are reduced by the figure eight polar curve.

It is easily demonstrated that the bass response of a pressure-gradient microphone rises considerably when it is used at close quarters to a relatively small source of sound such as a human voice. Microphones of this type are often called 'velocity' transducers because the velocity of the ribbon when operated on by the pressure gradient corresponds to the air particle velocity in the sound wave over a fair part of the frequency range.

The first points to be settled in the design of the microphone were the details of the ribbon itself. Its area, aspect ratio and the side clearance slits at the pole pieces were fixed. The slits are usually 0·007 in wide to allow safe clearance with a thin ribbon. A high sensitivity demands that the area of the ribbon should be large compared to that of the side slits, the final area being fixed at slightly over $1\,\mathrm{cm}^2$. The effects of diffraction and wave phase interferences on the vertical polar curve restrict the vertical length of the ribbon to 1 in, its width then being just under ¼in. A good sensitivity demands a magnetic gap flux density of 6000 Gs or 0.6 T. Research by various authorities has shown that the use of a part-toroidal ring form of magnet is an economical way of generating the flux across a wide (¼in) magnetic gap. Having fixed the general proportions of the microphone, a considerable amount of development is needed to ensure that the ribbon functions correctly; the frequency response of the microphone can then largely be controlled by the shape of the pole pieces and certain subsidiary parts such as the outer case and some gauze baffles, which are fitted around the pole pieces.

A microphone ribbon behaves like an elastic bar and its fundamental resonance is given by

$$f = \frac{3.56}{L^2} \sqrt{\left(\frac{E\kappa^2}{\rho}\right)}$$

where zero tension is assumed, and

  L = ribbon length, cm
  ρ = density of material
  E = modulus of elasticity of material
  κ = radius of gyration of the ribbon cross-section.

It is essential that all resonances from the fundamental upwards are critically damped, otherwise the transient response, and to a lesser extent the frequency response, of the microphone will be affected. Any lack of damping of the vibrational modes of the ribbon means that the modal frequencies are shock excited by transients and can persist as an unwanted 'tail' to the reproduced transient. Unless it is very carefully made and mounted between the poles without any tendency to twist or to move in any direction other than perpendicular to its plane, a ribbon may exhibit torsional or lateral modes which generally are not effectively damped by air loading and hence may cause large irregularities in response. As a large part of all programme matter is transient in nature, the seriousness of a poor transient response is apparent. Two microphones may have apparently similar frequency response; one will sound clean and smooth, while the other, with a poor transient response, will sound rough and harsh. Unfortunately, it is difficult to make a direct check on the transient response of a microphone, as an acoustic source or loudspeaker with a near-perfect transient response is usually unobtainable.

The easiest indirect general assessment of a ribbon-microphone response can be made by examining its electrical 'motional impedance'. This is obtained by examining the impedance looking into the microphone terminals with a special bridge which balances out the 'static' component of electrical impedance, leaving unbalanced only the dynamic component, which is a faithful replica of the net effective motion of the ribbon at the test frequency. Figure 1.23 shows motional-impedance frequency runs on an unsatisfactory ribbon and shows the impedance of a satisfactory ribbon. This ribbon is made from a soft aluminium leaf material $\frac{1}{40}$ mil (0.6 μm) thick. The radiation resistance load of the air and the coupled acoustic resistance of the protective wire gauzes on the poles give a smooth curve, indicating critical damping of all the vibrational modes of the ribbon. The response of a thicker ribbon made of harder material shows the number of undamped resonances which an unsuitable ribbon can exhibit.

However, it has been found possible to make satisfactory ribbons for particular classes of microphone using thicker rolled aluminium foil (5–10 μm thick) by forming the centre area by transverse curvature with a few corrugations at the ends, the system then approximating to a rigid bar flexibly mounted at the ends (Figure 1.25).

The ribbon is clamped to electrical terminal blocks at the top and bottom, its back and forth movement in the pole-gap magnetic field generating an e.m.f. Its electrical resistance is usually about 0.25 ohm, and thus a small input transformer has to be mounted closely in order to step up this e.m.f. and match the ribbon to the usual amplifier input impedance (30 ohm, 600 ohm etc.) The transformer must be fitted with a magnetic shield, and wiring between it and the ribbon must be arranged in a 'humbucking' layout to neutralize the effects of any stray electromagnetic fields originating from power transformers in the vicinity of the microphone. If these precautions are not taken, the ribbon and its wiring form a considerable loop and the microphone is likely to pick up hum.

Figure 1.24 gives a view of the high-quality ribbon microphone with part of the outer case removed so as to show the construction.

The ribbon is very light, its weight being 0·2 mg, and this, combined with the very effective acoustic damping, gives the microphone an excellent transient response. A flat frequency response and high efficiency have also been achieved by careful proportioning of the pole pieces and the magnet, the former being precision castings in vanadium permendur, while the latter is a high-energy-content magnet alloy. The driving force on the ribbon is approximately proportional to $d \cos \theta$ where d represents the effective distance corresponding to the sound pressure-gradient force, θ being the angle of sound incidence relative to the normal. A flat response results if d is small compared to the wavelength of sound, so that the pressure-gradient force is proportional to frequency, and if the ribbon acoustic or mechanical impedance is mass-controlled.

Several difficulties are encountered in practice. First it is difficult to assign one simple value to *d*, as it is impossible to make the pole-piece cross-section the same all the way down. The poles have to be enlarged at the bottom to meet the necessarily greater area of the magnet face. The net result is that the gradient distance *d* tends to have a larger value at the bottom of the poles than at the top. The larger distance gives a gradient driving force of larger magnitude, but with a high-frequency cut off at a lower frequency than is given by the smaller top

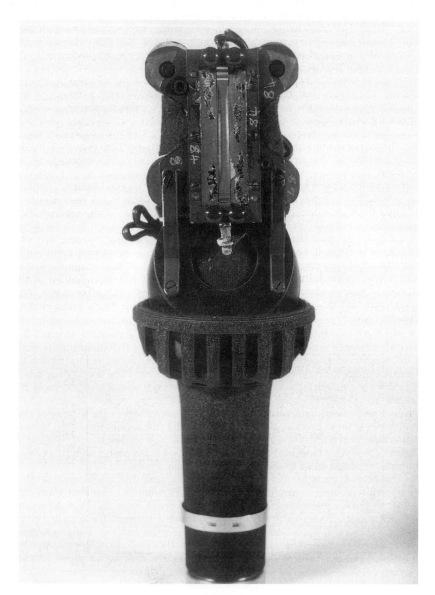

Figure 1.25   *Dual unit ribbon microphone using a thicker ribbon with laterally curved central portion and end corrugations*

distance. Such ribbon microphones are likely to have a range of several decibels in magnitude in their frequency response at middle or high frequencies. In the present design, the pole piece and magnet proportions are such that the effect is quite negligible. The response tends to fall at frequencies below about 150 Hz owing to the introduction of resistance control rather than mass control: a result of the critical damping of the natural resonances of the ribbon. A porous gauze baffle with a fairly low value of acoustic resistance is fitted around the pole pieces and magnet window in order to compensate for this loss. The baffle performs this function because the constant pressure difference built up between its front and rear surfaces at all frequencies is a significant addition to the small low-frequency pressure differences produced across the ribbon surfaces by the pole pieces, but is negligible compared to the much larger pressure differences produced by the pole pieces at higher frequencies. The natural tendency for the high frequencies to fall off as the distance d approaches a half-wavelength is compensated by a broad resonant cavity gain due to the pole-piece chamfer, and also by a standing-wave reflection gain due to the spacing of the case from the ribbon plane, the acoustic impedance of the holes in the fluted upper part of the case having a preferred value. The frequency response of a similar type of microphone is given in Figure 1.26. The total harmonic distortion has a maximum value of less than 0·1% for the vast majority of frequencies and sound levels normally encountered in operation.

This type of ribbon microphone is widely used as a general-purpose studio microphone for all types of music. The response is well-balanced, the linearity is good and the bidirectional polar response reduces excessive studio reverberation and noise.

Some more compact studio and general-purpose ribbon microphones are now available. It is difficult, however, to reduce the size and weight to any great extent without sacrificing a considerable amount of sensitivity, as large magnets need to produce the required flux density in a relatively wide gap. The availability of the latest high-energy permanent magnets such as neodymium iron boron can offer considerable advantages for ribbon microphones, as well as for other electro-magnetic types.[11]

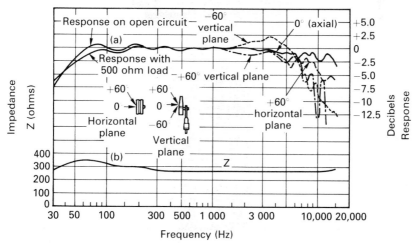

Figure 1.26 *The response and impedance characteristics of a studio ribbon microphone.*

High-quality noise-reducing ribbon microphones for close talking

It is often desired to broadcast or reproduce speech from noisy or highly reverberant surroundings where the conditions are so severe that the use of normal types of microphone is largely ruled out owing to the high level of speech masking, or by the annoyance caused by a high acoustic background noise level at the microphone position. This sort of condition is encountered in communication systems where orders have to be passed from flight decks or aerodromes in the vicinity of aircraft, from moving vehicles, from operators under conditions where other operators are talking independently alongside, etc. Outside broadcasts may demand commentaries of high speech quality from ringside seats or other positions surrounded by excited crowds, from persons interviewed in streets, where a high traffic noise level is present, or from moving vehicles.

The best solution is usually to employ microphones which are specially designed for very close talking and which confer a measure of background-noise reduction by means of a differential response to close and distant sounds. A specially designed type of gradient microphone is generally used. First- or second-order gradient microphones give improvements of 4·8 and 7·0 dB, respectively, in discrimination against randomly incident background noise by virtue of the cos θ and cos² θ respective polar responses, as compared to an omnidirectional microphone. The proximity effect can also be exploited in order to achieve a useful additional discrimination against distant noises. The plane-wave or distant-sound response of a first-order gradient microphone can be curtailed by 25 dB at 200 Hz (or 55 dB for a second-order gradient microphone), the close-talking frequency response then being level for a small source at ¾in (19 mm).

The general properties of first-order gradient microphones have already been described. Small gradient units have been designed to give a near approach to these theoretical noise-reduction figures with speech of acceptable communication quality. One of the major difficulties is the elimination of breath and explosive 'pop' noises from speech at very close range. It is also difficult to get natural-sounding speech unless the sound radiated from the nostrils is adequately reproduced. Generally, in small communication-type units, it is not possible to shield the microphone against breath noises and it is not possible to ensure that the microphone is always sufficiently accurately positioned with regard to the nose and mouth.

High-quality close-talking microphones have been designed so that broadcasting commentaries etc. can be given under conditions where the level of background noise is such that the microphone must be used very close to the mouth. As mentioned above, this involves special problems in securing natural-sounding speech reproduction, and any noise-reducing properties which can be conferred on the microphones are invaluable.

In studying close-range speech, strictly, account must be taken of the fact that there are four sound sources concerned: the mouth, the nose, the throat and the chest, the last two being predominantly low-frequency sources. Other problems are concerned with breath noises and the effects of explosive speech sounds e.g. certain consonants, such as 'P'. In addition, the frequency spectrum of sound energy in close-range speech differs from that in speech at normal distances, various subjective experiments indicating that a slight rise in the high-frequency response is desirable between 4 and 7 kHz.

Figures 1.27 and 1.28 show a close-talking noise-reducing commentator's microphone, which has been developed for broadcasting. It is a gradient ribbon unit with a high degree of acoustic damping in the form of silk gauzes mounted close to the ribbon. It is fitted into a special case equipped with nose and mouth

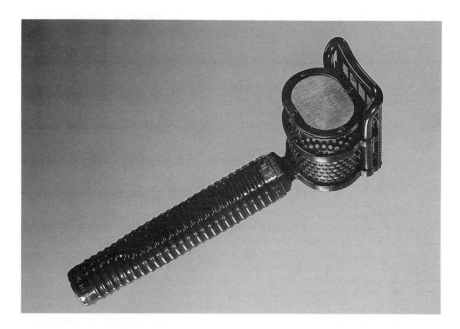

Figure 1.27 *A close talking noise reducing ribbon microphone. (Courtesy Coles Electroacoustics)*

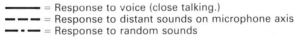

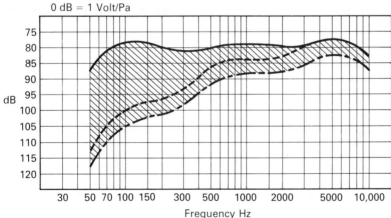

Shaded area gives discrimination against background noise (random)

Figure 1.28 *Response of the microphone in Figure 1.27. (Courtesy of Coles Electroacoustics)*

breath shields made from stainless-steel woven mesh, the talking distance with respect to the nose and mouth being accurately controlled by means of a positioning bar which contacts the talker's upper lip. Noise reduction occurs firstly because the bidirectional cosine polar response affords a theoretical reduction of about 66% in the pickup of random noise energy, and secondly because the proximity effect of a first-order gradient microphone of these proportions is such that, in order to get a flat frequency response at the correct talking distance of about 2 in (50 mm) between the lips and the ribbon, the free-field frequency response has to show a fall of about 30 dB between 50 Hz and 3 kHz, thus greatly attenuating low- and middle-frequency noises.

1.11 Acoustic boundary microphones

It has been known for a long time that a microphone placed in the vicinity of a large reflecting surface such as a table top has its response modified to a possibly serious extent by the interaction between the direct and reflected wave at the microphone pick-up point, producing a comb-filter effect.[1] Acoustically transparent studio table tops have been used to avoid this effect. A few years ago it was pointed out that if the microphone was mounted flush in the surface the deleterious effects are avoided, the microphone effectively responding over a full hemisphere upwards with a 6 dB gain in local sensitivity at the reflecting surface. If the microphone effective point is less than 0.4 in (1 cm) from the surface, comb filter effects will not start below 20 kHz. Small omnidirectional condenser microphones can be let into the surface of a wooden square about 0.5 in (12 mm) thick, as this type of unit is relatively insensitive to any mechanical vibrations that may occur on floors or table surfaces. Small unidirectional microphones have also been used in some cases, standing up slightly from the surface and angled horizontally rather than vertically, to give an unobtrusive and accurate sound pick-up of a speaker at a table, with a half-cardioid pick-up pattern. Alternatively, small microphones have been housed in a minimum-sized casing to give an unobtrusive 'mouse' microphone. A small diameter (< 1 cm) microphone, let in flush to the surface of a small flat baffle set on the boundary surface of a room, has a substantially uniform + 6 dB frequency response over the full hemisphere of frontal frequency sound incidence. This often gives a very good subjective impression of the sound ambience in a room, the free field and the diffuse field response being substantially the same for such a microphone. They may also be used as spaced microphones in an A − B stereo set-up, as much of the spatial impression is considered to be due to the horizontal wave propagation in a room.

An interesting variant of the boundary-layer microphone is the spherical-case microphone. This uses small omnidirectional condenser microphones let in flush to the surface of a sphere of about the size of a human head; in effect two boundary-layer microphones at 180°. Tests show that this spherical microphone may produce stereo signals with natural interaural differences similar to dummy head signals, which however combine at the listening loudspeakers to give auditory clues which are more compatible with natural hearing.[6]

Commercial boundary-layer microphones generally have to be housed in a rigid baffle or a wooden plate of a thickness suitable to accommodate a small omnidirectional condenser microphone cartridge let in flush with the surface. The surface of the plate then projects slightly above the floor or wall surface on which it is mounted. Acoustic diffraction wave effects occur at the discontinuity represented by the edges of the plate and these may interfere with the direct wave to cause irregularities in the response of the microphone. These effects are

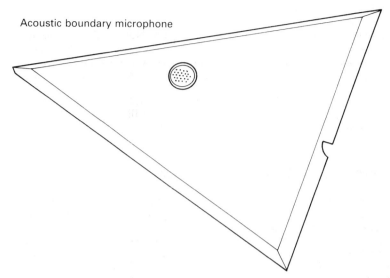

Acoustic boundary microphone

Figure 1.29   *The plate of the Neumann boundary-layer microphone is designed so that at the location of the electroacoustic transducer the secondary sound field resulting from diffraction at the edges of the plate has, in aggregate, a frequency independent phase relationship and a frequency independent level, aiding a smooth response.*
This is achieved by distributing the path lengths from each edge point to the centre point of the transducer evenly for all wavelengths. (Courtesy Neumann)

additive if the microphone is central in a symmetrical plate. Figure 1.29 shows a microphone cartridge mounted off-centre in a triangular plate so as to minimize diffraction effects and to give a smoother response.

It may be noted that spherical microphone cases have been used on some of the classical studio moving coil microphones evolved in the 1930s because this form gave less severe diffraction irregularities in the response than are encountered with the more usual cylindrical form of case. It is now considered worth while housing studio omnidirectional cartridges on the surface of a sphere so as to obtain a smoother response for sound incident in off-axis directions, in effect a smoother random incidence or reverberant sound pick-up (Bruel & Kjaer).

1.12 Lavalier microphones

These microphones cover a wide class of units that are mounted on a neck cord or tie/lapel clip, or otherwise attached to the person in the vicinity of the chest. This is convenient for many types of programme and a number of different designs of microphone are made for this purpose. Unless some care is taken in the design and mounting of the units speech quality may be inferior. Curves have been produced showing how the higher frequencies fall off both in the horizontal and vertical planes through the centre of the mouth. It is also noticed that pop effects are reduced at 30–45° off the central mouth axis. Another effect is that the surface of the chest can radiate appreciable amounts of middle frequency sound.

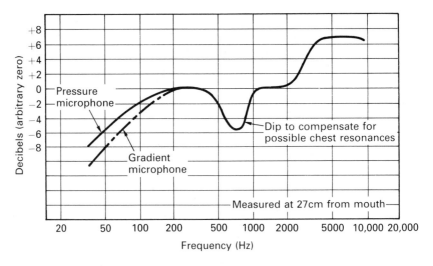

Figure 1.30 *The desired response for a pressure Lavalier microphone for average talkers. Note the bass roll-off, the dip at 800 Hz and the top rise.*

Curves have been produced to compensate for this by introducing a dip of about 6 dB between 500 and 1000 Hz (Figure 1.30). There is also bound to be considerable variation in response with inclination or turning of the head, as well as between individual male and female voices. Better class microphones designed for the purpose are given an optimum median type of response curve sometimes with an IRT dip and a high-frequency rise between about 1.5 and 10 kHz. It is also desirable to give a low-frequency roll-off to compensate for proximity effect if a gradient type of microphone is used. Bidirectional units have been used to reduce background noise by orienting the maximum pick-up axis towards the mouth and the dead axis towards any prominent or over-loud sounds in the vicinity. One example is that of a singer near to loud percussion. In view of the possibility of considerable individual variations it has often been found worth while to apply some equalization in the control desk etc. at rehearsal. With such measures and the use of good specialist microphones, it is possible to get good speech or vocal quality. This is often particularily valuable when a radio microphone is being used in a prestiguous TV set-up. The question of clothes rustle and other vibrational pick up is also important. Careful fixing and low-noise highly flexible cables and the use of small omnidirectional Lavalier microphones help in this respect. Some of the above considerations apply to boom-mounted microphones on headsets where good-quality speech is required.

1.13 Highly directional microphones

The original directional line microphone was evolved in the 1930s by WE and RCA in the USA as an alternative to earlier directional microphone arrays which achieved directionality through their planar size compared to the wavelength. These early arrangements consisted of a large number of omnidirectional microphone units arranged over a large flat disc or along a line to produce obstacle or diffraction type directionality in two or three dimensions, the curves being dependent on frequency to a serious extent. The parabolic reflector is an extension of this approach, using one microphone at the focal point, with a considerable increase in efficiency.[7]

Multitube units are bulky and expensive to make, and studies were made on the use of a single tube with holes spaced along the length. It was seen that acoustic damping was needed along the length of the tube, both covering the holes and in the bore of the tube, and that this resistance had to be graded so as to increase from the far end of the tube up to the microphone diaphragm, so as to avoid losses due to sound leaking out from later holes and to ensure equal contributions arriving at the diaphragm from all inlets. This was worked out in some detail by Tamm and Kurtz in Germany.[8,9] Production microphones of this type are made by Sennheiser and others. The basic action of line microphones can be seen: if sound waves incident along the microphone axis travel with the same velocity inside the tube as in free air, then all waves incident at angles near to the 0° axis will arrive at the diaphragm in phase, whilst waves arriving at angles up to 90° will enter the row of holes and will arrive substantially out of phase at the diaphragm, causing cancellation. Figure 1.31 shows the basic construction and vector relationships. The directivity is reduced at low frequencies, where the wavelength is comparable to or exceeds, the tube length. This effect is partially compensated by the use of a unidirectional microphone capsule. The longest units made were about 5 ft (1.5 m) in length, giving directionality down to 200 Hz. This is inconveniently long for operational use and most modern line microphones range from about 2 ft (0.6 m) to 10 in (0.25 m). These give a useful degree of directionality combined if necessary with some bass cut. Some short units use slender tubes and small cardioid capsules to give inconspicuous and convenient line microphones.[1]

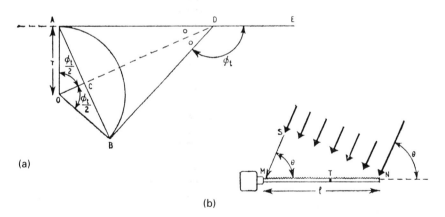

Figure 1.31 *The principle of a highly directional line microphone. (a) Shot-gun line microphone vector diagram showing the resultant pressure on the diaphragm at angle of incidence* θ; *(b) diagram showing how the phase angle between the pressure contributions from the near and distant end of the line varies with angle of incidence* θ.

At any angle of incidence θ *pressures enter the slot along its whole length and the resultant pressure* $p_θ$ *on the diaphragm is represented in (a) by the chord AB. AD and BD are tangents to the arc and indicate the phase of the pressure contributions which enter. The resultant pressure* $p_θ$ *is represented by the chord AB.*

Tubular line directional microphones are now made by several makers, including AKG, Beyer, EV, Sennheiser, Sony, RCA etc. The units range in length from about 28 in (700 mm) to 10 in (26 mm). The microphone used at the tube end is either a small dynamic or a miniature cardioid condenser microphone. Foam or mesh type overall wind shields are available which can give the microphones a fair degree of protection against wind and weather. The main forward response lobes tend to vary with frequency, being fairly sharp at higher frequencies and tending to a cardioid at low frequencies. Minor unwanted side and back lobes have been fairly well suppressed by careful design but may still not be entirely negligible. The shorter line microphones are often used on booms or as hand microphones in audience participation shows, etc. The broadening of the polar pick up at lower frequencies can be combatted by using a bass cut filter below about 200 Hz. The on-axis response is generally not adversely affected, due to carefully arranged damping in the tube.

We may summarize the present position by noting that there is quite a wide range of units available in the various categories described above. The strengths and limitations of each category are apparent. A careful choice of microphone for any particular application is required.

1.14 Integrated circuit microphones

Historically there have been a number of attempts to integrate a microphone transducer with an amplifying stage in a composite substrate or chip, so as to achieve an adequate output signal at a low impedance line level to suit modern cabling and mixers etc. In the 1950s Olson produced a diaphragm system driving a pivoted beam electrode set in the glass of a vacuum tube so as to modulate the electron flow. Efforts were also made to exploit the stress sensitivity of a transistor by a diaphragm and sapphire pin applied to the emitter region of a transistor chip (Sikorski)[5] and a similar contact loading of a tunnel diode (Rogers) in the 1960s. Later directly actuated strain gauge bridge systems have been used. In the late 1970s Sank designed a range of piezo-resistive silicon strain gauges on a flexible substrate which responded to direct pressure actuation. They were manufactured by National Semiconductor and claimed a flat response to 20 kHz. The sensitivity was adequate for close-range speech and acoustic measurements at similar sound levels. The chip was less than ⅛in (4 mm) in size and when supplied with a DC bias gave a low-impedance balanced electrical output. These gauges were environmentally robust.

There have also been proposals for making an electret condenser microphone on one side of a substrate with an FET transistor stage processed on the other side. This may be uneconomic because of the relatively large size (0.4 in – 10 mm) of diaphragm needed for general acoustic purposes. In view of the future need of highly automated mass production of commercial microphones it seems likely that attempts will continue to produce integrated designs – MSI concepts, possibly leading to A/D conversion effectively in the microphone to produce a direct digital output at a high enough frequency to enable optical cables to replace the present rather clumsy A/F cabling.

1.15 Specialist microphone types

The main types of microphone have been reviewed in the preceding pages. There are various microphones designed for specific specialized applications. These include measurements of sound pressure level (SPL), otological hearing measurements, hearing aids, communications of all kinds including telephone sets, headsets and booms, inside mask microphones, underwater microphones, units for extreme climatic conditions, hostile environments, etc.

Other sections of this book and selected references will give much information on many of the aspects mentioned above.

1.15.1 Bone conduction hearing and microphones

Audible bone conduction processes occur when the cochlear end organs in the ear are acted on by vibrations of the skull bones, particularly around the mastoid area. The result is to excite the auditory nerves and to communicate a sensation of sound to the brain. Vibrations of the skull occur when normal sound waves act on it and also when vibrations are set up by the talker's normal speech. This is why we hear a modified version of our own voice which is particularly noticeable if we close both ear canals while speaking. Bone conduction transducers are useful for communication in high-noise conditions where very effective ear defenders are mandatory to avoid ear damage and ear canal receivers are not desirable. Bone conduction audio vibrators have been found useful here and also in cases where damage to the middle ear ossicles etc. precludes any type of normal acoustic hearing aid. Bone conduction microphones have also been used to pick up the voice vibrations produced in the skull and teeth, by employing miniature piezoelectric accelerometers secured to the teeth or skull. Accelerometers are insensitive to normal airborne sound waves and may be very small in size to constitute a tooth microphone. An 'ear-com' transducer device has also been made for application to the mastoid bone area to give two-way communication. The mechanical impedance in the thinner areas of the skull such as the mastoid shows a minimum around 1–3 kHz but considerable overall equalization may be required. It has been suggested that a highly miniaturized system of telemetry might be used with a tooth microphone to avoid wires coming out of the mouth.

1.16 Electrical and radio interference

There are few spaces where microphones operate with complete freedom from interference of some sort. Electromagnetic and electrostatic fields are present from lighting, electric motors, equipment power supplies, control equipment. Radio and TV transmitters, monitoring equipment, recorders and other auxiliaries are all liable to be around. Microphones, cables and associated auxiliaries should be shielded and interference-suppressed in accordance with the best practice. Electromagnetic types such as moving coil and ribbon microphones should have 'hum-bucking' windings disposed around the sensitive axes, wired in series with the output. Electrostatic microphones must have continuous earthed casing, free from chinks and significant entry holes or cable sockets which could allow electrostatic induction to penetrate. Microphone cables should be balanced to earth and shielded or double shielded, as well as treated against internal static generation on being moved. Careful balancing of cable circuits with regard to earth cancels 'longitudinals' induced due to electromagnetic fields.

Standard induction loops of wiring are specified to enable known electromagnetic fields to be set up to check that microphones meet the specified standards for interference.

When microphones are fitted with switches, these must be of good quality and quiet in action without possibility of inadvertent action. Lines should be terminated in the 'off' position to avoid interference in the open-circuit position.

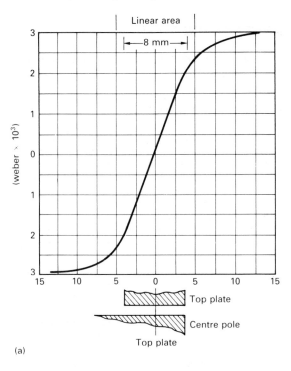

(a)

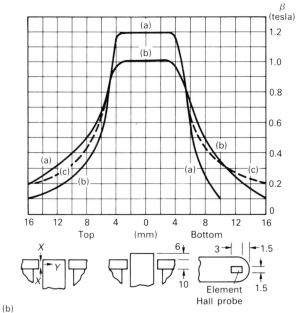

(b)

Figure 1.32   *Measurements by two methods of the flux density in a moving coil microphone gap. (a) Swept flux using a single search coil; (b) magnetic measurements using a Hall probe.*

1.17 Microphone appearance and finish

The design should generally be visually good, avoiding sharp edges and anything suggesting a forbidding or intrusive appearance, without light reflections.

1.18 Microphone responses and linearity

The salient characteristics of microphones from a fidelity point of view are frequency response, phase response and freedom from non-linearity, as these are the factors which define the ability to preserve the sound waveform. The first two are governed by the main microphone design and layout, in the manner which we have outlined. Non-linearity is a subtle factor which may involve a detailed study of the transducer elements concerned.

The maximum excursion of the diaphragm system due to peak signal levels must be estimated, to determine the linearity of the restoring stiffness and to check that any overloads due to surges, wind, etc. will not result in permanent elastic deformation or rupture.

The mounting tension of a diaphragm and the effect of any corrugations, dimples, or other rolls, ridges, flutes etc. have to be considered. For instance, studies of thin plates or foils as used in condenser microphone diaphragms can exhibit changes in the deflection curvature and linearity, according to whether the edge is clamped or guided. Moving coil units have roll or fluted surrounds for good linearity and centring, with a stiffening dome over the centre area. The height of the dome can have an acoustic effect,[10] whilst any undue tension or stretching of the surround due to faulty assembly or overload can produce drastic nonlinearity due to an 'oil-can bottom' effect, so that the diaphragm can flop between two unstable height configurations. Ribbons are corrugated along their length or at the ends with a centre cylindrical or fluted stiffening area (Figures 1.24, 1.25). The corrugations are stretched slightly so as to mount the ribbon with a suitable amount of tension. In some cases the maximum excursion of the diaphragm is limited by hitting a back plate, pole piece, or porous gauze front and rear damping covers, in the case of ribbons. These limits represent stops or clipping conditions. Except in cases where some degree of non-linearity occurs around the zero point, e.g. carbon microphones, magnetic hysteresis, etc., the systems are substantially linear up to almost the highest levels; shows that a stretched condenser microphone diaphragm deflected in a linear manner until a deflection corresponding to double the normal polarizing voltage caused the diaphragm to collapse on to the back plate. The normal output of a condenser microphone at an SPL of 130 dB would be only about 1 V compared to 100 V polarization.

Dynamic or ribbon microphones operate between magnetically polarized pole faces. The flux distribution around the zero position becomes non-linear at each extreme and represents a distortion limit (Figure 1.32).

References

1. Borwick, J. (1991) *Microphones: Technology & Techniques*. Focal Press, Oxford.
2. Hikaru (1985) A variable directional microphone on the Doppler principle. *Proc. 5th International Congress on Acoustics*, Liege.
3. Larsen, P. (1965) Comments on the zoom microphone. *J.AES*, **33**, p. 969, December.
4. Borwick, J. (ed.) (1988) *Loudspeaker and Headphone Handbook*. Butterworth-Heinemann, Oxford.

5. Sank, J.R. (1985) Microphone Survey. *J.AES*. **33,** July.
6. Thiele, G. (1991) Study of naturalness of 2 channel stereo sound. *J.AES*, **39,** p. 761, October.
7. Walstrom, S. (1985) Improved parabolic microphone reflectors for microphone directionality. *J.AES*, **33,** June.
8. Gayford, M.L. (1971) *Acoustical Techniques & Transducers*. Macdonald & Evans.
9. Boré, G. (1989) *Microphones for Professional Use*. G. Neumann, Berlin.
10. Knoppow, R. (1985) A bibliography of microphones. *J.AES*, **33,** July.
11. Magnetic Developments Ltd, Swindon, SN6 7NA, UK.

Further reading

Abbagnaro, L.A. (ed) (1979) *Microphones*. AES anthology series. AES.
Bauer, B.B. (1987) A Century of Microphones. *J.AES*, **35,** April.
Beranek, L.L. (1949) *Acoustic Measurements*. Wiley/Chapman & Hall, London.
Borwick, J. (ed.) (1987) *Sound Recording Practice*. 3rd edn. OUP/APRS, Oxford.
Brand, W. (1955) Non-linearity of condenser microphones. *B & K Technical Review*, No. 3, July.
Burnett, E.D. Free field reciprocity calibration of microphones. *Journal of Research*. National Bureau of Standards, USA.
Ellithorn, H.E. and Wiggins, A.M. (1964) Anti-noise differential microphones. *Proc. IRE*. February.
Fahy, F.J. and Holland, K.R. (1968) End fire acoustic radiator. Acoustic dual of the shotgun microphone. Calculation and modelling enable optimum size, spacing and number of holes to be specified. *J.AES*, **39,**
Falkus, A.E. (1960, 1963) Design of permanent magnet systems. *Wireless World*. January, July.
Flanagan, J.L. (1972) *Speech Analysis Synthesis & Perception*. Springer-Verlag.
Fletcher, H. (1953) *Speech & Hearing in Communications*. Van Nostrand.
Gayford, M.L. (1950) Application of dynamic analogies to microphones. *Akustica*. **4.**
Gayford, M.L. (1970) *Electroacoustics, Microphones, Earphones, Loudspeakers*. Butterworth-Heinemann, Oxford.
Gayford, M.L. (1986) Studio and noise reducing ribbon microphones. *Studio Sound,* September, October.
Gayford, M.L. (in press) Electro-mechanical analogies. In *The Audio Engineers Reference Book*. Focal Press, Oxford.
Gerlach, H. (1989) Stereo recording with shotgun microphones. *J.AES*, **37,** October.
Hunt, F.V. (1954) *Electro-Acoustics*. Wiley, Chichester. (Reprinted by JASA 1982)
Julstrom, S. (1991) Coincident stereo microphone patterns: ms and XY in 3 dimensions. *J.AES*, **39,** September.
Julstrom, S. and Tichy, T. (1984) Direction sensitive gating in multi-microphone systems. *J.AES*, **32,** July, August.
Lipshitz, S. (1986) Stereo microphone techniques; a review. *J.AES*, **34,** September.
Marshall, R.N. and Romanov, F.F. (1935) A non-directional moving coil microphone. *Bell System Technical Journal*, October.
Mason, W.F. *Piezoelectric Crystals & Ultrasonics*. Van Nostrand.
Massa, F. (1945) A working standard measuring microphone *JASA*, **17.**
Meyer, J. (1993) The sound of the orchestra. (Recent research on subjective blending and other effects in orchestral halls and on microphone techniques). *J.AES*, **41,** (4) April.
Negishi, H. (1993) Wide imaging stereo (WIS) technology. Audio mirror loudspeaker technique extends central listening area. *IEE Electronics & Communication*, **5** (1) p. 37, February.
Nisbett, A. (1993) *The Sound Studio*. Focal Press, Oxford.
O'Hara, G. (1967) Mechanical impedance & mobility concepts. *JASA*, **41.**
Olson, H.F. (1957) *Acoustical Engineering*. Van Nostrand.
Preese, K.E. (1965) Point contacts on silicon junctions. *Proc. IEEE*, **13.**
Preston, V. and Freeman, W. (1968) Electret microphones of high quality & Reliability. *J.AES*, **16.**
Robertson, A.E. (9163) *Microphones*. Illiffe Hayden.
Rumsey, F. (1989) *Stereo Sound for Television*. Focal Press, Oxford.

Shorter, D.E.L. (1959) Lavalier microphones. *Proc. 3rd International Congress on Acoustics,* Stuttgart.

Shulman, Y. (1987) Reduction of off-axis comb effects in shotgun microphones. *J.AES,* **35,** June.

Shultz, T.J. (1956) An air stiffness controlled condenser microphone. *JASA,* May.

Steinke, G. (1980, 1985) Sound delay system for large halls. *65 th AES Convention.* (Preprint No 1599). Developments using the Delta Stereo System. *77th AES Convention.* (Preprint No 2187).

Streicher, R. and Dooley, L. (1985) Basic stereo microphone perspectives. *J.AES,* **33,** July/ August.

Talbot-Smith, M. (1993) *Broadcast Sound Technology* Focal Press, Oxford.

Weingartner, B. (1966) Two-way cardioid microphones. *J.AES,* **14,** July.

Weinhofer, W. and Sennheiser, J. (1988) The measurement of microphone pop sensitivity. *J.AES,* **36,** June.

Woszczyk, W.R. (1992) Study of microphone arrays optimised for music stereo recording. *J.AES,* **40,** (11) p. 926, November.

Wuttke, J. (1992) Microphones and wind. Techniques and measurements on current windshielding structures and materials. *J.AES,* **40,** (10) p. 809, October.

2 Precision microphones for measurements and sound reproduction

Torben Nielsen

2.1 Introduction

After many years of intermittent scientific experiment in many countries various methods of measuring the magnitude of sounds were evolved. Some of these, such as the Rayleigh disc and particle movement detectors, may still survive, but more robust and convenient instruments have now become universal.[1,2]

For the measurement of sound wave pressure, the advantages of condenser microphones and, to some extent, piezoelectric microphones, have now become apparent, in that they can be made mechanically stable, they are amenable to various forms of absolute calibration, and can cover a wide frequency range.[3,4,5]

2.1.2

In the post-war period the development and use of electroacoustic measuring equipment advanced rapidly and the use of stable calibrated microphones became of central importance. The advantages of condenser microphones can be summarized as follows:

- Condenser microphones can be designed and made using durable metals and stable insulators by precise and repeatable engineering.
- Condenser capsules are made under controlled clean conditions and virtually sealed against the entry of foreign matter.
- The required characteristics can be realized to close tolerances and be maintained indefinitely.
- They are readily calibrated and checked in use. The effects of moisture and climatic changes can be guarded against and corrections applied if necessary.
- Condenser microphones can be designed to offer other useful attributes. They offer a constant electrical source impedance with a substantially constant charge from a polarizing or electret source, they are normally made in compact cylindrical form with virtually flush diaphragms, suitable for different forms of front cover, and offer convenient response corrections when required.
- Different sizes and constructional details can provide a wide frequency range. Low frequencies down to 0.01 Hz are possible by adjustment of the necessary small rear atmospheric leak, whilst small diameter microphones with a high diaphragm resonance can respond up to 150 kHz.
- They are largely insensitive to external electrical interference as they offer good electrical shielding and low electromagnetic coupling. They have a low seismic response to unwanted mechanical shocks and vibrations transmitted via mountings etc.

Table 2.1 *Microphone types*

Dimension	B & K microphone type				
	4144, 4145	4133, 4134, 4149	4147, 4148 4165, 4166	4135, 4136	4138
A	23.11	12.7	12.7	6.35	3.175
thread A	60 UNS-2	60 UNS-2	60 UNS-2	60 UNS-2	M 3.175 × 0.2
B	19	12.6	16.3	10.5	6.7
C	17	11.5	15.2	9	6
D	23.77 ± 0.02	12.7 ± 0.03	12.7 ± 0.03	6.35 ± 0.03	$3.175^{+0.05; \, -0}$
E	23.77 ± 0.02	13.2 ± 0.02	13.2 ± 0.02	7 ± 0.03	3.5
thread F	60 UNS-2	60 UNS-2	60 UNS-2	60 UNS-2	M 3 × 0.25
F	23.22	11.7	11.7	5.7	3
G	8.5	6	6	5	2.75
H	1.5	3.6	3.6	1	0.5
J	3.3	2.8	2.8	1.8	1.75
K	22.70	12.2	12.2	5.95	3.0

Figure 2.1 shows the construction of various sizes of capsule. Where national and international Standards exist, it may be noted that particular cartridge types will meet the requirements. Table 2.3, ANSI Standards, lists B & K cartridges which apply.

2.1.3

Modern conditions and developments such as digital sound recording have led to a demand for microphones of high precision for studio and concert hall broadcasting and recording etc. Condenser microphones have long been used and have great potential for these wider applications, and many improved types have been developed. Omnidirectional condenser microphones such as those described so far have often been used in 'dry' acoustic conditions where there is no excessive reverberation. Bruel & Kjaer have developed specialized omni and directional condenser microphones for these uses, with a specialized range of accessories, feed units and connectors to the audio systems in use. Figure 2.2 illustrates a range of studio condenser microphones. These are described in detail in Section 2.8 and illustrated in Figure 2.3, 2.4 and 2.5.

2.1.4

Piezoelectric microphones have long been used for sound measurement purposes, as well as for general sound reproduction. Natural piezoelectric crystals and piezo ceramics have the advantage that in suitable configurations their absolute sensitivity may be obtained from a knowledge of the mechanical piezo constants of the material, and they have been used to provide standards. However, when directly actuated in slab form, their sensitivity to sound waves in air is rather low, largely restricting their use to high sound levels. They are very suitable for use as hydrophones, as the material is a much better match to the medium and they have the advantage that they can be completely sealed.

Table 2.2 Microphone specifications

Cartridge type no.	Diameter mm (in)	Sensitivity (mV/Pa)	Pol. voltage (V)	Frequency range (± 2dB)	Frequency response characteristic	Associated preamplifier No.	Dynamic range* (dB)	Application areas
4138	3 (⅛)	1		6.5 Hz–140 kHz –	Pressure and random	2633 + UA0160 / 2639/45/60 + UA0036	2633: 56–168 / 2639: 55–168	Very high-frequency, high-level and pulse measurements
4135	6(¼)	4	200	4 Hz–100 kHz –	Free field and random	2633 or	2633: 39–164 / 2639: 36–164	High-level, high-frequency measurements and model work
4136		1.6		4 Hz–70 kHz	Pressure	2639/45/60 + UA 0035	2633: 47–172 / 2639: 43–172	High-level, high-frequency and coupler measurements
4129		50	0†	6.5 kHz–8 kHz	Free field and random	2639/45/60	2639: 15–142	General SPL measurements – IEC/ANSI – Type 2
4130		10	28	–		2642 or 2639/45/60	2642: 26–142 / 2639: 22–142	
4133, 4149		12.5	200	4 Hz–40 kHz –	Free field	2639 / 2645 / 2660	2639: 22–160	4133: Electroacoustic measurements. 4149: (Semi) permanent outdoor installations. 4134: Lab. & coupler measurements
4134				4 Hz–20 kHz –	Pressure and random		2639: 21–160	
4147	13 (½)	1.6	28	0.4 Hz–18 kHz –	Pressure and random	2639	2639: 38–160	Low-frequency and sonic-boom measurements
4155			0†	4 Hz–16 kHz	Free field		2639: 15–146	SPL measurements IEC–Type 1
4165		50	200	2.6 Hz–20 kHz –	Free field	2639 / 2645 / 2660	2639: 15–146	SPL measurements ANSI–Type 1
4166				2.6 Hz–10 kHz –	Pressure and random		2639: 15–146	
4176			0†	6.5 Hz–12.5 kHz –	Free field and random		2639: 14–142	SPL measurements IEC/ANSI–Type 1
4180		12.5		4 Hz–20 kHz	Pressure	2645	2645: 21–160	Lab. std., precision calibration
4144				2.6 Hz–8 kHz	Pressure	2639/45/60	2645: 10–146	Lab. std., low-level calibration General lab. use, low-level measurements
4145		50		2.6 Hz–18 kHz	Free field		2645: 11–146	
4160	25 (1)	47	200	4 Hz–8 kHz‡	Pressure	2645 + UA0786	2645: 10–146	Lab. standard, precision calibration
4179		100		10 Hz–10 kHz	Free field (compensated)	2660	2660: 2.5–102	System for very low SPL measurements (–2.5 dB(A))

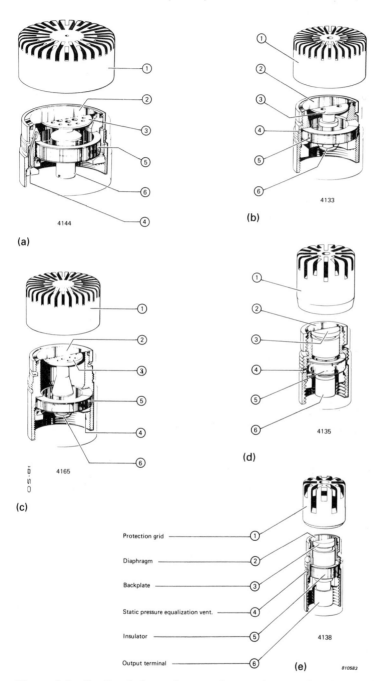

(a)

(b)

(c)

(d)

(e)

Protection grid ———————— ①

Diaphragm ———————— ②

Backplate ———————— ③

Static pressure equalization vent. ———— ④

Insulator ———————— ⑤

Output terminal ———————— ⑥

Figure 2.1 *Sectional views of measuring condenser microphones. (a) Types 4144, 4145 (1 "diameter) (b) Types 4133, 4134, 4149 (½" diameter) (c) Types 4147, 4148, 4165, 4166 (1/2 " diameter) (d) Types 4135, 4136 (¼" diameter) (e) Type 4138 (1/8 " diameter) (Courtesy Bruel & Kjaer)*

Table 2.3

ANSI type	B & K Type
XL	4144, 45, 47, 48, 33, 34, 49, 60, 65, 66
L	4144, 45, 60
M	4133, 34, 35, 36, 47, 49
H	4135, 36, 38

Individual Calibration

The microphones fulfil the American standard ANSI S1.12-1967 "Specifications for Laboratory Standard Microphones" as indicated

Piezoelectric materials in the form of cantilever 'Bi-Morph' piezoelectric sandwich elements connected to a conventional microphone diaphragm directly or by an actuating rod can offer a much better sensitivity in air and have been widely used for general sound reproduction and in sound level meters etc. Their relatively large capacitative source impedance gave a good match to the valve input amplifiers used in the past. Figure 2.6 shows the construction of one of the above types of piezoelectric microphone.

Modern piezoelectric air microphones and hydrophones are described in Section 2.7.

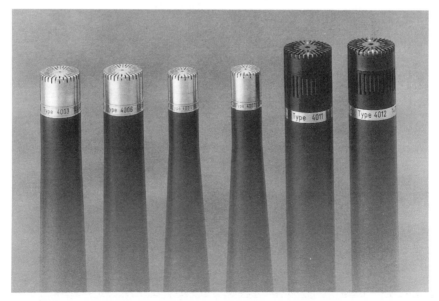

Figure 2.2 *Precision microphones for sound reproduction (a) Low noise omnidirectional; (b) high intensity omnidirectional; (c) unidirectional cardioid. (Courtesy Bruel & Kjaer)*

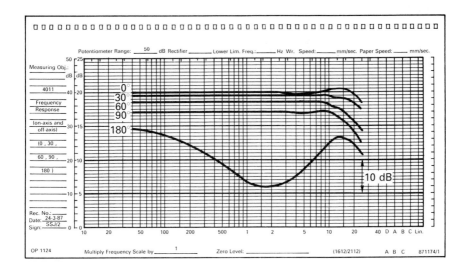

(a)

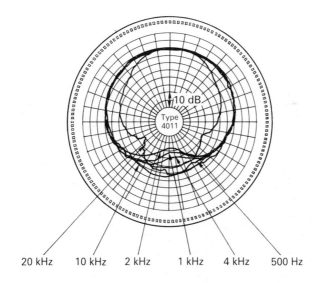

| 20 kHz | 10 kHz | 2 kHz | 1 kHz | 4 kHz | 500 Hz |

(b)

Figure 2.3   *Cardioid condenser microphone responses. (a) The on- and off-axis frequency responses of Type 4011 (measured at a distance of 30 cm); (b) The directional characteristics of Type 4011 (1 m distance). (Courtesy Bruel & Kjaer)*

Figure 2.4 *An omnidirectional studio condenser microphone. (Courtesy of Bruel & Kjaer.)*

Figure 2.5 *Cardioid studio microphone. (Courtesy of Bruel & Kjaer.)*

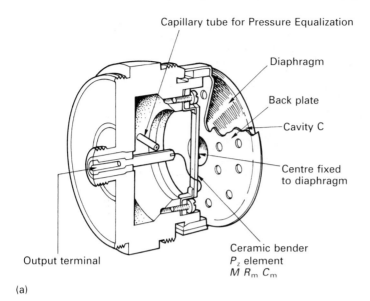

(a)

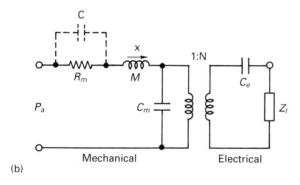

(b)

Figure 2.6 *(a) Cantilever bimorph piezoelectric microphone; (b) analogous circuit of (a). (Courtesy of Bruel & Kjaer.)*

2.2 Condenser microphone theory

The basic design and theory of measuring condenser microphones has been studied. They are primarily sound pressure operated transducers characterized by a highly tensioned conductive diaphragm stretched before an insulated backplate. A flat back plate carries a relatively high static charge, due to E_0. The equation of motion is

$$F_a = M_m \ddot{x} + R_m \dot{x} + \frac{x}{C} + Q^2/2\Sigma S \tag{2.1}$$

$F_a \sim F_e$ is the net force on the diaphragm system. F_e is constant. The mechanical system is thus effectively a simple resonant circuit, as shown in the equivalent circuit of Figure 2.7, which is completed by adding the rear vent components, etc.

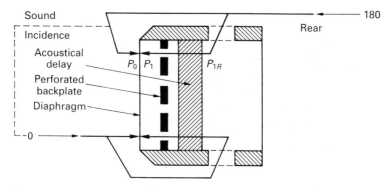

P_0 = pressure on the front of the diaphragm
P_1 = pressure on the back of the diaphragm
P_{1R} = pressure at rear of microphone

(a)

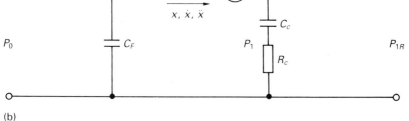

(b)

Figure 2.7   *(a) Mechanical layout of a condenser microphone with rear access; (b) analogous circuit. (Note. the rear access may be for barometric equalization or for phase shift.)*

The system is basically stiffness-controlled up to the diaphragm resonant frequency at the upper end of the response, giving a flat response.[6]

$$F_e = \text{the electrostatic force} = Q^2/2\Sigma S \qquad (2.2)$$

Q = the electrostatic charge between the diaphragm and the backplate. Q is effectively constant due to high insulation

F_a = the acoustic driving force. x, $\dot{x}$, $\ddot{x}$ are displacement, velocity and acceleration, respectively

C_m = capacitance

e = instantaneous voltage across plates

Σ = permittivity of dielectric

E_0 = polarizing voltage

S = effective area of the condenser plates

M_m, R_m, C_m = the mechanical constants of the diaphragm system

The Q^2 term in equation (2.1) shows it as a non-linear equation. It is apparent that if the charge Q can be kept substantially constant and at a suitably high level, the system can be made linear for all but extreme excursions. This is confirmed in practice. A stretched diaphragm exhibits a fundamental resonance and a series of higher resonant modes. The back plate is designed to cover as much as possible of the diaphragm rear area and applies a suitable amount of viscous air damping in the narrow slit and vent holes, thus critically damping all the resonances. For smaller microphones, the diaphragm resonances are higher in frequency and a correspondingly higher frequency range is covered.

The sections on microphone design show that a high and constant charge can be maintained by high values of insulation in the microphone and preamplifier. High precision assembly is also described.

An inspection of the equation of motion (2.1) shows the solutions for the displacement x and the velocity $\dot{x}$ given to the diaphragm by the sound pressure, as well as the air particle motions in the various passages and enclosed volumes within the microphone structure. These relationships show that the system can be represented by an equivalent or analogous electrical circuit, which is easier to interpret and to analyse than the original mechanical system, due to our greater familiarity with electrical circuits. Figure 2.7 shows the equivalent circuit of a

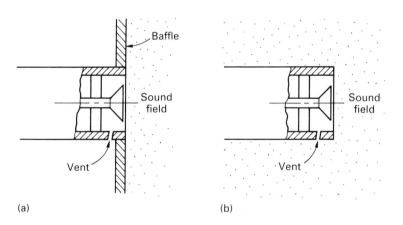

Figure 2.8 *Equalization vent (a) outside and (b) inside the sound field.*

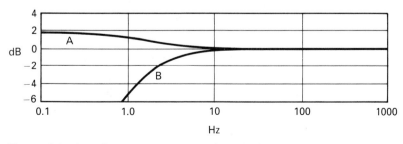

Figure 2.9 *Low frequency response of one-inch microphone Type 4144 (−3 dB point at 1.46 Hz) with static pressure equalization vent outside (curve A) and inside (curve B) the sound field.*

Table 2.4 *Maximum sound pressure level to which the microphones may be exposed*

Microphone type		Maximum sound pressure level dB re 20 µPa
One-inch	4144, 4145	160
Half-inch	4133, 4134 4147, 4149	174
	4165, 4166	160
	4148	154
Quarter-inch	4135	178
	4136	186
Eighth-inch	4138	178

condenser capsule with the various parts and air motions identified. The frontal sound access to the diaphragm is via the elements of the front cover, but some access occurs to the rear of the diaphragm by means of the narrow vent passage necessary for atmospheric pressure equalization. The acoustic mass effect of the mass of air in this small tube restricts any air flow into the microphone rear cavity at all but the lowest frequencies.

The response of the microphone at frequencies below a few herz can be calculated from the time constant of the vent and cavity constants, as shown in Figure 2.10. To reproduce very low frequencies down to 0.01 Hz approximately, it is necessary to specify a microphone with a higher than average value of acoustic resistance in the vent tube, or alternatively to vent the rear opening into a virtually sealed preamplifier or other space which is sealed against access from the sound wave. For cases where barometric pressure varies much faster than usual, e.g. when in an aircraft, the vent constants may have to be reduced at the expense of the VLF response, so as to retain the proper operation of the microphone. For microphones with normal vent settings a rate of climb of 500 m/s at ground level affects sensitivity at that level by less than 1 dB.

It is apparent that the analogy given by the equivalent circuit of an acoustical system such as the above microphone may be of only limited validity at higher frequencies where the dimensions of the various passages and cavities become comparable to the wavelength of the sound. Standing wave effects may then occur, possibly in more than one dimension. This means that the element concerned cannot be accurately represented by a single electrical component in the equivalent circuit. An intelligent study of the equivalent circuit will still give a useful idea of the effect on the response of the various elements. An interesting example may be observed from the construction of 6 and 3 mm (¼ and ⅛ in) capsules, shown in section in Figure 2.1. It was originally found that an irregularity in the response occurred at higher frequencies due to standing wave effects in the rear cavity (7) of the capsules. The cavity was accordingly reduced in volume by the addition of an insulating ring.

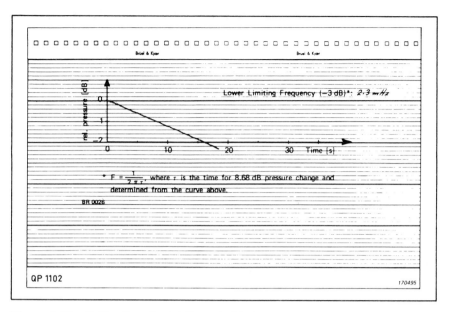

Figure 2.10 *Calibration chart delivered with the condenser microphone cartridge enabling calculation of the low frequency −3 dB point by extrapolation.*

2.3 Cartridge sizes and applications

Standard sizes of cartridge are made with a range of different characteristics to suit a wide field of specific applications. For instance, the 25 mm (1 in) cartridge is made the US ANSI Standard S1 12–1967. It can also be provided with a special front so as to make it exactly equivalent to the original WE standard microphone, which is still in use. The 13 mm (½ in) cartridges are the most widely used size and are made to cover a wide range of measurements and applications. The good long-term stability of these microphones is a special feature, as it is extrapolated to be better than 1 dB for an indefinite period (100 years). The various sizes of cartridges have much the same general form, the smaller sizes generally covering a wider frequency band at the expense of reduced sensitivity.

The high order of stability has been achieved by careful design and the selection of the most suitable materials and forms of construction. Special care also has to be devoted to the precise realization and setting of the rear vent which provides equalization of the static air pressure between the inside and outside of the cartridge. The presence of this leak also provides a small amount of pressure gradient operation at extreme low frequencies. The exact setting of the leak has to be precisely adjusted so as to give a well-defined lower limiting frequency characteristic to meet various specified applications. Figure 2.11 shows examples of side-vented and back-vented arrangements, the latter being particularly suitable for use with de-humidifier arrangements within the rear enclosure. The lower limiting response may be precisely determined as a function of the low-frequency time constant of the vent, as shown in Figure 2.10. For long-term exposure to out-door, damp or corrosive atmospheres the backplate and diaphragm may be coated with a thin layer of quartz, in addition to a rear de-

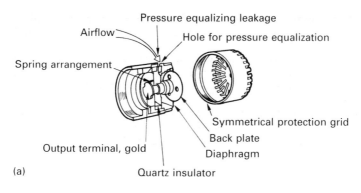

Pressure equalizing leakage

Airflow

Hole for pressure equalization

Spring arrangement

Symmetrical protection grid

Back plate

Output terminal, gold

Diaphragm

(a)

Quartz insulator

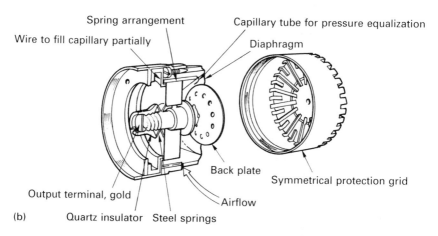

Spring arrangement

Capillary tube for pressure equalization

Wire to fill capillary partially

Diaphragm

Back plate

Symmetrical protection grid

Output terminal, gold

Airflow

(b)   Quartz insulator   Steel springs

Figure 2.11   *Vent constructions. (a) side vented microphone cartridge; (b) cross-sectional view of the quartz-coated microphone Type 4149 back vented. (Courtesy of Bruel & Kjaer.)*

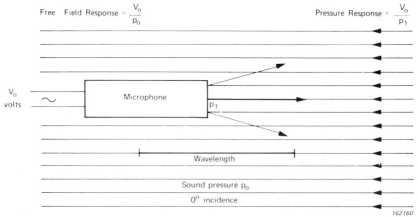

Free Field Response = $\dfrac{V_o}{p_o}$

Pressure Response = $\dfrac{V_o}{p_1}$

V_o
volts

Microphone

p_1

Wavelength

Sound pressure p_o

0° incidence

162160

Figure 2.12   *Definitions of free-field response and pressure response.*

humidifier system. Figure 2.13 shows a weather-proof kit which also adds a wind-shield, a rain cover and a preamplifier and power supply in a stainless steel housing.

Some cartridges are also made with low (28 V) polarization or stable electret polarization on the backplate, for use in portable or battery-operated equipment. A wide range of special accessories is described in Section 2.8, together with the relevant response corrections.

Where a minimum disturbance of the sound field is desired the use of 6 or 3 mm (¼ or ⅛ in) cartridges is recommended. An example is the use in acoustic model auditoria, etc., where scaled-up frequencies are used. It is also worth

(a)

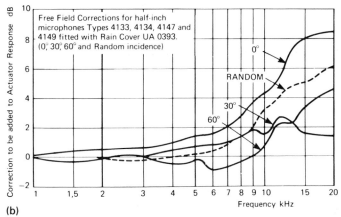

(b)

Figure 2.13 *(a) Rain cover and (b) corrections. An electrostatic actuator is built in which can be excited for remote calibration. It is important to mount the unit upright, the microphone diaphragm facing straight up. The rain cover is particularly recommended for permanent outdoor use), factory calibrated to give an equivalent sound pressure level of 90 ±1 dB when 215 V AC is applied to the actuator terminal (80 ±1 dB for 121 V AC).*
Free-field corrections of the ½ inch microphones fitted with the rain cover are given in (b). From these curves, the response of the ½ dp microphone can be shown. (Courtesy of Bruel & Kjaer.)

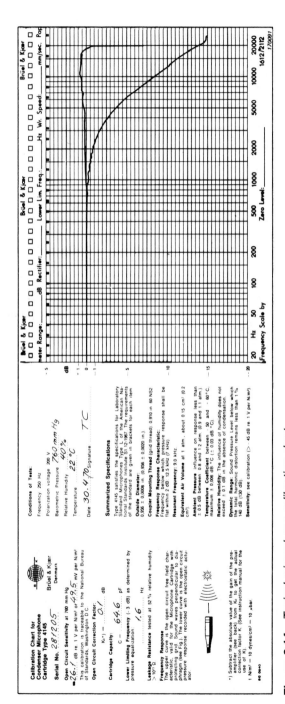

Figure 2.14 *Typical microphone calibration chart.*

noting that a free-field-calibrated microphone should generally be aimed towards the main sound of interest, whilst a pressure-calibrated capsule should be set for 90° incidence.

2.4 Factors affecting cartridge response

The free field correction is a function of frequency, and the figures give curves for correcting the electrostatic actuator response given on the supplied calibration chart for various angles of sound incidence. The Random Incidence obtained using these curves characterizes the frequency response of the microphone in a diffuse sound field. It is an averaging of the microphone's response at all angle of incidence. The corrections given are calculated as defined in IEC 651 1979 'Sound Level Meters', Appendix B:

$$S^2 = 0{,}018\,(S_0^2 + S_{180}^2) + 0{,}129\,(S_{30}^2 + S_{150}^2) + 0{,}244\,(S_{60}^2 + S_{120}^2) + 0{,}258\,S_{90}^2$$

where S_0, S_{30}, S_{60}, S_{90}, S_{120}, S_{150}, S_{180} are microphone sensitivities at the respective angles of incidence (mV/Pa).

The free-field corrections should be added to the pressure response characteristics of the microphone given on the electrostatic actuator chart to give the free-field characteristic for the respective angle of incidence. The corrections are functions of the geometry of the microphones and to a very small degree the acoustical impedance of their diaphragms. Therefore, the corrections are similar for a given microphone type.

Measuring microphone cartridges (B & K) are provided with an individual calibration chart showing a continuous pressure response curve produced by an electrostatic actuator. For cartridges specified as free-field measuring types, free-field responses derived from the free-field corrections for that type are given (Figures 2.14 and 2.15). The diffuse field response is given for certain specified

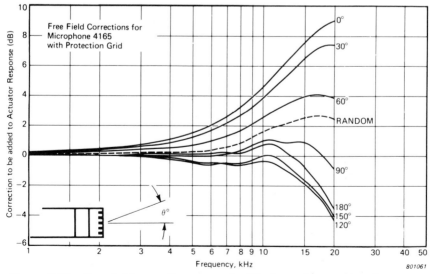

Figure 2.15 *Free-field correction curves for ½ inch condenser microphone Type 4165 fitted with normal protection grid. (Courtesy Bruel & Kjaer.)*

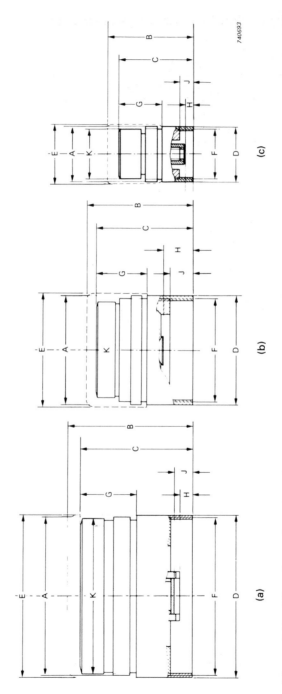

(a)   (b)   (c)

Figure 2.16 Microphone dimensions (see also Table 2.1). (a) 1 inch; (b) ½ inch; (c) ¼ inch and ⅛ inch. (Courtesy Bruel & Kjaer.)

cartridges. The time constant of the rear vent is also given, with a chart giving the precise VLF response for cartridges specified for low-frequency use.

It is seen that the random incidence diffuse field corrections are relatively small at most audio frequencies of interest for cartridges of 13 mm (1/2 in) or less diameter. Thus microphones with a flat pressure response are often preferred for diffuse field indoor measurements. It may be noted that specially designed nose cones (B & K) can be fitted to free-field cartridges so as to make the response independent of angle over a wide frequency range (Figure 2.18). Data given with condenser microphone capsules normally refers to open-circuit operation. In practice, connections and preamplifier input circuits have some effect, though this is often small enough to be neglected in many cases. The total response is found by adding the open circuit response of the capsule to the preamplifier response curve for that particular type (Figure 2.19).

It may be noted that stabilized 200 V polarizing supplies for cartridges are normally available from preamplifiers and complete measuring equipments (B & K, etc). For battery-operated equipment, such as sound level meters, prepolarized electret cartridges are normally used.

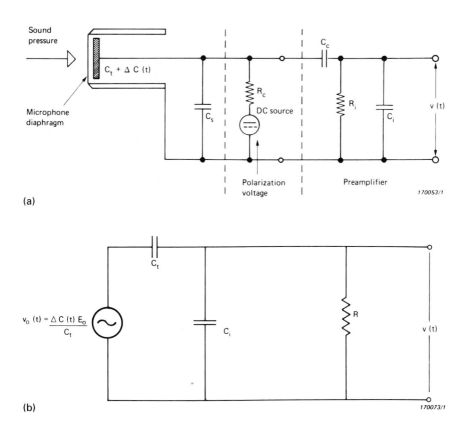

Figure 2.17 *(a) Equivalent circuit of a condenser microphone and microphone preamplifier; (b) simplified equivalent circuit of a condenser microphone and microphone preamplifier.*

4133/49 + 2619   UA 0386

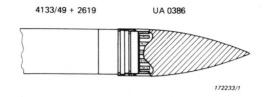

(a)

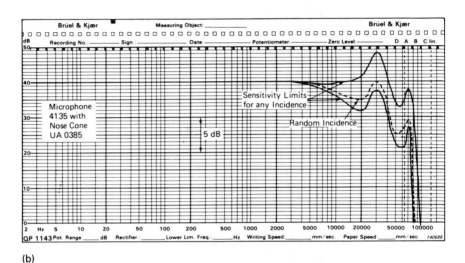

(b)

Figure 2.18 *(a) The nose cone is designed to substitute the normal protection grid of the microphone and is intended for use in air flows of high speed and well defined direction. Its streamlined shape gives the least possible resistance to air flow, thereby reducing turbulence. A fine wire mesh around the nose cone permits sound pressure transmission to the microphone diaphragm while a truncated cone behind the mesh reduces the air volume in front of the diaphragm.*
(b) Sensitivity limits for any incidence for ¼ inch microphone type 4135 fitted with Nose Cone UA 0385. (Courtesy Bruel & Kjaer.)

The lower limit for measurement of sound is ultimately set by the inherent thermal noise in the cartridge diaphragm given by:

$$\overline{P} = (4\pi TR\Delta f)^{1/2} \tag{2.3}$$

where $\overline{P}$ = sound pressure (Table 2.5)

2.5 Cartridge and preamplifier response

These preamplifiers are primarily designed for use with condenser microphone cartridges, but they may also be useful in other electronic circuits, where a very high input impedance is required, e.g. for special voltmeters, etc. Preamplifiers

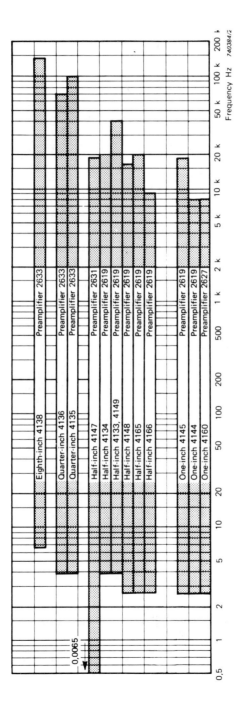

Figure 2.19 Comparison of the frequency response ranges (±2 dB) of recommended microphone and microphone preamplifier combinations. (Courtesy Bruel & Kjaer)

Table 2.5 *Cartridge inherent noise specifications*

Microphone type		Normalised noise pressure (μPa/$\sqrt{Hz}$)	Inherent noise level (dB(A) re 20 μPa)	Broad-band noise level dB re 20 μPa
One-inch	4144	0.6	9.5	10
	4145	0.9	10	10.5
Half-inch	4133	2.2	20	21.5
	4134	1.3	18	21.5
	4147			
	4148	1.0	12.5	12.5
	4149	2.2	20	21.5
	4165	1.2	14.5	15.5
	4166	1.1	15	15.5
Quarter-inch	4135	5.7	29.5	34
	4136		30.5	37.5

must be compact to match up and must be capable of operating over a wide range of temperature and other possible environmental hazards. They must have a very low inherent noise level and a wide dynamic range. A low output impedance allows long output cables to be used without appreciable losses. The need for low-noise DC power and polarizing supplies has already been mentioned.

Figure 2.20 shows a range of Bruel & Kjaer preamplifiers for various cartridges, adapting couplers being used where necessary. De-humidifier elements and current supplies for microphone heating elements are provided in special cases. The 2627 preamplifier for 25 mm (1 in) cartridges also allows for a series insert voltage to be injected for overall calibration by reciprocity or other methods of calibration. One must note that polarizing supplies must be disconnected if pre-polarized electret capsules are to be used.

A system of double screening of capsules allows very low cross-talk between systems to be obtained, figures better than – 60 dB being obtainable. The shield immediately around the input contact on the preamplifier goes to the amplifier output line or separately to ground, thus allowing either a 'driven' or a grounded shield. Connecting cables for preamplifiers must be designed for low capacitance and very good shielding and anti-noise performance, lengths up to 30 m or more being usable. There may then be some slight effects on noise and distortion at higher frequencies, as shown in Figure 2.22 (a–c).

Some very special connections may have to be made: for example, flat flexible cables may be needed to go under doors and windows etc. Bruel & Kjaer Type AR 0001 flat 7-core cable is only 0.2 mm thick $\times$ 300 mm long.

The frequency response of the preamplifiers depends on the transducer capacitance connected to the preamplifier input and the capacitance load (extension cables) connected to the output. The curves show the frequency response of the preamplifiers with different B & K condenser microphone cartridges and no extension cables connected.

1″ Cartridge 4179 + DB 0375 + 2660

1″ Cartridge + UA 0786 + 2645

1″ Cartridge + DB 0375 + 2639

1/2″ Cartridge + 2639

1/2″ Cartridge 4149 + UA 0308 + 2639

1/4″ Cartridge + UA 0035 + 2639

1/4″ Cartridge + 2633

1/8″ Cartridge + UA 0036 + 2639

1/8″ Cartridge + UA 0160 + 2633

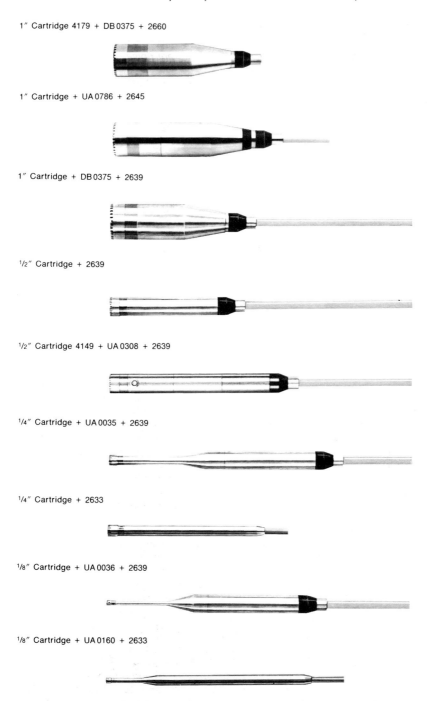

Figure 2.20 *The mounting of various sizes of capsule on different preamplifiers using adaptors where necessary. (Courtesy of Bruel & Kjaer.)*

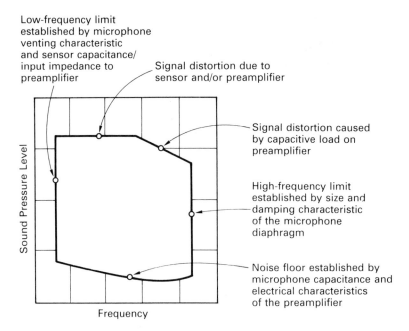

Low-frequency limit established by microphone venting characteristic and sensor capacitance/ input impedance to preamplifier

Signal distortion due to sensor and/or preamplifier

Signal distortion caused by capacitive load on preamplifier

High-frequency limit established by size and damping characteristic of the microphone diaphragm

Noise floor established by microphone capacitance and electrical characteristics of the preamplifier

Sound Pressure Level

Frequency

Figure 2.21 *Factors defining the operating limits of a microphone system. (Courtesy Bruel & Kjaer)*

Typical B & K microphone capacitances are: Type 4144, 4145, and 4160, 50 to 66 pF (1" dia); Type 4125, 4133, 4134, 4147, 4148, 4149, 4155, 4165, 4166 and 4175, 17 pF (½" diameter); Type 4135 and 4136, 6 pF (¼" diameter); Type 4138, 3 pF (⅛" diameter).

Table 2.6 gives the full specifications and performance data for the pre-amplifiers. Figure 2.23 gives the response limits and the dynamic range of the preamplifiers when fitted with various capsules. Figures 2.25 and 2.22 give the overall response for various inputs and outputs for various cable capacitance loads and distortion limits for various preamplifiers and capsules. We have seen that the study of the overall performance of the capsule and preamplifier combination must involve the analysis of a combined equivalent circuit. We have also seen that the condenser capsule consists basically of a thin tensioned diaphragm stretched before a rigid metal back plate, constituting an air dielectric capacitor, to which a constant polarizing charge is applied. The static pressure on the diaphragm is equalised by the vent leak; sound pressure causes voltage variations. The stray capacitance due to the condenser capsule itself, wiring capacitance and that of the preamplifier input stage must be kept as low as possible to preserve sensitivity and the input load resistance must be as high as possible to ensure a good lower limiting frequency response. The preamplifier stage is designed as an impedance converter to give a suitably low output impedance. Figure 2.17a, b show a simplified overall equivalent circuit of the capsule and preamplifier. A constant charge Q is applied to the diaphragm-back plate capacitor C by the polarizing voltage E_0 via the high resistor R_c:

$$E_0 = Q/C_t \tag{2.4}$$

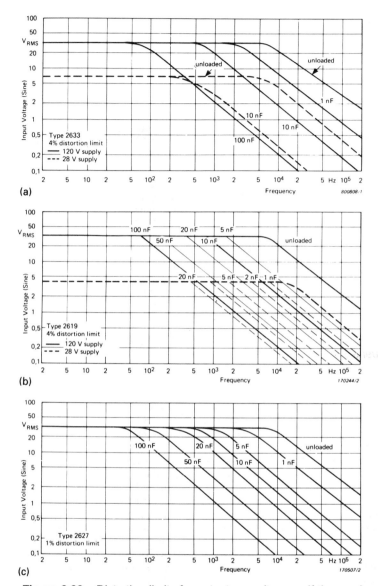

Figure 2.22 *Distortion limits for output capacitances. If the maximum output current of the preamplifier is exceeded the signal will be distorted. These curves give the upper distortion limit (4%) as a function of output voltage and load capacitance at the output of the preamplifier. For normal use and cable lengths the distortion will be lower than 1%. (a) Preamplifier Type 2633. Upper limit of dynamic range (4% distortion) due to capacitive loading as a function of input voltage and frequency. (b) Preamplifier Type 2619. Upper limit of dynamic range (4% distortion) due to capacitive loading as a function of input voltage and frequency. (c) Preamplifier 2627. Upper limit of dynamic range (1% distortion) due to capacitive loading (extension cables) as a function of input voltage and frequency. (Courtesy Bruel & Kjaer)*

Table 2.6 *Preamplifier specifications. (Courtesy of Bruel & Kjaer.)*

B & K type preamplifiers	2633	2619	2627
DC power supply	120 V/2 mA 28 V/2 mA	120 V/2 mA, Heater 12V/80 mA 28 V/0.5 mA	120 V/2 mA Heater 12 V/60 mA
Polarization voltage	Transmitted through preamplifier to microphone cartridge from power supply		
Input impedance	>50 GΩ//0.25 pF at 60°C (approx. 5 GΩ// 0.25 pF at 100°C) 10 GΩ//0.4 pF (typical)	>10 GΩ//0.8 pF >7 GΩ//1 pF	grounded shield: <5 pF driven shield: >10 GΩ//<0.5 pF
Output impedance	<100 Ω 600 Ω (typical)	<25 Ω <70 Ω	<50 Ω
Max. output current	1.4 mA peak 0.25 mA peak (typical)	1.5 mA peak 0.5 mA peak	1.4 mA peak
Temp. range	$-20°$ to $+60°$C ($-4°$ to $140°$F)	$-20°$ to $+60°$C ($-4°$ to $+140°$F)	$-10°$ to $+60°$C ($14°$ to $140°$F)
Attenuation (preamplifier alone)	<0.06 dB <0.1 dB	<0.03 dB <0.1 dB	<0.08 dB
Phase linearity (20 Hz to 20 kHz) (2 Hz to 200 kHz)	$\pm 2.0°$ $+8° - 13°$ (6pF)	$\pm 2.5°$ $+20° - 13°$ (20pF)	$\pm 2.0°$ $+2° - 12°$ (50pF)

Preamplifier noise (dummy microphone): Lin. 20 Hz to 200 kHz and A-weighted. Typical values are given in parentheses

60 pF (1" microphone)	—	<10 μV (7 μV) Lin. <2 μV (1.7 μV) A-weighted	<15 μV (9 μV) Lin <3 μV (2.3 μV) A-weighted
17 pF (½" microphone)	—	<25 μV (12 μV) Lin. <4.5 μV (3.3 μV) A-weighted	—
6 pF (¼" microphone)	<30 μV (19 μV) Lin. <7 μV (5.8 μV) A-weighted Approx. 150 μV, 34 μV at 100°C	<50 μV (23 μV) Lin. <15 μV (7.5 μV) A-weighted	—
3.5 pF (⅛" microphone)	<50 μV (23 μV) Lin. <12 μV (7 μV) A-weighted	<70 μV (50 μV) Lin. <25 μV (14 μV) A-weighted	—
Dimensions: Diameter Length	6.35 mm (0.25 in) 88 mm (3.46 in)	12.7 mm (0.5 in) 83 mm (3.25 in)	23.8 mm (0.936 in) 99 mm (3.9 in)
Cable length	3.4 mm (11.2 ft)	2 m (6.6 ft)	2 m (6.6 ft)

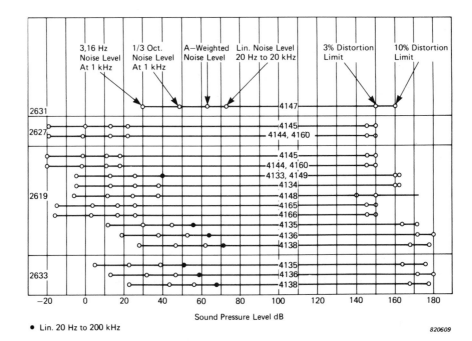

Figure 2.23   *The dynamic range of various preamplifiers and microphone combinations. The upper limit is shown for 3 % and 10 % THD. (Courtesy Bruel & Kjaer)*

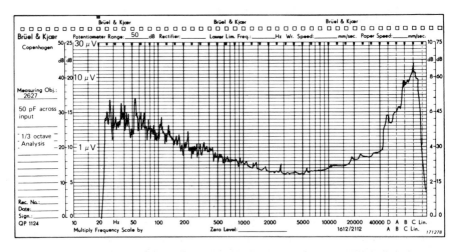

Figure 2.24   *Inherent noise of preamplifier Type 2627 (third-octave analysis) with input capacitance equivalent to 1" microphone. (Courtesy of Bruel & Kjaer.)*

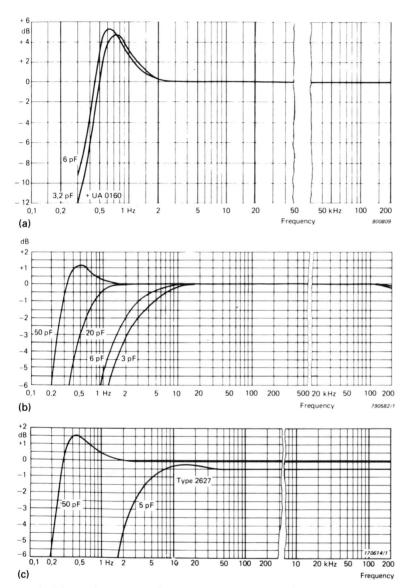

Figure 2.25   *Responses with different microphone capacitances*
The capacitance of the adaptors necessary to connect the cartridges to the preamplifiers has been included as indicated on the curves. (a) Preamplifier Type 2633. Frequency response curves with different transducer capacitances connected at the input. (b) Preamplifier Type 2619. Frequency response curves with different transducer capacitances connected at the input. (c) Preamplifier Type 2627. Frequency response curve of the preamplifier unloaded and an input capacitance of 50 pF corresponding to 1" microphone. Lower curve with 5 pF input capacitance indicates the influence on frequency response of varying input capacitances. With 50 pF input capacitance the response is linear from 2 Hz to 200 kHz ±0,5 dB. (Courtesy Bruel & Kjaer.)

When unloaded, the cartridge open circuit voltage is given by

$$v(t) = \frac{\Delta C(t)E_0}{C_t} \times \frac{C_t}{C} \times \frac{j\omega RC}{1 + j\omega RC} = \frac{\Delta C(t) \cdot E_0}{C_t} \cdot \frac{j\omega RC_t}{1 + j\omega RC}$$

where

$C = C_t + C_i$ the sensitivity, S is given by

$$S = \frac{v(t)}{\Delta p(t)} = \frac{\Delta C(t)E_0}{C_t \times \Delta p(t)} \times \frac{j\omega RC_t}{1 + j\omega RC}$$

where

$\Delta p(t)$ = sound pressure variation with time

and

$\Delta p(t) = K_1 \Delta C(t)$ where K_1 is a constant

$$S = K_2 \frac{E_0}{C_t} \times \frac{j\omega RC_t}{1 + j\omega RC} \qquad \text{where } K_2 \text{ is a constant}$$

From the expression for the microphone sensitivity S two basic cases can be considered. First, the high-frequency response, when $\omega RC \gg 1$. In this case

$$S = K_2 \frac{E_0}{C}$$

It is seen that the sensitivity is proportional to the polarization voltage but inversely proportional to the total capacitance C. Any capacitance additional to the transducer capacitance, such as that of the preamplifier, would load the transducer and decrease the overall sensitivity. This means that the total capacitance in the circuit should not greatly exceed the capacitance of the transducer, hence the usual close proximity of the cartridge and its associated preamplifier.

The second case is at low frequencies, when $\omega RC \ll 1$, thus giving

$$S = K_2 \frac{E_0}{C_t} j\omega RC_t$$

This expression is frequency dependent, and associated with the transition to frequency dependence of sensitivity a cut-off frequency (f_c) can be defined when $\omega RC = 1$ and the sensitivity is reduced by 3 dB. For this case

$$f_c = \frac{1}{2\pi RC} \tag{2.5}$$

Preamplifier input circuits can be made to present a very high effective resistance in the DC polarizing circuit to the microphone and a high shunt resistance to the microphone signal. Effective values in excess of the passive resistors fitted to the amplifier are obtained by the use of feedback in an FET source follower.

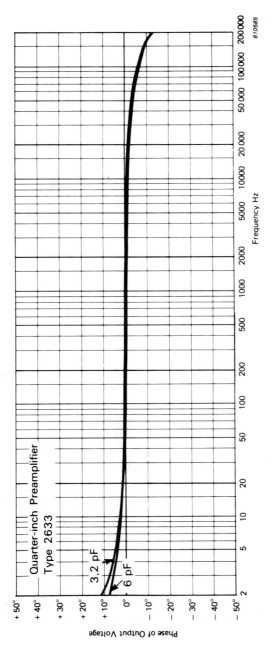

Figure 2.26 *Phase response of preamplifier Type 2633 with input capacitance equivalent to (a) ¼ inch microphone; (b) ⅛ inch microphone. (Courtesy Bruel & Kjaer.)*

The preamplifier feedback circuits are primarily designed to reduce input capacitance and to increase the input resistance loading on the capsule. They produce some secondary effects such as capacitative secondary loading being fed back to the input. This can produce some additional attenuation for low-capacitance capsules and some phase changes. Figure 2.19 gives the frequence response of preamplifiers without any additional output extension cables. Figure 2.22 shows the effect of additional capacitance loads due to output cables. Distortion tolerances limit the upper level of the dynamic range imposed by the maximum output current or voltage. Figure 2.22 shows the 4% distortion limit as a function of the output capacitance load.

Figure 2.26 show the phase response of the overall microphone and amplifier combination, the phase response of the capsule being the dominant factor. Figure 2.28 shows the transient pulse response. Figure 2.24 give the ⅓ octave noise spectra for various capsule capacitances.

Microphone pulse responses are compared in (a) Figure 24 one-inch, half-inch and quarter-inch microphones are excited by a rectangular pulse at a frequency of 2 kHz. The crest factor of the pulse was varied by altering the pulse duration, crest factors of 10, 5, 2, and $\sqrt{2}$ being produced by pulse durations of 5, 19, 100, 250 μs respectively. The 19 μs pulse is repeated with an expanded time base, which is also the time base corresponding to the 5 μs pulse. Fig 24 (b) compares the response of the two quarter-inch microphones Types 4135 and 4136 and the eighth-inch microphone Type 4138 to a 20 μs pulse. The marked difference between the responses of Types 4135 and 4136 is due to the difference in diaphragm thickness of the two microphones and the interaction of the two major resonance modes of the diaphragm.

The effect of humidity is generally less important for the preamplifiers than for the microphone capsules. Most preamplifiers dissipate enough power to avoid condensation in the combination for the majority of normal measuring environments. For more humid conditions heating elements built into the 2619/2627 preamplifiers raise the temperature of the combination.

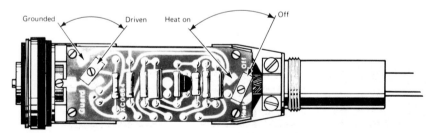

Figure 2.27   *A preamplifier showing a switched heater circuit and a switch for disconnecting the earthing shield. Use of the driven shield will minimize the stray capacitance, and hence reduce the input capacitance of the preamplifier. If the driven shield arrangement is used, the open circuit voltage produced from a microphone by a known sound pressure level differs by less than 0,02 dB from the open circuit voltage which would be obtained if the shield were grounded (ANSI S1 ·10–1966). The choice of configuration needs to be in agreement with different methods used in standards laboratories. Input impedance of the 2627 is >10 GΩ and output impedance <50 Ω. Input capacitance is <5 pF (grounded shield); is <0·5 pF with driven shield. (Courtesy Bruel & Kjaer)*

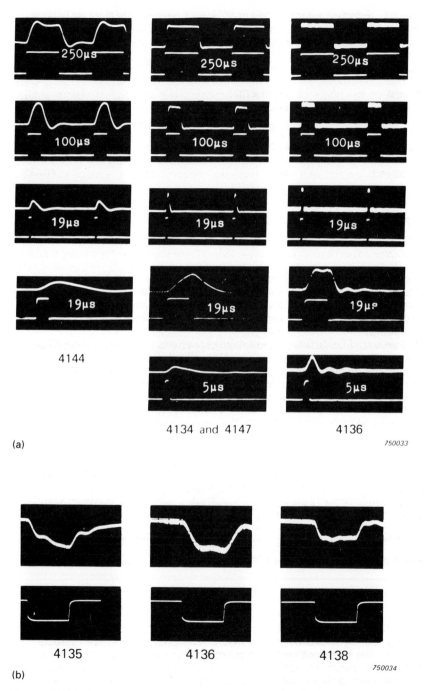

(a)

4144

4134 and 4147   4136

750033

(b)

4135   4136   4138

750034

Figure 2.28 *Comparative pulse responses of microphones. (a) 1 inch, ½ inch and ¼ inch microphones. (b) The response of the two ¼ inch microphones and the ⅛ inch microphone to a 20 μ s pulse. (Courtesy Bruel & Kjaer.)*

Table 2.7   *Effect of magnetic field of 80 A/m (1 oersted) at 50 Hz on various capsules. Equivalent SPL for 80 A/M 50 Hz field*

		dB	range dB
Half-inch	4133, 4134 4147, 4149	20	6–34
	4148	28	10–38
	4165, 4166	30	10–40
Quarter-inch	4135	30	10–42
	4136	38	18–46
Eighth-inch	4138	40	30–50

These preamplifiers usually meet the specified limits for ambient temperature limits of –20°C to +60°C. For much higher temperatures up to 150°C the preamplifier may be placed in a cooler environment than the capsule by the use of extension cables, the capsule usually being less affected than the preamplifier.

Intense sound fields usually have negligible effects on solid-state amplifiers of good construction, as can be confirmed by substituting a dummy capsule and clamping the preamplifier to a calibrated vibration table and finding the axis of maximum sensitivity for a frequency sweep from 10 Hz to about 2 kHz. The response to magnetic fields is usually greatest when the field direction is perpendicular to the preamplifier axis. Normal operation is usually possible for fields up to 100 A/m (see Table 2.7). The response of condenser capsules to EM fields is usually low and can be investigated separately. IEC and ANSI standards apply. Very high fields are normally only encountered close to large electrical power transformers or machines.

The inherent noise level of a microphone is defined as the equivalent A weighted SPL due to the cartridge noise. Table 2.5 gives noise levels for 25 mm (1 in) to 6 mm (¼ in) cartridges, the noise being worse for smaller cartridges. Figure 2.24 shows the one-third octave noise spectra for the cartridge and

Table 2.8   *Example of magnetic fields measured in a power plant*

Measurement location	Max. magnetic field strength (A/m)
13 kV to 120 kV transformer 1, at 1 m	110
1 m from 120 kV output cable, transformer 1	200
13 kV to 5 kV transformer at 1 m	14
2 m from coils of 30 kV output	350
1 m from generator 1, machine room	56
1 m from 13 kV output cable of generator 1	1250

preamplifier combinations. The preamplifier noise predominates for smaller cartridges of lower sensitivity because the smaller cartridge capacitance increases the preamplifier noise level. The capacitance of cartridges varies with the polarization voltage. The microphone cartridge capacitance with 200 V polarization is measured for each individual cartridge at 250 Hz and is quoted on its calibration chart. Owing to the effects of diaphragm movement, however, capacitance of the polarized cartridge is slightly variable with frequency. This also affects the sensitivity of the microphone and the high-frequency response. Figures 2.30 and 2.31a,b show the effects for 25 mm (1 in) and 13 mm (½in) cartridges.

The leakage resistance of a capsule should be 1000 times that of the polarizing circuit in the amplifier, which should be of the order of 1000 megohms. The

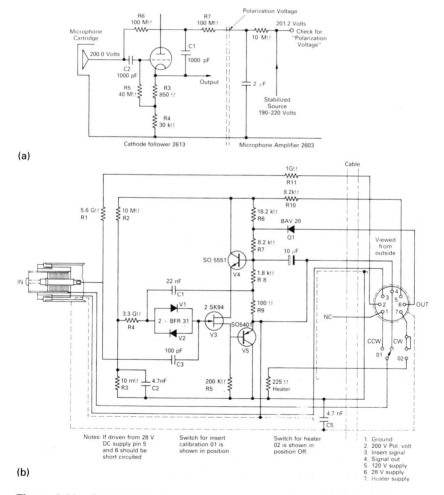

(a)

(b)

Figure 2.29 *Preamplifier circuits using (a) valve cathode follower. (b) The more modern semiconductor microphone which uses a low noise FET input stage with feedback circuits to increase the input impedance to a high value. (Courtesy Bruel & Kjaer)*

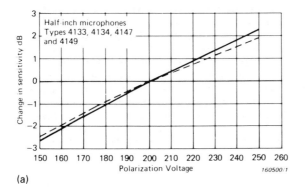

(a)

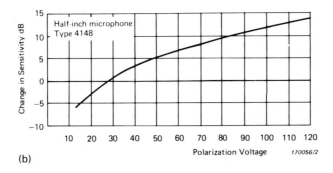

(b)

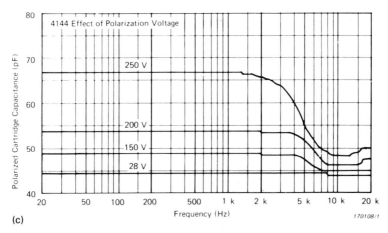

(c)

Figure 2.30 *Polarization effects. (a) Sensitivity variation of ½ inch microphones Types 4133, 4134, 4147 and 4149 at 250 Hz as a function of polarization voltage. The dashed line indicates the theoretical relationship, not allowing for electrostatic attraction.*
(b) Sensitivity variation of ½ inch microphone Type 4148 at 250 Hz as a function of polarization voltage
(c) Variation of cartridge capacitance with polarization voltage and frequency for 1" pressure response microphone Type 4144.

insulators in B & K capsules are quartz, ruby or sapphire treated with silicon. Ionic current across the diaphragm-to-back-plate gap may also cause leakage. Special cleaning and polishing methods are applied during production to minimize these effects. The effect of high humidity is usually small, provided no condensation occurs. Quartz coating helps, but may cause small reversible changes of about 0.2 dB for relative humidity up to 80%. The long-term changes of the capsule diaphragm structure due to high-temperature cycles and mechanical shocks are usually very small, being due to minute creep affecting the diaphragm tension. The materials for capsules are chosen to give the best possible thermal match. B & K 25 mm (1 in) microphones and the associated calibration equipment when kept at 8–10°C typically show reproducibility better than ± 0.2 dB in overall calibration. See Figure 2.31a,b. Ambient atmospheric pressure changes cause some usually small changes in microphone response and sensitivity as the coupled air stiffness and damping resistance on the diaphragm

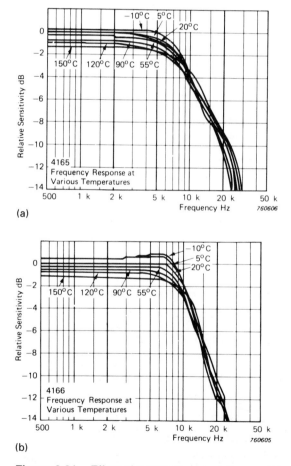

Figure 2.31 *Effect of temperature variation on ½ inch microphones Types 4165 and 4166. (a) Effect on frequency response of Type 4165; (b) effect on frequency response of Type 4166. (Courtesy of Bruel & Kjaer.)*

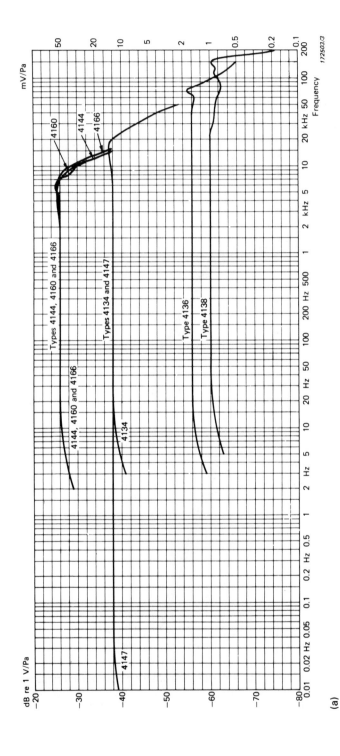

(a)

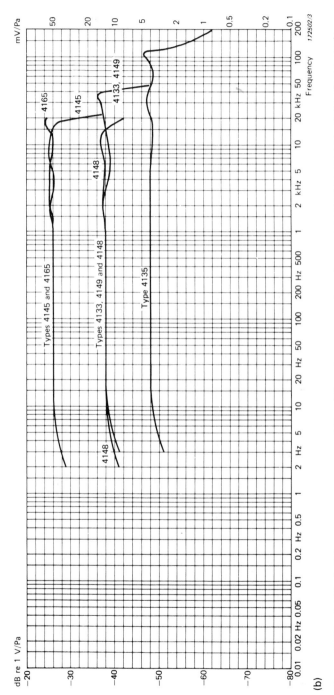

Figure 2.32 (a) Typical frequency responses of the different pressure microphones recorded by means of the electrostatic actuator method. (b) Typical 0° incidence frequency response of the different free-field microphones recorded by means of the electrostatic actuator method and corrected according to the curves shown in Figure 2.35. (Courtesy of Bruel & Kjaer.)

Table 2.9   *Equivalent volumes at 250 Hz (Bruel & Kjaer)*

Microphone type		Typical equivalent volume at 250 Hz (mm³)
One-inch	4144	148
	4145	130
Half-inch	4148	80
	4165, 4166	40
	4133, 4134	10
	4147, 4149	
Quarter-inch	4135	0.6
	4136	0.25
Eighth-inch	4138	0.1

change. Makers' handbooks show the effect on response of ambient pressure changes.[7] In aircraft, a rate of climb of 500 m/s near ground level typically causes changes of 1 dB in sensitivity for normal back vent values.

Direct sunlight on microphone diaphragms causes some heating but the effect is usually <0.1 dB for normal diaphragm fronts.[8] For measurements in small cavities where the microphone diaphragm forms one wall, the diaphragm compliance changes the effective volume of the cavity, the increment being termed the equivalent volume of the capsule (Table 2.9).

2.6 Field calibration equipment and accessories

Accurate calibrations of microphones and complete sound measuring set-ups can be made in laboratories or in the field using accessories such as the B & K 4220 pistonphone or the sound level calibrator. These units are battery operated and produce 124 dB ± 0.15 dB and 94 dB ± 0.3 dB SPL respectively (see Figures 2.33 and 2.34).

The B & K high-pressure microphone calibrator measures 25 to 3 mm (1 to ⅛ in) capsules up to 164 dB SPL 3 to 1000 Hz in order to check the sensitivity at specific frequencies.

Electrostatic actuators are made to fit cartridges of 25 to 3 mm (1 to ⅛ in) and give accurate pressure calibrations and periodic checks to a high level of accuracy of the response.

Nose cones improve omnidirectionality, are highly polished and so reduce aerodynamically induced noise when exposed to high wind speeds. They replace the B & K microphone protective grid and have a fine wire mesh-covered aperture to allow sound access. They are particularly suitable for measurements in ducts etc. where wind flow is in a specified direction. The B & K turbulence-reducing tubular screen fits on to 13 mm (½ in) microphones for 16 dB extra reduction of duct wind noise from 70 to 1500 Hz.

Windscreens for general use are made by B & K in special polyurethane sponge material. They are made to fit 13 mm and larger microphones and attenuate moderate outdoor wind noise by 10 dB or more. Useful extra moisture protection is obtained.

Rain covers are made to fit 13 mm microphones, to allow permanent outdoor use and incorporate an electrostatic actuator for periodic checks. It also improves the omnidirectional characteristics (Figure 2.18).

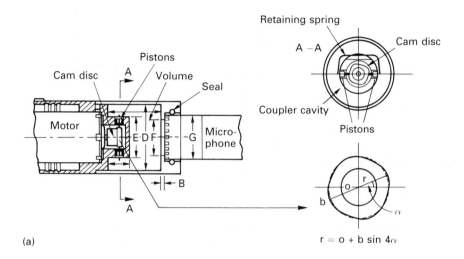

(a) $r = o + b \sin 4\alpha$

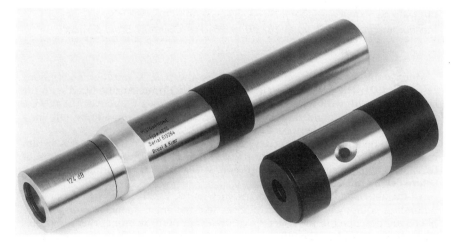

(b)

Figure 2.33 *(a) Cross-section of a double piston pistonphone with principles and dimensions. The cam disk is shown on the right. The high degree of precision is obtained by incorporating two symmetrically mounted pistons with a floating retaining spring arrangement which eliminates shaft eccentricity and backlash from bearings. (b) Pistonphone Type 4220; (c) sound level calibrator type 4230. (Courtesy Bruel & Kjaer)*

Figure 2.34 *High pressure microphone calibrator Type 4221 which allows measurements at levels up to 162 dB or 20 μPa (Courtesy Bruel & Kjaer)*

More ambitious protection is afforded by the B & K outdoor microphone assembly 4921. This consists of a quartz-protected 13 mm capsule fitted with the large windshield, with wires to prevent bird perching, a rain cover UA 0393, a preamplifier, de-humidification unit, calibrator and power supply, all in a stainless tube and weather-proof case (Figure 2.13).

A 13 mm dehumidifier unit is designed to be mounted between a B & K 13 mm back-vented cartridge and its preamplifier. A small window allows the colour of the silica gel to be checked for saturation. It lasts about 1 month at 100% R H, before drying out is needed.

All the above and other instruments such as the B & K reciprocity laboratory calibrating equipment, and other devices like the small-diameter probe microphone are described in more detail with relevant performance curves in Section 2.6.1, which also describes the use of the microphone K factor in setting up measuring equipment connected to the preamplifier in order to calibrate the system to read sound-pressure levels.

2.6.1 Laboratory calibration of microphones

Microphone field calibration and checking methods have been outlined. Laboratory methods involve thorough calibrations of sensitivity at specified frequencies, and also over complete frequency ranges.[9,10]

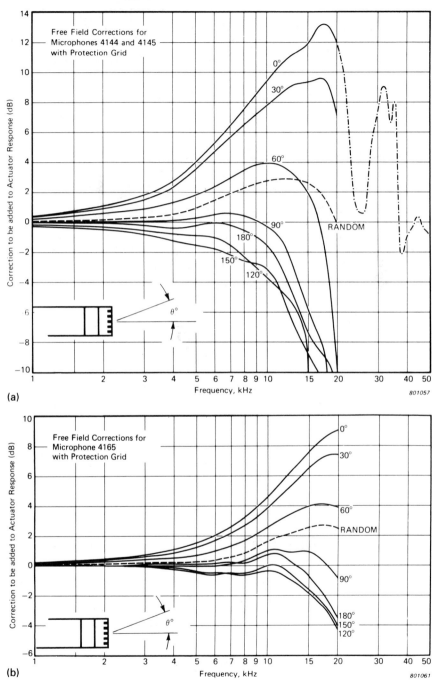

Figure 2.35   *Free-field correction curves for various condenser microphone cartridges (a) 1 inch; (b) ½ inch; (c) ½ inch; (d) ½ inch; (e) ½ inch; (f) ¼ inch; (g) ⅛ inch. (Courtesy Bruel & Kjaer.)*

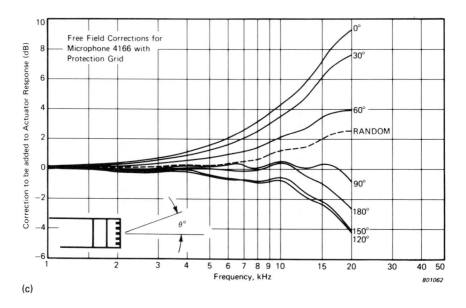

(c)

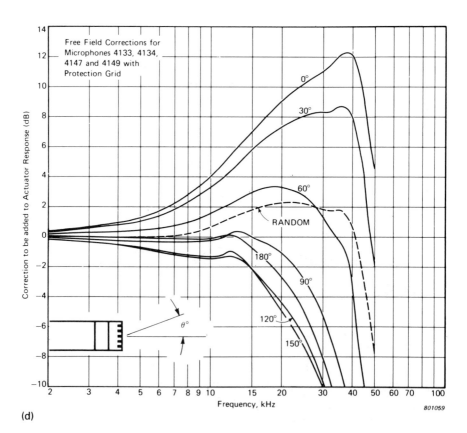

(d)

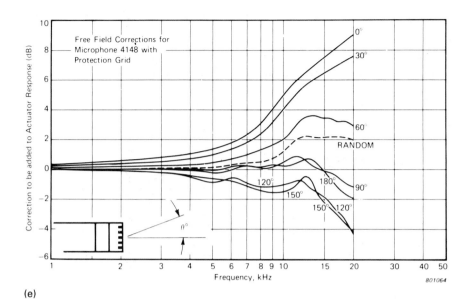

(e)

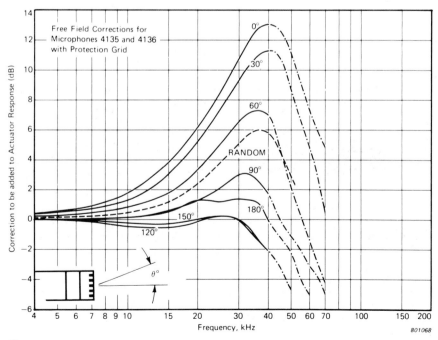

(f)

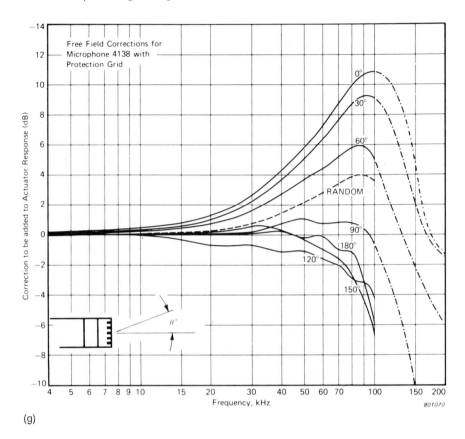

(g)

(a)

Figure 2.36   *(a) Reciprocity calibration apparatus Type 4143; (b) couplers and electrostatic actuators; (c) For laboratory calibration of standard microphones in accordance with IEC R 327 and IEC R 402 (reciprocity method) and for measurement of the frequency response of 1 inch, ½ inch, ¼ inch and ⅛ inch condenser microphones by the electrostatic actuator method, the reciprocity calibration uses current measurement in a standard capacitor. (Courtesy Bruel & Kjaer.)*

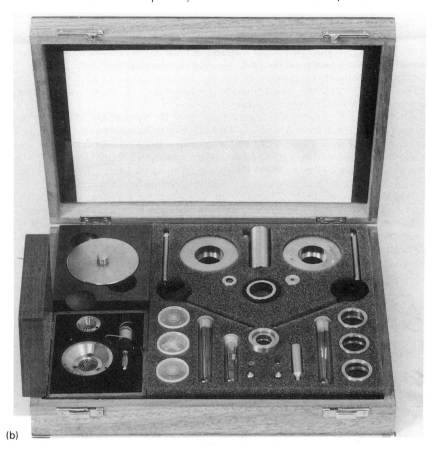

(b)

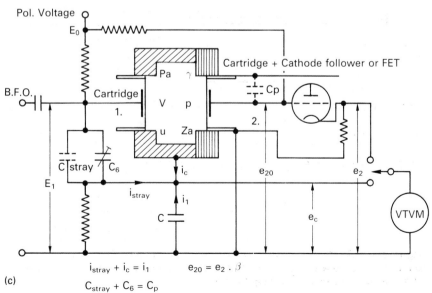

(c)

Reciprocity calibration is an absolute calibration method and is capable of giving the highest laboratory standards of accuracy at specified fixed frequencies. The IEC standard publication 327 applies, as regards tolerances on the accuracy. Figures 2.36 a–c show a reciprocity calibration instrument and a range of accessories, couplers etc, which enable absolute calibration of 25 mm (1 in) and 13 mm (½ in) condenser microphones, measurement of equivalent volumes, comparison of accelerometers and reference sound sources.

Electrostatic actuators of various sizes and high-pressure calibrators and pistonphones used for microphone calibration have been mentioned for field calibration, but these devices can be sufficiently refined for laboratory use.

Figure 2.37 shows direct comparisons of microphone calibration results for the main calibration methods. These tables are reproduced from a two-part article which describes calibration methods including reciprocity in detail.[9,10]

IEC Recommendations 327 specify an absolute accuracy of ± 0.05 dB with a reproducibility of better than ± 0.02 dB for laboratory standard microphones.

Laboratory calibration methods have to be considered in the context of the microphones concerned, as well as the frequencies and sound levels concerned. Accurate standardization of microphone sensitivity is usually carried out at a convenient middle frequency such as 250 Hz, where the dimensions of the

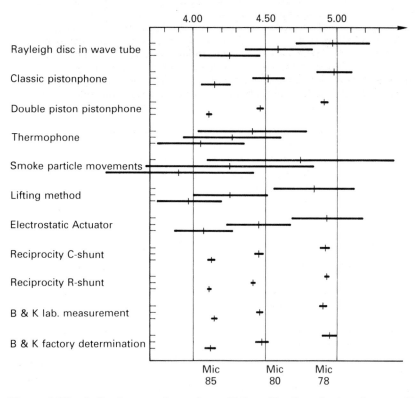

Figure 2.37   *A direct comparison of sensitivity calibration of microphones with nine different calibration methods. The length of the bars indicate uncertainties found for each method. (Courtesy Bruel & Kjaer.)*

microphones and associated equipment, such as couplers etc, are small compared to the wavelength of the sound. This particularly applies to the most accurate methods such as reciprocity and pistonphones. Calibrations of the full frequency range involve considerable difficulties when high accuracy is required. Continuous curves of the response are needed and the limitations of reciprocity and other comparison methods are obvious when completely accurate free field acoustic conditions are required.

The electrostatic actuator is a convenient and accurate method for the pressure calibration of accurately defined capsules with accessible metal diaphragms such as condenser-measuring microphones. These also lend themselves to accurate extension of the pressure response to free field responses for various angles of incidence, by using well-established curves relating to the geometry of a cylinder with a sensor on one end. This is a convenient factory method, as shown on the B & K microphone calibration in Figure 2.11.

Electrostatic actuators can be designed to apply a known evenly distributed force to a plane metal diaphragm at a controlled distance, the resultant mechanical force being analogous to sound pressure. The actuators have to be accurately dimensioned for the particular cartridges involved. Figure 2.41 shows the layout of actuators for 13 mm (½ in) and 25 mm (1 in) B & K capsules. They are basically a coupler applied to the diaphragm of the capsule with an acoustically transparent insulated flat plate accurately spaced from the diaphragm, holes or slots in the driving plate allowing a substantially even driving force and free field acoustic loading on the diaphragm.

Alternating current applied to the actuator would produce frequency doubling, so a high DC polarizing voltage (800 V) is applied with an AC component of 10–100 V maximum, which applies a substantially accurate sinusoidal drive. The actuator cannot take account of the back or side vent access on a capsule and thus independent corrections must be applied below the lower limiting frequency of the microphone. This is applied in terms of the capsule vent time constant at frequencies below the lower limiting frequency of the response (see Figure 2.10).

It is easier to measure the LF time constant of a capsule vent system than to obtain a direct calibration. The capsule diaphragm is placed in a sealed cavity. The pressure therein is increased by a suddenly applied increment and the time taken for the pressure to fall to the previous static value noted. The –3 dB point is then obtained from the formula relating to the time constant of the equivalent RC circuit of the vent, where

$$f = 1/(2\pi t) \tag{2.6}$$

f = frequency of the –3 dB point

t = time for an 8.68 dB pressure fall, as above.

The B & K type 4147 extra-low-frequency 13 mm (½ in) cartridge has a time constant between 30 and 160 s and a lower limiting frequency of 0.001 to 0.005 Hz.

A VLF constant-pressure electrodynamic form of pistonphone has a freely suspended large-area piston comprising one wall of a thin annular cavity, microphones under calibration being let into the opposite wall. The system operates down to 0.001 Hz at an atmospheric pressure of 1 bar.

Insert voltage calibration is a technique which can be used for two purposes:

1. In calibration laboratories it is used to assess the open circuit sensitivity of microphone cartridges.

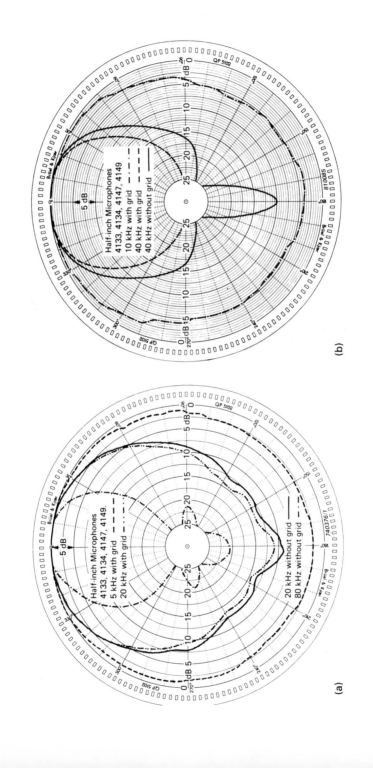

(b)

(a)

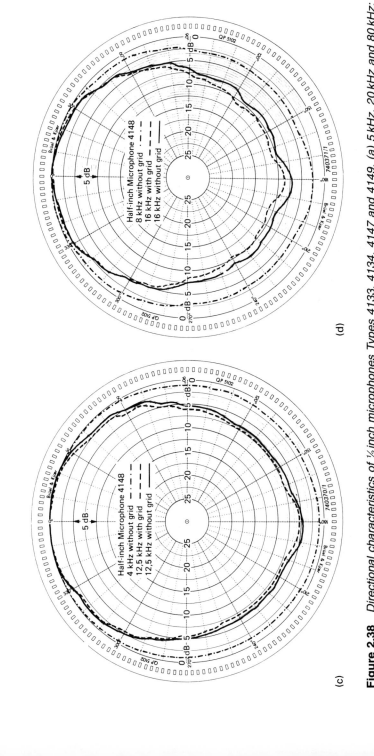

Figure 2.38 *Directional characteristics of ½ inch microphones Types 4133, 4134, 4147 and 4149. (a) 5 kHz, 20 kHz and 80 kHz; (b) 10 kHz and 40 kHz. Directional characteristics of ½ inch microphone Type 4148. (c) 4 kHz and 12.5 kHz; (d) 8 kHz and 16 kHz. Directional characteristics of ⅛ inch microphone Type 4138*

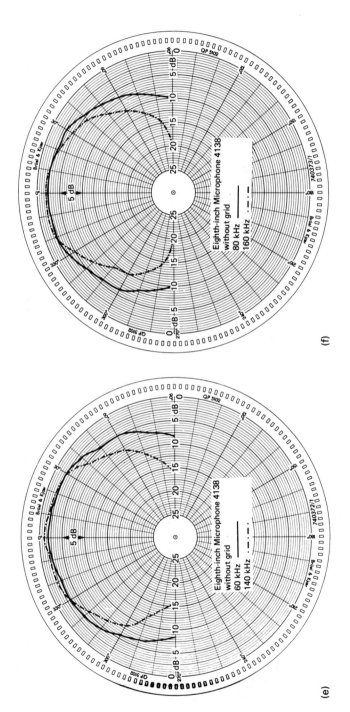

Eighth-inch Microphone 4138
without grid
60 kHz ————
140 kHz —·—·—

QP 5102
740372/1
QP 5102

(e)

Eighth-inch Microphone 4138
without grid
80 kHz ————
160 kHz —·—·—

QP 5102
740373/1
QP 5102

(f)

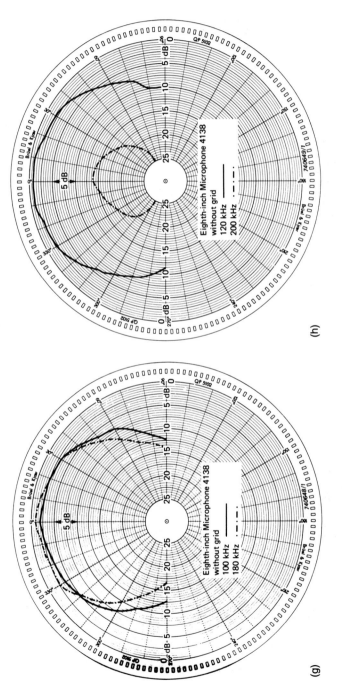

(g)

(h)

Figure 2.38 (e) 60 kHz and 140 kHz; (f) 80 khz and 160 khz; (g) 100 khz and 180 khz; (h) 120 khz and 200 khz. (Courtesy Bruel & Kjaer)

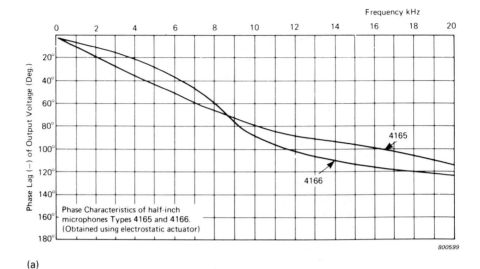

(a)

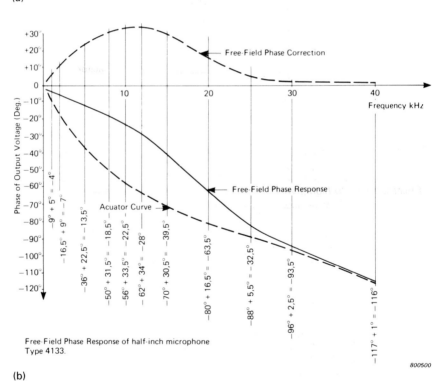

(b)

Figure 2.39 *(a) Pressure phase response of ½ inch microphones Types 4165 and 4166;*
(b) Free-field phase response of ½ inch microphone Type 4133 obtained using the pressure phase response curve and the estimated free-field phase correction. (Courtesy Bruel & Kjaer.)

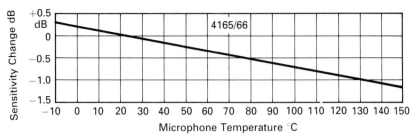

Figure 2.40 *Effect on sensitivity at 250 Hz of temperature. (Courtesy Bruel & Kjaer)*

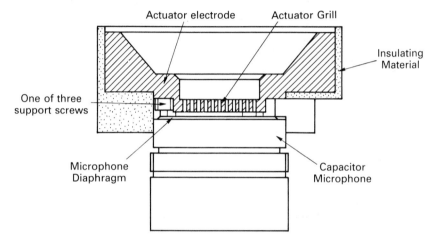

Figure 2.41 *Electrostatic actuator. Actuators are designed for measurement of the pressure frequency response of condenser microphone cartridges. Electrostatic actuators not only enable laboratory frequency response calibration of microphones to high levels of accuracy but also allow users without extensive laboratory facilities to obtain calibrations. (Courtesy of Bruel & Kjaer.)*

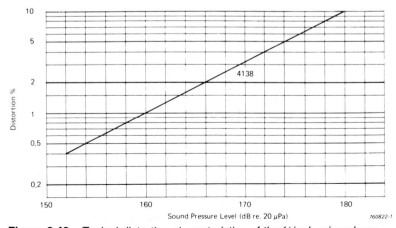

Figure 2.42 *Typical distortion characteristics of the ⅛ inch microphone.*

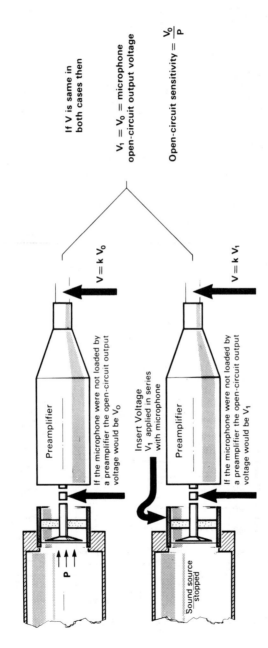

If V is same in
both cases then

$V_1 = V_0 =$ microphone
open-circuit output voltage

Open-circuit sensitivity $= \dfrac{V_0}{P}$

$V = k\,V_0$

$V = k\,V_1$

If the microphone were not loaded by
a preamplifier the open-circuit output
voltage would be V_0

Preamplifier

Insert Voltage
V_1 applied in series
with microphone

Preamplifier

If the microphone were not loaded by
a preamplifier the open-circuit output
voltage would be V_1

P

Sound source
stopped

Figure 2.43 *Principle of the insert voltage calibration.*

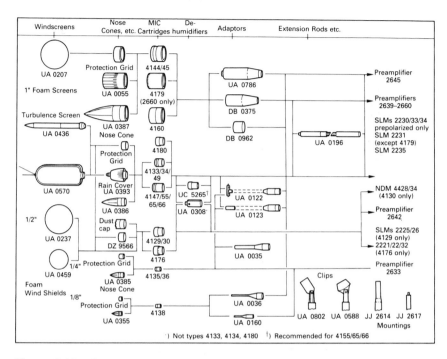

Figure 2.44   *Chart showing condenser microphone accessories. (Courtesy Bruel & Kjaer)*

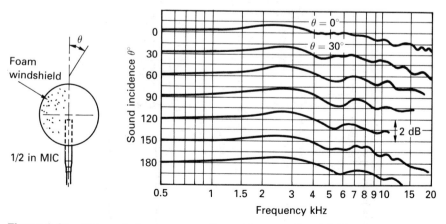

Figure 2.45   *Foam windscreen corrections. (Courtesy Bruel & Kjaer)*

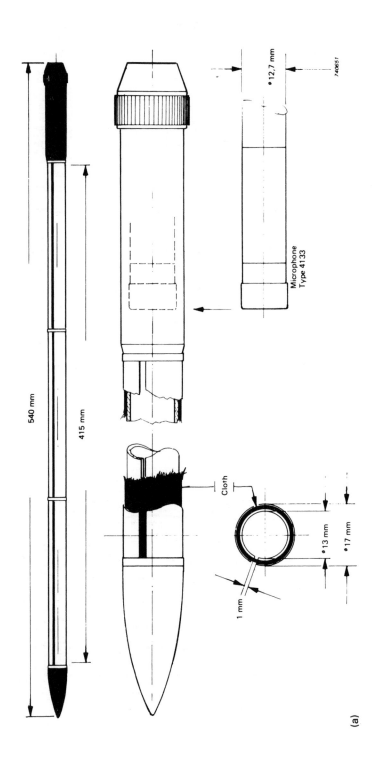

540 mm

415 mm

Microphone
Type 4133

∅12.7 mm

740651

Cloth

∅13 mm

∅17 mm

1 mm

(a)

2. It can provide a convenient means for checking in the field the electrical sensitivity of a complete sound measuring system, including preamplifier and cables. However, the method does not account for the mechanical parameters which determine the acoustical properties of the microphone cartridge itself.

For routine testing of microphones against standard microphones of the same type the insert voltage method may be used (Figure 2.43).

Pistonphones have been refined in design to give accurate pressure calibrations of complete cartridges from 25 mm to 3 mm (1 in to ⅛ in) by using suitable couplers. Fixed frequencies of 250–1000 Hz can be used at sound levels>120 dB SPL. Figure 2.33 shows a double-piston pistonphone which obtains high precision by driving two symmetrically mounted pistons from a rotating cam disc. A very high pressure pistonphone (B & K type 4221) allows levels of up to 170 dB with a 15% pulse gating. VLF couplers facilitate measurements down to 0.01 Hz. A convenient small battery-operated calibrator operates at 1000 Hz at 94 dB SPL with an accuracy of ± 0.3 to 0.5 dB. It is particularily suitable for calibrations in the field.

Some subsidiary microphone constants that are calibrated are capacitance at the specified polarization voltage and the equivalent volume of a cartridge at 250 Hz, as given on B & K calibration charts. This is done in a special coupler which takes three microphones: a 6 mm (¼in) driver, a 6 mm (¼in) pressure monitor and the capsule under measurement, which is then replaced with a blank dummy with an adjustable volume-increasing screw, which is set to bring the cavity pressure to the former level; the equivalent volume of the measured microphone is then read off from a scale.

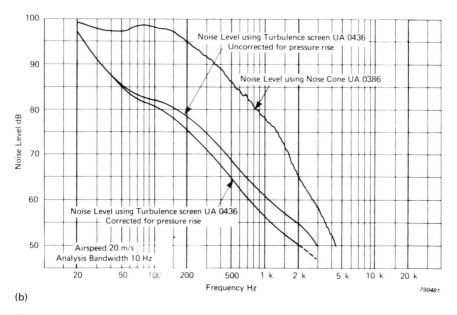

(b)

Figure 2.46   *(a) Anti-turbulence screen UA 0436; (b) comparison of turbulence induced noise levels using Turbulence Screen UA 0436 and half-inch Nose Cone UA 0386. (Courtesy Bruel & Kjaer)*

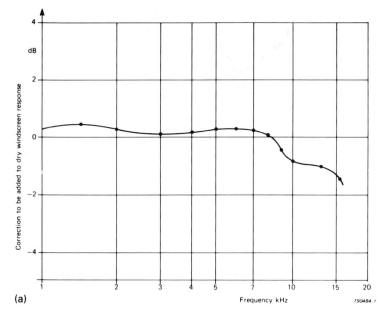

(a)

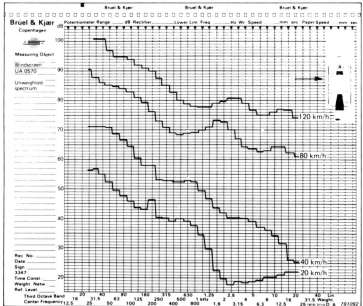

(b)

Figure 2.47 *(a) Effect of wet screen on the attenuation of Windscreen UA 0570 (The large foam screen for weatherproof microphone).*
(b) 90° incidence wind-induced noise levels for half-inch microphone Type 4133/49 fitted with Rain Cover UA 0393 and Dehumidifier UA 0308 mounted inside Windscreen UA 0570. (Un-weighted 1/3 octave bands up to 20 khz). (Courtesy Bruel & Kjaer)

B & K microphone cartridges are calibrated at the factory and this calibration will normally be valid indefinitely. However, recalibration may be desirable in certain cases, such as after lengthy operation at very high temperatures or after mechanical shocks. A complete calibration involves measuring sensitivity at a fixed frequency and a complete frequency response, a combination of calibrating methods being required. For most routine purposes a simple check of sensitivity in the 250 Hz to 1 kHz region is considered adequate, as it is unlikely that the frequency characteristics will have altered.

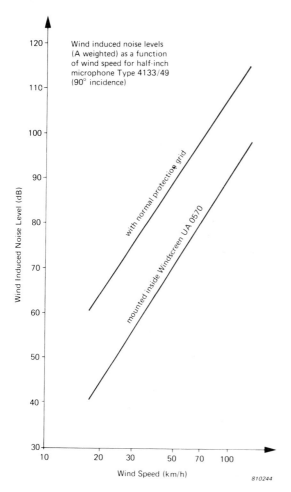

Figure 2.48 *A-weighted wind-induced noise level as a function of wind speed for half-inch microphone Type 4133/49 fitted with Rain Cover UA 0393 and Dehumidifier UA 0308 mounted inside Windscreen UA 0570. (Courtesy Bruel & Kjaer)*

(a)

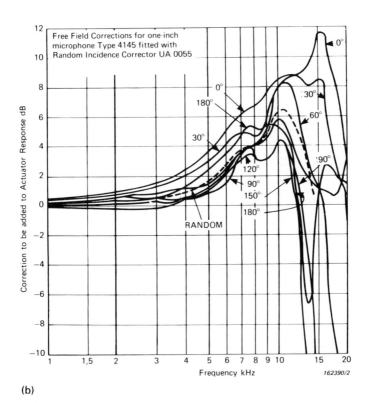

(b)

Figure 2.49   *(a) Random incidence corrector UA 0055*
(b) Free-field correction curves for 1" microphone Type 4145 fitted with random incidence corrector UA 0055. (Courtesy Bruel & Kjaer)

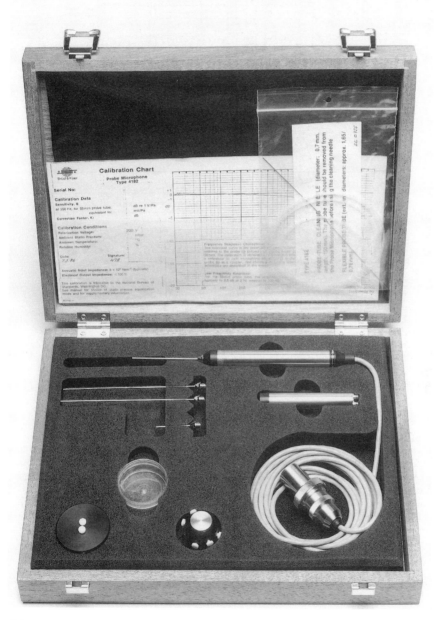

(a)

Figure 2.50 *(a) Probe microphone kit*

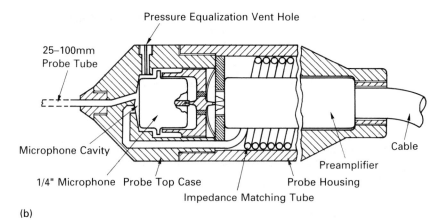

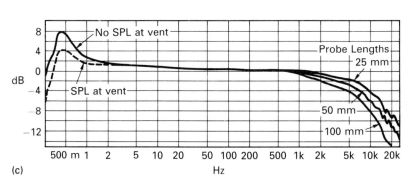

(c)

Figure 2.50 *(b) Schematic representation of the internal construction of Probe Microphone Type 4182 showing the microphone cavity, equalization vent, and impedance matching tube. (c) Typical frequency responses for various stiff probe tube lengths. (Courtesy of Bruel & Kjaer.)*

2.62 Amplifier reference voltage – Microphone K-factor

B & K measuring amplifiers and analysers have a built-in reference voltage, which is usually 50 mV at 1 kHz. This reference voltage can be used to calibrate the measuring system after the preamplifier, assuming accuracy of the microphone open circuit sensitivity and nominal gain of the preamplifier, which generally is quite justified. The sensitivity of the microphone and preamplifier combination is given by:

$$S_{\mathrm{MP}} = S_0 + g + 20\log_{10}\left(\frac{C_t}{C_t + C_i}\right) \text{ dB with reference to } 1\,\text{V per Pa}$$

where: S_0 = microphone open-circuit sensitivity (dB with reference to 1 V per Pa)

g = preamplifier gain (dB)
C_t = microphone capacitance (pF)
C_i = preamplifier capacitance (pF)

Values of S_0 and C_t are given on the microphone calibration chart. Values of g and C_i are given with sufficient accuracy for general-purpose use in tables.

In practice, use of the equation is made simpler by information on the microphone calibration chart. The calibration chart gives a correction factor K_0 which is the dB difference between a reference sensitivity of –26 dB with reference to 1 V/Pa (50 mV) and the microphone open-circuit sensitivity:

$$K_0 = -26 - S_0 \text{ dB}$$

Typical overall gain of the preamplifier in use is given on the back of the microphone calibration chart. The figure given on the back of the calibration chart, G, includes the effect of microphone and preamplifier capacitance, so

$$G = g + 20 \log_{10} \left(\frac{C_t}{C_t + C_i} \right) \text{ dB}$$

Hence, an overall correction factor K can be defined:

$$K = K_0 - G \text{ dB} \tag{2.7}$$

For example, the open-circuit correction factor K_0 is given as + 12.3 dB on the calibration chart of a 13 mm ($\frac{1}{2}$ in) microphone Type 4133. When used with Preamplifier Type 2619, the back of the calibration chart gives an overall gain factor G of –0.2 dB. Therefore, the overall correction factor K is + 12.3 – (–0.2 dB) = + 12.5 dB.

This is very simply used in practice by first setting up the measuring instrumentation to the reference sensitivity of –26 dB with reference to 1 V per Pa and then correcting this by adding K dB to the readings obtained. Alternatively, the measuring instrumentation can be set up for a microphone sensitivity of –26 dB – K dB with reference to 1 V per Pa.

2.7 Piezoelectric microphones and hydrophones

Some earlier laboratory piezoelectric transducers used directly actuated expander bars. A stack of bars is exposed to end sound pressure by a cemented-on diaphragm, the sides being electroded and polarized to be in series in order to increase output. The bottom of the bars rests on a massive base structure which also extends up to seal off side access to sound.

Modern hydrophones/high-level sound transducers may use suitably polarized piezoelectric rings or plates, which may be stacked to increase output, or hemispherical cups as shown in Figures 2.51 and 2.52.

The use of these piezoelectric structures means that they are responsive to the surrounding sound pressure and can thus conveniently be completely sealed in an acoustically transparent protective sheath, which may also isolate the sensors from any mechanical vibration of the cable and mountings.[11] Figure: 2.53 and Table 2.10 give data.

Units of this type may also be used for sound-pressure measurements in air where very high sound levels and extreme conditions of humidity and pollution may be encountered. Some units are sensitive to mechanical vibration in certain planes and may have to be suitably mounted.

For regions in the response where the curve is flat, the sensitivity in terms of sound pressure is the same in air as in water. The much higher characteristic impedance of water means that, for a given energy in the sound wave, the pressure in water and hence the output is much higher.

Some of these hydrophones may be used as senders as well as microphones and hence lend themselves to reciprocity calibrations in tanks, when electronic gating is used to avoid reflections.

B & K make a special multi-piston pistonphone calibrator which can be coupled to the hydrophones for sensitivity at 250 Hz.

A number of standards apply to hydrophone calibrations and methods of measurement.

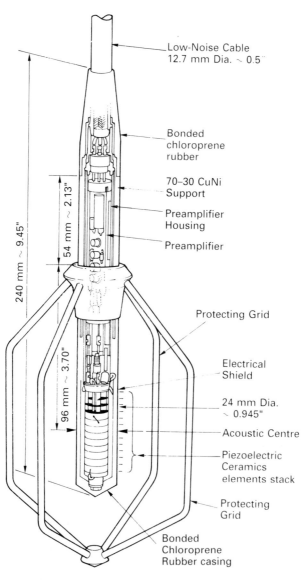

Figure 2.51   *Piezoelectric Hydrophone Type 8101. (Courtesy Bruel & Kjaer)*

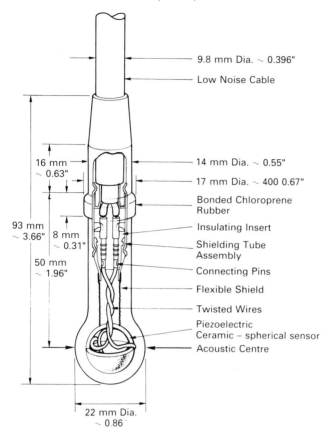

Figure 2.52 *Spherical Hydrophone Type 8105*
*The ceramic elements are specially shaped and polarized to produce a voltage
corresponding to the sound wave pressure. It may also be used as a sender.*

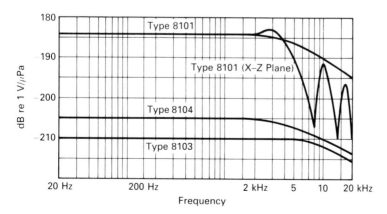

Figure 2.53 *Hydrophones. Typical receiving characteristics in air: 8103 and
8104 in the x-y plane, and of the 8101 in the x-y and x-z planes. (Courtesy
Bruel & Kjaer)*

Table 2.10 *Hydrophone specifications 8181, 8103, 8104 and 8105. (Courtesy of Bruel & Kjaer.)*

Type	8101	8103	8104	8105
Voltage sensitivity, nominal: (with integral cable)	630 µV/Pa* (−184 dB re 1 V/uPa)	30 µV/Pa* (−211 dB re 1 V/µPa)	56 µV/Pa* (−205 dB re 1 V/µPa)	
Charge sensitivity, nominal:	—	0.12 pC/Pa*	0.42 pC/Pa*	0.40 pC/Pa*
Capacitance, typical: (with integral cable)	—	3850 pF	7800 pF	7200 pF
Frequency range:	(+1.5 dB) 0.1 Hz to 60 kHz (−2.5 dB) / (+2 dB) 1 Hz to 120 kHz (−10 dB)	(+0.5 dB) 0.1 Hz to 100 kHz (−3.5 dB) / (+2 dB) 0.1 Hz to 180 kHz (−10 dB)	(±2 dB) 0.1 Hz to 60 kHz / (+2 dB) 0.1 Hz to 120 kHz (−10 dB)	(+0.5 dB) 0.1 Hz to 100 kHz (−4 dB) / (+2 dB) 0.1 Hz to 160 kHz (−10 dB)
Horizontal directivity† (at 100 kHz)		± 2 dB (typical)	± 2 dB (typical)	
Vertical directivity:†	± 2 dB (typical) at 15 kHz	± 4 dB (typical at 100 kHz)	± 2 dB (typical) at 50 kHz	± 2 dB (typical) over 270° at 100 kHz
Leakage resistance: (at 20°C)	—		> 2500 MΩ	
Operating temperature range Short-term continuous	−10°C to +60°C (14°F to 149°F)		−40°C to +120°C (−40°F to +248°F) −40°C to + 80°C (−40°F to 176°F)	
Sensitivity change with temperature Charge:		0 to +0.03 dB/°C (0 to 0.017 dB/°F)	0 to +0.03 dB/°C (0 to 0.017 dB/°F)	0 to +0.03 dB/°C (0 to 0.017 dB/°F)
Voltage:	0 to −0.03 dB/°C (0 to −0.017 dB/°F)	0 to −0.03 dB/°C (0 to −0.017 dB/°F)	0 to −0.04 dB/°C (0 to −0.022 dB/°F)	0 to −0.03 dB/°C (0 to −0.017 dB/°F)
Max. operating static pressure	252 dB = 4 × 10⁶ Pa = 40 atm. = 400 m ocean depth			260 dB = 9.8 × 10⁶ Pa = 100 atm. = 1000 m ocean depth

Sensitivity change with static pressure:	0 to -3×10^{-7} dB/Pa (0 to -0.03 dB/atm.)			
Allowable total radiation dose:	5×10^{7} Rad.			
Dimensions: Length: Body dia: Width across the cage:	248 mm (9.76") 24 mm (0.95") 132 mm (5.21")	50 mm (1.97") 9.5 mm (0.37")	120 mm (4.73") 21 mm (0.83")	93 mm (3.66") 22 mm (0.87")
Weight: (including integral cable)	3 kg (6.6 lb)	170 g (0.37 lb)	1.6 kg (3.5 lb)	
Integral cable:	10 m waterblocked low-noise shielded cable to MIL-C-915 with 7-pin B & K plug	6 m waterproof low-noise double-shielded teflon cable with standard miniature coaxial plug	10 m waterblocked low-noise shielded cable to MIL-C-915 with BNC plug	
	Individual calibration chart and calibration data			
Accessories included:	Removable protective cage 1 Adaptor .... DB 2609	Mahogany case		
Accessories available:	Waterblocked low-noise shielded extension cable to MIL-C-915 available to any length up to 300 m .... AC 0038 Underwater connectors: Male .... JP 0415 Female .... JJ 0415		Waterblocked low-noise shielded extension cable to MIL-C-915 available to any length up to 300 m .... AC 0034 Underwater connectors: Male .... JP 0415 Female .... JJ 0415	

2.8 Precision condenser microphones for studio and broadcasting use

Condenser microphones are in demand for high-quality sound work because of their good response characteristics and general suitability. Specialized condenser microphones have been developed to suit the situations and equipment involved in studio, concert hall and outside broadcasts.

Modern practice is exemplified by the Bruel & Kjaer range of professional condenser microphones, together with specially developed accessories, such as

Table 2.11   *Microphone specifications (omnidirectional). (Courtesy of Bruel & Kjaer.)*

On-axis frequency response:
20 Hz to 40 kHz ± 2 dB

Phase response:
See figure. Phase matching between any two microphones ± 5° (50 Hz to 20 kHz)

Sensitivity at 250 Hz:
Type 4004: 10 mV/Pa, unloaded
Type 4007; 2.5 mV/Pa, unloaded

Equivalent noise level:
Typically 24 dB(A) (max 26 dB(A))

Harmonic distortion:
Type 4004: ≤1% at 148 dB peak SPL, f < 20 kHz
            ≤0.002% at 94 dB (extrapolated)
Type 4007: ≤1% at 148 dB peak SPL f > 100 Hz
            ≤0.002% at 94 dB (extrapolated)

Difference frequency distortion:
Type 4004: ≤1% at 153 dB peak SPL
Type 4007: 1% at 153 dB peak SPL, f > 500 Hz

Maximum sound pressure level:
Type 4004: 168 dB peak SPL, f ≤ 4 kHz
Type 4007: 155 dB peak SPL, f > 200 Hz

Weight: 150 g

Dimensions:
Length: 165 mm excluding connector
Cartridge diameter: 12 mm

Accessories included (Types 4004 S & 4007 S):
Cable AO0261 (type 4004 S)
Cable AO0182 (type 4007 S)
Windscreen: UA0658
Microphone clamp: UA0639

Table 2.12 Cardioid microphone specifications. (Courtesy of Bruel & Kjaer.)

CARTRIDGE TYPE:
Prepolarized condenser-microphone cartridge

PRINCIPLE OF OPERATION:
Pressure gradient

POWERING:
48 V phantom powering (in accordance with DIN 45 596). Transformerless

DIRECTIONAL CHARACTERISTICS:
Directional – first-order cardioid pattern (see Fig. 3)

FREQUENCY RESPONSE:
(See also Fig. 2)
On-axis: 40 Hz to 20 kHz ± ½ dB (0.3 m distance)

NOMINAL SENSITIVITY:
10 mV/Pa (−40 dB re 1 V/Pa) (individually calibrated)

POLARITY:
Positively increasing sound pressure produces positive-going voltage at Pin 2.
XLR Output Connector: Pin 1: Shield; Pin 2: Signal (+); Pin 3: Signal return

PHASE MATCHING:
Phase matching between any two microphones ± 5 degrees (100 Hz to 20 kHz)

EQUIVALENT NOISE LEVEL:
A-weighted: typically 19 dB(A) re 20 μPa (Max. 20 dB(A))
CCIR 468-1: typically 25 dB

MAXIMUM SOUND PRESSURE LEVEL:
158 dB SPL peak before clipping (attenuator 0 dB or −20 dB)

TOTAL HARMONIC DISTORTION:
<0.5% at 110 dB SPL peak (worst case in the specified frequency range)

DIFFERENCE FREQUENCY DISTORTION (DF2, DF3, ΔF = 70 Hz):
<0.5% at 110 db SPL (worst case in the specified frequency range)

INFLUENCE OF VIBRATION:
<71 dB equivalent SPL for 1 m/s² in direction of greatest sensitivity

INFLUENCE OF MAGNETIC FIELD:
45 dB equivalent SPL for 80 A/m, 50 Hz in direction of greatest sensitivity

PREAMPLIFIER:
Frequency Range (−3dB):
20 Hz to 40 kHz
Output Impedance: 180 Ω

OPERATING TEMPERATURE RANGE:
−10 to +55°C (+14 to +131°F)

WEIGHT:
165 g (0.36 lb)

DIMENSIONS:
Overall length: 175 mm (6.75 in)
Cartridge diameter: 19 mm (0.75 in)

ACCESSORIES INCLUDED, TYPE 4011S:
Windscreen .... UA 0896
Microphone holder .... UA 0961
Cable .... AO 0182

ACCESSORIES INCLUDED, TYPE 4011P:
4 × Windscreen .... UA 0896
4 × Microphone Holder .... UA 0961

ACCESSORIES AVAILABLE:
Set of 4 Windscreens (UA 0896) ...... UA 0964
Set of 4 Microphone Holders
(UA 0961) .... UA 0962
Set of 4 Microphone Clips
(UA 0639) .... UA 0963
Shock Mount .... UA 0897

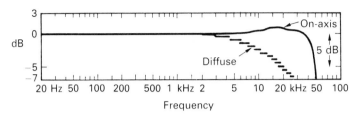

(a)   Frequency responses of types 4004 and 4007

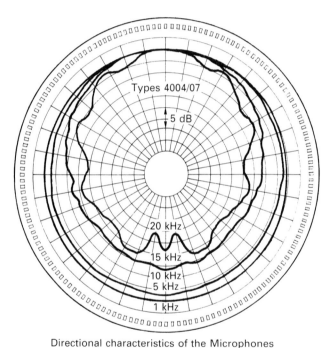

Directional characteristics of the Microphones

(b)

Figure 2.54   *Responses of omnidirectional studio condenser microphones.*
(a) Types 4004 and 4007 are 12 mm diameter, prepolarized condenser
microphones of essentially the same design and construction as Types 4003
and 4006. The smaller cartridge diameter results in extended frequency and
phase responses and a considerably higher degree of omnidirectivity.
(b) Type 4006 is specifically designed for powering from standard P48 Phantom
supplies and is distinguished from Type 4003 by the silver ring at the base of
the main body housing. It is acoustically identical to the 4003 and has the same
signal-to-noise ratio, although the sensitivity is a factor of four lower (nominally
12,5mV/Pa unloaded) due to the integral transformer circuitry. Like the 4003,
Type 4006 is ideally suited for low noise applications. (Courtesy Bruel & Kjaer)

alternative fronts, nose cones, clips, holders, anti-vibration mounts, windshields, transformerless interfaces and power supplies, phantom feeds etc.[12]

Individual types of microphones of 12 and 16 mm diameters with unidirectional and cardioid polar response, for especially low background noise situations and for high sound levels, are produced with high-stability electret polarization. The range, comprising six types, is illustrated in Figure 2.2. Tables 2.11 and 2.12 summarize their characteristics. Full specifications and response curves follow for the various microphones (Tables as above, and Figures 2.54, 2.3).

The above microphone types, which are fitted with a bright bottom ring, are suitable for standard 48 V DIN 45596 power-supply systems. The other microphones without bright bottom rings are designed to be powered from the

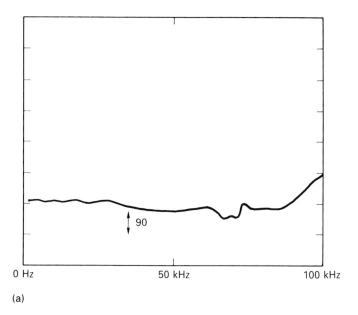

(a)

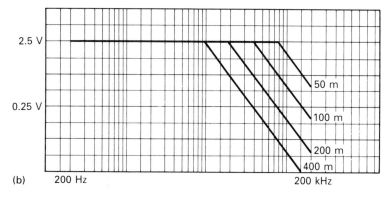

(b)

Figure 2.55   *(a) Phase responses; (b) output characteristics. (Courtesy Bruel & Kjaer)*

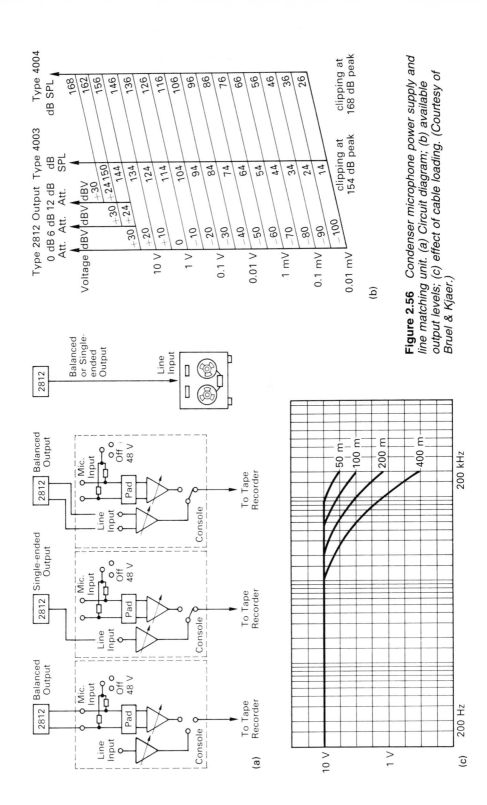

Figure 2.56 Condenser microphone power supply and line matching unit. (a) Circuit diagram; (b) available output levels; (c) effect of cable loading. (Courtesy of Bruel & Kjaer.)

Table 2.13 *Power supply unit specifications (Bruel & Kjaer)*

FREQUENCY RANGE:
15 Hz to 200 kHz ± 0.5 dB

DYNAMIC RANGE:
1 µV TO 10 V (140 DB)

INPUT:
Via modified 4-pin XLR connector.
Accepts cable AO 0261 supplied with
Microphones Types 4003 S and 4004 S

INPUT ATTENUATOR:
Switchable 0; −6; −12 dB

MAXIMUM INPUT VOLTAGE:
16; 32; 64 V peak

INPUT IMPEDANCE:
10 kΩ

MAXIMUM OUTPUT VOLTAGE:
32 V peak

MAXIMUM DC OFFSET:
± 50 mV

MINIMUM LOAD IMPEDANCE:
600 Ω

TOTAL HARMONIC DISTORTION:
<−75 dB (20 Hz to 40 kHz)

CHANNEL CROSS-TALK:
<−90 dB (0 Hz to 20 kHz)

EQUIVALENT INPUT NOISE:
A-weighted: <2 µV
C-weighted: <2.5 µV
CCIR468-1:   <10 µV

POWERING:
100 to 127 V and 200 to 250 V, 50 to 60 Hz
AC mains supply

OPERATING TEMPERATURE RANGE:
−10 to +70°C (+14 to +158°F)

WEIGHT:
1.75 kg (3.85 lb)

DIMENSIONS:
200 × 126 × 46 mm (7.9 × 5.0 × 1.8 in)

ACCESSORIES INCLUDED:
Power Cable AN0027
Two 100 mA slow blow fuses VF 0026

Bruel & Kjaer two-channel power supply type 2812. This unit provides a high-level 130 V with single-ended or balanced outputs for symmetrical or asymmetrical standard microphone or line inputs in use. Figure 2.56 and Table 2.13 give the specification of this unit, which is transformerless in the signal circuit. The output may also be fed to circuits already equipped with phantom power if required. The mains input voltage selection is automatic and needs no manual tap changing.

The use of a 130 V high-level power supply to the microphones and preamplifiers gives increased headroom for handling high sound levels.

The B & K 4003 and 4006 condenser microphones are low-noise omnidirectional types with preamplifiers incorporated. The 4006 is equipped with a standard XLR socket for 48 V phantom supply.

The 4004/7 are similar omnidirectional microphones designed to accept higher sound levels for close-source use. The 4004 powered from the B & K 130 V supply unit will accept up to 168 dB SPL before clipping. Alternative fronts can give HF boost, or a nose cone can improve omnidirectional response.

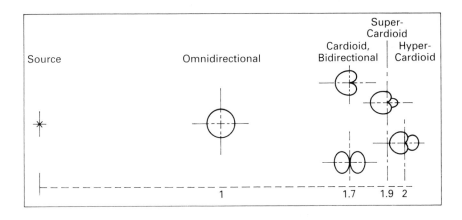

(a)

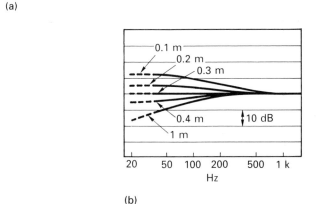

(b)

Figure 2.57 *(a) Source-to-microphone distances for various directional characteristics giving the same proportions of direct and diffuse sound. (b) Proximity effect with Type 4011. (Courtesy of Bruel & Kjaer.)*

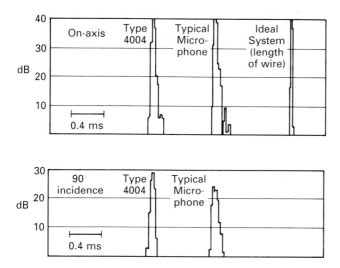

Figure 2.58 *Transient response of microphones in the form of energy-time response. Energy-time curves illustrate the transient response of the microphones. The narrower the ETC, the better the response. (Courtesy of Bruel & Kjaer.)*

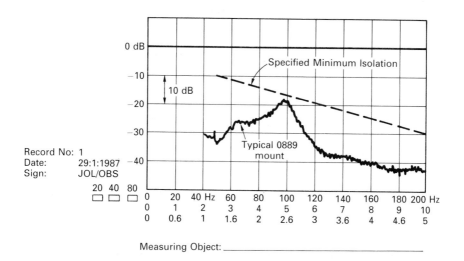

Figure 2.59 *Characteristics of a microphone anti-shock mounting. (Courtesy Bruel & Kjaer.)*

The 4011 and 4012 microphones are first-order gradient types with cardioid polar response (see Figure 2.5). These units incorporate openings in the sides of the tubular casing which lead to the rear of the diaphragm via phase-shifting networks acting on the sound wave inside the case. These microphones are transformerless and may be supplied from a standard P 48 phantom or from the B & K 130 V supply unit. The 4011 has a recessed –20 dB attenuator switch located between the XLR pins in the microphone base, in order to allow use of close to high-level sources.

The usual low-frequency gradient proximity boost is shown in Figure 2.3. Output cables up to 200 m in length may be used but may cause a small reduction of high frequencies and a small reduction in the upper clipping level.

Figure 2.5 and Table 2.13 show the responses, polar curves and other performance data for these microphones.

A complete range of alternative fronts, mounting clips, anti-shock mounts and windscreens has been produced for these microphones. It may be noted that the amount of information given in the curves and data relating to these microphones is unusually full and indicates the amount of research and development that has been devoted to these microphones.

As a matter of interest, it may be noted that the theoretical equivalent circuit for the above microphones is of the same basic form as for the measuring microphones (see Figure 2.7). The form and values assigned to the rear vent passages are modified in the case of the cardioid microphone so as to give the desired phase shift to the wave reaching the rear of the diaphragm, in contrast to the higher resistance leak value required for the atmospheric vent in omnidirectional microphones.

References

1. Hunt, F.V. (1954) *Electroacoustics*. John Wiley, Chichester. Reprinted by the Acoustical Society of America, 1982.
2. Rassmussen, G. and Bruel, P.V. (1972) *Selected Reprints on Measuring Microphones*. Bruel & Kjaer, Naerum, Denmark.
3. Rasmussen, G. (1959) A new condenser microphone. *B & K Technical Review*, No. 1.
4. Rasmussen, G. (1963) Miniature pressure microphones. *B & K Technical Review*, No. 1.
5. Frederiksen, E. (1969) Long term stability of condenser microphones *B & K Technical Review*, No. 2.
6. Gayford, M.L. (1970) *Electroacoustics*. pp. 168, 241. Butterworth-Heinemann Newnes.
7. Brüel & Kjaer (1982/92) *Condenser Microphone Data Handbook*.
8. Frederiksen, E. (1974) Influence of sunbeams striking the diaphragms of measuring microphones. *B & K Technical Review*, No. 1.
9. Brüel, P.V. (1964) The accuracy of condenser microphone calibration methods. Part I. *B & K Technical Review*, No. 1.
10. Brüel, P.V. (1965) The accuracy of condenser microphone calibration methods. Part II. *B & K Technical Review*, No. 1.
11. Bruel & Kjaer (1990) *Instruction Manual on Hydrophones 8101–8105*.
12. Bruel & Kjaer (1989) *Instruction Manual on Professional Microphones*.
13. Brüel, P.V., and Rasmussen, G. (1959) Free field response of condenser microphones. *B & K Technical Review*, Nos 1 and 2.

Further reading

Bernard, P. Microphone intermodulation distortion measurements using the high pressure microphone calibrator 4221. *Bruel & Kjaer Application Note*, 083–80.
Bernard, P. and Fredericksen, E. High pressure measurements with the Mic Calibrator Type 4221. *Bruel & Kjaer Application Note*, 17–231.

Bruel & Kjaer. (1990/91/92) *Short Form Catalogue.*

Bruel & Kjaer (1970) *National and International Standards.* (Brochure.)

Frederiksen, E. (1977) Condenser microphones used as sound sources. *B & K Technical Review*, No. 3.

Fredericksen, E. (1977) The low impedance microphone calibrator advantages. *Bruel & Kjaer Technical Review*, No 4.

Frederiksen, E. (1979) Prepolarised microphones for measurements. *Bruel & Kjaer Technical Review*, No. 4.

Rasmussen, G. (1960) Pressure equalization of condenser microphones and performance at varying altitudes. *B & K Technical Review*, No. 1.

Rasmussen, G. (1969) The free field and pressure calibration of condenser microphones using electrostatic actuator *B & K Technical Review*, No. 2.

Skøde, F. (1966) Windscreening of outdoor microphones. *B & K Technical Review*, No. 1.

Taniguchi, H.H. and Rasmussen, G. (1980) Microphones for aircraft noise measurements. *Bruel & Kjaer Technical Review*, No 4.

3 Optical microphones

David Keating

3.1 Introduction

For many years there have been no viable alternatives to the electrostatic (capacitor or electret) or electromagnetic (dynamic or moving coil) methods of converting diaphragm movements into electrical signals. Recent research (Keating[1]), however, has shown that optical techniques may be capable, not only of offering alternatives of similar performance to conventional methods, but also of offering an improvement in performance. Optical techniques have in the past been employed to examine the detailed behaviour of microphone diaphragms especially when excited at high frequencies (e.g. Ohashi *et al.*[2]). These techniques have, however, been restricted to very costly precision laboratory set-ups which are quite unsuitable for use as part of a commercial microphone system. In recent years the communications industry has developed integrated optical components (Weardon[3]) which offer the possibility of constructing inexpensive small-scale optical measuring systems. One example of this is the system used to read the information stored on the domestic compact disc. This system consists essentially of a miniature interferometer designed to give a '1' or '0' output depending on the presence or absence of bumps of a specific height which have been pressed on to the internal surface of the disc.

This chapter gives an introduction into optical components and techniques, and examines several optical microphone configurations. The use of force-feedback to improve linearity and increase dynamic range is discussed at the end of the chapter.

3.2 Optical components

This section is intended to give a brief introduction to optical components for those who are not familiar with the field of optics. Readers requiring a more detailed description of optical sources and sensors will find useful information in Senior[4] whilst Dyson[5] contains details of the operation of more general components such as beam splitters and quarter- and half-wave plates.

3.2.1 Sources

The light sources used in interferometry must be coherent, that is all the photons must have a constant phase shift with respect to one another or more simply they must all have precisely the same wavelength. It is usually convenient if the source produces a narrow beam of light, the beam being of low divergence. The most convenient way of producing such a beam is to use a laser. Traditionally gas lasers such as the helium – neon type would have been used, but more recently semiconductor laser-diodes have gained favour, primarily due to their lower cost and simpler drive requirements. Laser diodes have the disadvantage that they produce divergent elliptical beams and are prone to 'mode hopping' (small but sudden changes in wavelength). The output power and wavelength are also temperature dependent. Due to their small size, however, they may be temperature stabilized and further due to their simple drive requirements the output power may also be stabilized. Lenses may be used to convert the divergent elliptical beams into parallel circular ones, thus giving a source which meets the requirements of interferometry.

If the interferometer is of the optical heterodyne type, the source must produce two physically separable beams of slightly different wavelength. This may be achieved by using two different modes of one laser or by splitting the beam of a single mode laser and changing the wavelength of one or both beams by acousto-optic modulators or acousto-optic tunable filters.

3.2.2 Beam splitters

The simplest and perhaps most common beam splitter is the partial mirror. When a beam strikes a partial mirror some of the incident light is reflected whilst some passes through, thus splitting the beam. Another common beam splitter is the polarizing beam splitter. This allows light of a certain polarization to pass through whilst light with a polarization at 90° to that is reflected. An advantage of polarizing beam splitters is that they may be used in conjunction with quarter-wave plates to ensure no light is reflected back into the laser. This is important to ensure correct operation of the laser.

3.2.3 Quarter-wave plates

The function of a quarter-wave plate is to turn linearly polarized light into circularly polarized light and vice versa. If a linearly polarized beam is passed through a quarter-wave plate and is then reflected from a mirror so as to pass back through the plate, then the polarization of the beam will have undergone a 90° rotation.

3.2.4 Half-wave plates

The function of a half-wave plate is the same as that of two cascaded quarter-wave plates, the polarization of the incident beam is rotated by 90° in the output beam.

3.2.5 Optical isolators

If the configuration of the optics could allow some reflected light to pass back towards the laser, then this must be prevented from occurring as it may affect the operation of the laser or even damage it. The reflected light may be isolated by means of a polarizer and a quarter-wave plate. The polarizer allows linearly polarized light to pass through and on to the quarter-wave plate where the linearly polarized light becomes circularly polarized. On subsequent reflection the direction of rotation is effectively reversed and once the light has passed back through the quarter-wave plate it produces linear polarization once more, but at 90° to that of the incident beam. The polarizer prevents light polarised in this plane from passing back through and hence stops it from reaching the laser.

3.2.6 Mirrors

Mirrors are usually of glass 'silvered' on its front surface. A metal or metallized plastic microphone diaphragm may act as a mirror if it is sufficiently smooth.

Partial mirrors are possible by varying the amount of silvering giving reflectances anywhere between close to 0% to close to 100%. The actual silvering is often a metal such as aluminium but may also be a dielectric coating.

3.2.7 Sensors

The most common sensor for the visible or near infrared region is the PIN photodiode. This is a device that has low dark current and hence low noise but also has high responsivity. Avalanche photodiodes give higher responsivities but suffer from proportionally higher noise and so are not often used. Photo-multipliers may also be used but are usually impractical.

3.3 Optical measurement techniques

3.3 1 Mirror galvanometer

This is the only non-interferometric technique which will be considered in this chapter as it is the only technique capable of the detectivity required. The principle of operation is simple but effective; a beam of light striking the microphone diaphragm near its edge will be deflected if there is any curvature of the diaphragm. The deflection of the beam can be detected by a pair of sensors. The principle of operation is shown in Figure 3.1.

The peak-to-peak deflection in the centre of a typical capacitor microphone diaphragm is $8.5\,\mu m$ for a sound pressure of $140\,dB$ SPL (ref. 2×10^{-5} Pa r.m.s.). This results in a displacement angle at the edge of the diaphragm of $0.13°$ or $2.2\,mrad$. The displacement (d) of the beam at a given distance (s) from the diaphragm is given approximately by

$$d = 2.2 \times 10^{-3}\,s \tag{3.1}$$

if the displacement of the diaphragm is small.

For a sensor to diaphragm separation of $25\,mm$ the deflection of the beam at a sound pressure of $140\,dB$ is $55\,\mu m$. The optimum diameter of the sensors is thus

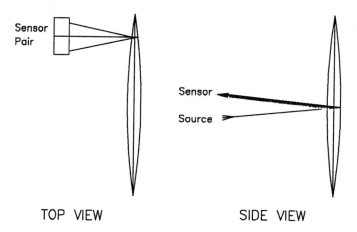

TOP VIEW   SIDE VIEW

Figure 3.1   *Principle of operation of mirror galvanometer.*

55 μm. This is most easily achieved by using a pair of optic fibres of about 50 μm diameter connected remotely to photodiodes. If the full power of the laser source is to be utilized, then the beamwidth must also be 50 μm.

Noise

Assuming a 50 μm diameter beam of low divergence is available (and this may be difficult to achieve in practice) the standing light power at each sensor and at zero displacement of the diaphragm will be approximately one third of the peak power. The shot noise in the photodiode sensor will be the dominant source of noise in the sensor system, and this is given by

$$i_n = \sqrt{(2eIB)} \qquad\qquad (3.2)$$

where i_n is the r.m.s. noise current in amperes, e is the electron charge in coulombs $(1.6 \times 10^{-19}\text{C})$, I is the diode current in amperes and B is the bandwidth in hertz.

A typical laser diode output power is 4 mW and a typical photodiode responsivity is 0.5 A/W. The standing current in each photodiode is thus approximately 670 μA, which results in a noise current in a bandwidth of 20 kHz of 2 nA rms. The peak output current will be 2 mA and the dynamic range, taken here as

$$\text{dynamic range} = 20 \log_{10} (\text{max r.m.s. signal current/r.m.s. noise}$$
$$\text{current in 20 kHz}) \qquad\qquad (3.3)$$

will be 114 dB. This figure takes into account the fact that there are two photodiodes but takes no account of source noise. The equivalent noise level is 26 dB SPL, which is somewhat worse than a typical capacitor microphone with 48 V polarization and FET amplification.

There are many problems with this method of measuring diaphragm movement, but the main problem is probably that of generating and measuring a 50 μm wide light beam. The alternative to this is to use a focusing lens at the sensor, although this will reduce the displacement of the beam slightly. To ease alignment problems and also allow the beam to enter and leave on the same path (this would allow the beam to be passed through a small hole in an existing back plate) the configuration shown in Figure 3.2 may be used. This configuration uses optic fibres to give a small effective sensor width. The photodiodes are placed a small distance from the end of the fibre to allow the beam to diverge slightly. This prevents the light concentration from becoming too high at the photodiode.

Other problems with this technique are that the response is non-linear, although this is significant only for large sound pressures, and more importantly that the technique depends on the edge of the diaphragm being well behaved, which it seldom is. The edge of the diaphragm is sometimes crinkled and once the diaphragm breaks up into resonant modes the edge may respond in a totally different fashion to that of the diaphragm as a whole. If the diaphragm is carefully made, however, and the resonant modes well damped or out of band, then the only remaining problem is that of limited dynamic range. The source power of 4 mW quoted earlier is typical of cheap laser diodes but others are available with powers up to 100 mW. An increase in laser power of this amount would result in an increase in dynamic range of 14 dB. In the preceding calculation, however, the source noise was not taken into account and if this is done the dynamic range will be reduced by at least 3 dB. The dynamic range would thus become 125 dB, which could give an equivalent self noise of 15 dB and a maximum capability of 140 dB.

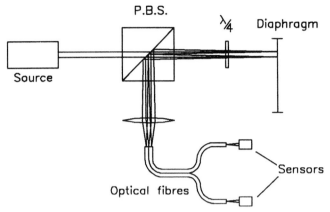

Figure 3.2   *Practical mirror galvanometer.*

These figures are comparable with those of existing microphones but the optical technique is more complicated and hence more expensive.

3.3.2 Fundamentals of interferometry

There are two distinct classes of interferometric technique which may be used in optical microphones. These are single wavelength interferometry and dual wavelength interferometry. In all interferometry, however, the basic principle is the same; the incoming beam of light is split into two beams which are then passed along different paths, the interference takes place once the two beams are recombined, having travelled different distances. In single-wavelength inter- ferometry the beams all have the same wavelength and the interferometer responds only to path-length difference. If one of the paths includes a reflection from a mirror, then any movement of the mirror along the beam axis will change the path length and hence the output of the interferometer. In dual-wavelength interferometry beams of slightly differing wavelength are combined, producing a beat frequency (usually at RF); this technique is known as optical heterodyning. In this case if one of the paths includes a reflection from a mirror moving along the beam axis, then the resulting beat frequency will be frequency modulated by the velocity of the moving mirror. The phenomenon is similar to that exploited in Doppler radar systems. An output proportional to the velocity of the mirror may be obtained by frequency demodulation of the beat frequency. This technique does not respond to the absolute position of the mirror and so the output has no DC component. This can be both an advantage and a disadvantage depending on the application.

3.3.3 Single-wavelength interferometers

The Michelson interferometer

The basic principle of the Michelson interferometer is shown in Figure 3.3. The incident beam is split by the beam splitter producing a reference beam and an object beam. The path-length difference is achieved by reflecting the reference beam from a fixed mirror and the object beam from a movable mirror. The two

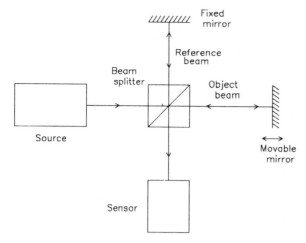

Figure 3.3 *Michelson interferometer.*

beams are recombined by the beam splitter and interference takes place. The resultant beam is then detected by the sensor. If the difference between the distance from the beam splitter to the fixed mirror and back and that from the beam splitter to the movable mirror and back is an exact number of wavelengths, then the reflected beams will arrive at the beam splitter in phase and constructive interference will take place. If, however, the difference in distance is an odd number of half-wavelengths, then the two beams will arrive out of phase and destructive interference will take place. At difference in distance between these two extremes partial interference will take place. These three cases are shown in Figure 3.4.

If the optical field of the reference beam is described by the equation

$$E_r = A_r \exp j(2\pi f t) \tag{3.4}$$

where f is the frequency of the source, then the field of the object beam will be phase shifted by an angle of ϕ, where ϕ is given by

$$\phi = 2\pi d/\lambda \tag{3.5}$$

where d is the path length difference and λ the wavelength of the source.

The optical field of the object beam can thus be described by the equation:

$$E_0 = A_0 \exp j(2\pi f t + \phi) \tag{3.6}$$

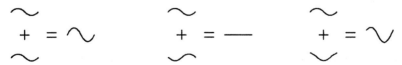

Constructive Interference   Destructive Interference   Partial Interference

Figure 3.4 *The results of different types of interference.*

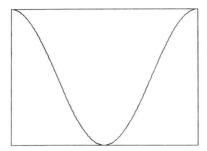

Figure 3.5   *Variation of photodiode current for path-length variation of one wavelength for Michelson interferometer.*

These two beams are combined at the beam splitter to produce an output beam described by

$$k(E_r + E_0) = k(A_r \exp j(2\pi ft) + A_0 \exp j(2\pi ft + \phi)) \qquad (3.7)$$

If the amplitudes of the beams A_r and A_o are equal (which is the optimal situation) then the resultant current in the photodiode sensor will be given by

$$I = k_d kS (1 + \cos \phi) \qquad (3.8)$$

where S is the source power, k_d is the photodiode responsivity and k depends on the losses in the optical system and has a maximum value of 0.5.

   This is shown graphically in Figure 3.5. A useful output is only achieved over a range of d/λ of $\frac{1}{4}$, which corresponds to a mirror movement of $\lambda/8$. The wavelengths of light typically used range from 500 nm to 1 μm, thus giving a typical maximum range (with λ = 800 nm) of 100 nm. Even within this limited range the relationship between the movement of the mirror and the output of the sensor is highly non-linear. Typical capacitor microphone diaphragms move several micrometres at the highest sound pressures. For this reason single-wavelength interferometers are not suitable for direct use with existing microphone capsules.

Half-mirror beam splitter

Figure 3.6 shows an implementation of the Michelson interferometer using a half-mirror beam splitter. In this implementation two output beams are generated, one passing to the sensor and one back to the laser. The second beam is prevented from reaching the laser by an optical isolator consisting of a polarizer and quarter-wave plate. Assuming the source is plane polarized the maximum power at the sensor will be half the source power, the remaining power being lost in the polarizer.

Polarising beam splitter

Figure 3.7 shows an alternative Michelson arrangement. In this case a polarizing beam splitter (PBS) is used in place of a half mirror. The beam from the source is split into two orthogonally polarized beams. These are passed through quarter-wave plates ($\lambda/4$) and onto the two mirrors. After reflection they are passed back through the quarter-wave plates, at which point each has had its plane of polarization rotated by 90°. The beam that was originally reflected by the beam splitter is now passed through whilst that which was originally passed through is now reflected. The two beams, which are still orthogonally polarized, are combined by the polarizer (P) and interference takes place. The resultant beam is

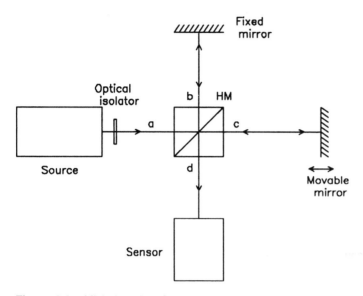

Figure 3.6   *Michelson interferometer using half-mirror.*

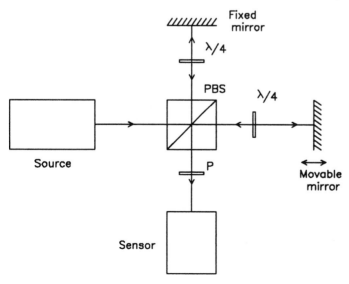

Figure 3.7   *Michelson interferometer using polarizing beam splitter.*

detected by the sensor. In this configuration the maximum power at the sensor is equal to the source power, no power is lost in the system, unlike the half-mirror arrangement.

Noise

To enable us to compare optical techniques with existing transducer techniques we must calculate the equivalent noise level due to the noise introduced by the opto-electronic components. A fuller description of noise in opto-electronic components can be found in Senior.[4]

Source noise

The noise in laser diodes is quite high if they are not stabilized by an external feedback loop. Many laser diodes, however, have photodiode sensors incorporated into the package which are subject to a proportion of the output of the laser. These sensors are intended for use in closed-loop stabilization of the laser output power. If the control loop works not only at DC but also over the whole of the audio bandwidth, then the noise in the laser will be greatly reduced over the audio bandwidth. The noise of the source can thus be made dependent on the noise in the photodiode sensor rather than the laser diode itself. External beam splitters and photodiodes may be used if the laser does not incorporate its own sensor or if the existing sensor is itself too noisy. The optimum power fed to this sensor is equal to the standing power which falls on the interferometer sensor. In the case of the polarizing beam splitter this will be one third of the laser output. Thus, of the laser output, one third is used to stabilize the laser, and two thirds to drive the interferometer.

Sensor noise

In most, if not all, interferometer configurations the standing light power at the sensor is a fair proportion of the source power. The current in the photodiode sensor is thus several orders of magnitude higher than the dark or leakage current. The major source of noise is thus the shot noise generated due to the standing current. The shot noise (i_n) in a diode was given in equation (3.2). A typical laser power might be 4 mW. The standing power at the sensor of a polarizing beam-splitter Michelson interferometer with external stabilization of laser output would be 1.33 mW. For a typical photodiode sensitivity of 0.5 A/W the standing current in the photodiode would be 667 μA. This results in a noise current of 2.06 nA r.m.s. in a bandwidth of 20 kHz. The responsivity of the interferometer is obtained by differentiating equation (3.8):

$$dI/dd = kSk_d([-2\pi/\lambda](\sin 2\pi d/\lambda)) \tag{3.9}$$

The value of this responsivity (R_b) at the bias point of $d/\lambda = \frac{1}{4}$ is given by

$$R_b = (kSk_d 4\pi)/\lambda \text{ A/m} \tag{3.10}$$

For a source power (S) of 4 mW, a photodiode responsivity (k_d) of 0.5 A/W; a wavelength (λ) of 800 nm, and a polarizing beam splitter externally stabilized Michelson, $k = 0.333$, this gives a responsivity of 10.46×10^3 A/m. The extra factor of 2 comes about because the responsivity is with respect to mirror movement and not path-length difference. The noise current of 2.06 nA thus corresponds to a mirror displacement of 1.97×10^{-13} m r.m.s. The movement of a typical capacitor microphone diaphragm at a 0 dB SPL (2×10^{-5} Pa r.m.s.) is about 10^{-13} m r.m.s. The shot noise current in the sensor is thus equivalent to a sound pressure of approximately 7 dB SPL. This figure will be worsened by 3 dB once the source noise is also considered, thus giving an overall figure of 10 dB SPL.

Dynamic range

The maximum output of the interferometer is achieved for a mirror movement of one quarter of the source wavelength. This has a value of 200 nm if the source wavelength is 800 nm. This is equivalent to a sound pressure of 117 dB SPL. As the noise level is 10 dB SPL this gives a dynamic range of only 107 dB. This should be compared with typical dynamic ranges of 120 dB for capacitor microphones. An optical microphone using the Michelson interferometer would thus have lower dynamic range than its electrostatic counterpart, but would have a lower self-noise. The distortion at high sound pressures would also be considerably worse due to the non-linearity of the response. Equations (3.2) and (3.10) show that the noise is proportional to the square root of the source power whilst the responsivity, and hence the signal, is directly proportional to the source power. The signal-to-noise ratio is thus proportional to the square root of the source power. If the source power were to be increased to 100 mW, then this would result in an increase in dynamic range of 14 dB. This would make the figure 121.5 dB, which is comparable with existing electrostatic techniques.

Direct analogue-to-digital conversion

A technique suggested by Mada and Koide[6] and further explored by Mada and Muramatsu[7] involves the direct digitization of the flexure of a tense diaphragm. The technique relies on the fact that, under constant pressure, different parts of the diaphragm deflect by different amounts, the deflection being greatest at the centre. By using an array of sensors, the displacement of different parts of the diaphragm may be observed. The principle, which is a derivative of the Michelson interferometer, is shown in Figure 3.8.

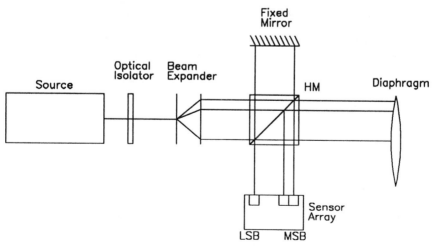

Figure 3.8 *Principle of direct A/D conversion.*

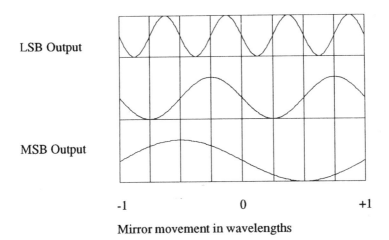

LSB Output

MSB Output

-1   0   +1

Mirror movement in wavelengths

Figure 3.9 *Photodiode outputs for a mirror movement of ± one wavelength for direct A/D conversion.*

The sensors are positioned such that they observe parts of the diaphragm that move by binary weighted amounts, that is the deflection doubles from one position to the next. The figure shows a three-bit system. As the diaphragm curvature increases the least significant bit (LSB) output will pass through several cycles of the repeated function. The next significant bit will pass through half as many cycles, and so on. This is shown, again for a three-bit system in Figure 3.9. The outputs, when passed through suitable thresholding circuits, would give a signed binary representation of the diaphragm displacement.

There are a number of problems associated with this technique, the most important of which are outlined here. The resolution of the system is limited by the size of sensors. In fact this is not as bad as it might seem, as the sensors may be optic fibres remotely connected to photodiodes. The minimum detectable deflection is half a wavelength thus for a twenty-bit system (which is the resolution required for a dynamic range of 120 dB) the deflection would have to be 2^{19}. This has a value of 0.4 m for a wavelength of 800 nm; thus the resolution is severely limited for realistic deflections. The linearity of the system depends wholly on the accuracy to which the sensors are positioned and the uniformity of curvature of the diaphragm. Both of these factors are difficult to control accurately. Another problem is that the part of the beam that produces the MSB will be deflected by the curvature of the diaphragm, once again this limits the possible resolution. Although some of these problems may be overcome, the technique could never give the resolution required for a high-quality microphone.

Multiple beam interferometers

Another class of single-wavelength interferometers is that of the multiple beam types. These rely on interference between multiple reflections of the same beam and hence avoid the need for a reference beam. Figure 3.10 shows a Fabry–Perot interferometer using two partial mirrors (see also Park and Diebold[8]).

The output beam (b) is the sum of a proportion of multiple reflections of the incident beam (a). If the separation of the two partial mirrors is an integer number

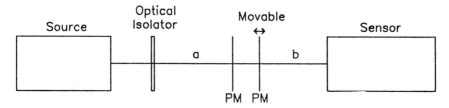

Figure 3.10 *Fabry-Perot interferometer.*

of half-wavelengths, then each of the reflections will be of the same phase. The interference at the movable partial mirror will thus be constructive and the amplitude of the output beam will be maximal. At other separations each successive reflection will be of a different phase and the interference will only be partial. The amplitude of the output beam is thus dependent on the separation of the two mirrors.

A disadvantage of this technique is that the light is required to pass through the movable mirror. This would be impractical in a microphone as the front of the diaphragm is usually open to the air. There is also a high optical loss associated with this technique as a lot of light is reflected from the fixed partial mirror. A variation of the Fabry – Perot interferometer is to make the movable mirror totally reflective and observe the light reflected from the fixed partial mirror by means of a beam splitter. This arrangement is shown in Figure 3.11 (See also Keating[1]).

The polarizing beam splitter (PBS) splits the output of the source into two beams. The proportion of light in each beam may be set by setting the plane of polarization of the source relative to that of the beam splitter. The reflected beam is passed to sensor 1, where it may be used to stabilize the source. The other beam passes through a quarter-wave plate ($\lambda/4$) and on to the Fabry–Perot cell

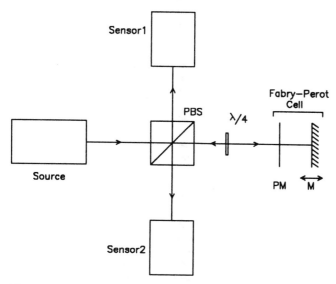

Figure 3.11 *Modified Fabry – Perot interferometer.*

consisting of the fixed partial mirror (PM) and the movable mirror (M). Interference takes place between the first reflection of the incident beam from the partial mirror and a proportion of the subsequent reflections from the movable mirror. The resultant beam passes back through the quarter-wave plate, is reflected by the beam splitter and passes on to sensor 2. The output current of sensor 2 can be described by the equation

$$I = kSk_d(\{R1 + R2 \exp (j2\pi d/\lambda)\}/\{1 + R1R2 \exp(j2\pi d/\lambda)\})^2 \qquad (3.11)$$

where d is the separation of the plates, λ is the wavelength of the source, R1 is the reflectivity of the partial mirror, R2 is the reflectivity of the moveable mirror, S is the source power, k_d is the responsivity of the photodiode and k depends on the losses in the optical system and has an optimal value of 0.8.

The response of this configuration with R1 = R2 = 0.7071 (which is optimal) is shown in Figure 3.12.

If the laser power is 4 mW, the photodiode responsivity 0.5 A/W and the wavelength 800 nm, then the slope of the output is 2.5×10^4 A/m at a bias point of 400 μA. The shot noise is 1.6 nA r.m.s., which is equivalent to a displacement of 6.5×10^{-14} m r.m.s. This, in turn, is equivalent to a sound pressure of −3.8 dB SPL. Once again this figure will be worsened by 3 dB due to the source noise, giving an overall noise level of −0.8 dB SPL. This is slightly better than that of the Michelson interferometer and the linearity of the Fabry–Perot is better. The maximum displacement is identical to that of the Michelson, hence the dynamic range of the Fabry–Perot is 11 dB better than that of the Michelson. This configuration also has the advantage that the most critical alignment is that the two mirrors must be parallel. This can be achieved by use of a precision spacing ring, which can also act as the supporting ring for the diaphragm. The whole assembly must also be aligned perpendicular to the incident beam. The alignment is no more difficult than that required for the Michelson, and has the advantage that the system is not susceptible to vibration perpendicular to the incident beam. This is important in a microphone where susceptibility to vibration is a disadvantage. The Fabry–Perot interferometer, in common with the Michelson, must have the standing path-length difference accurately set to ensure linearity of the output.

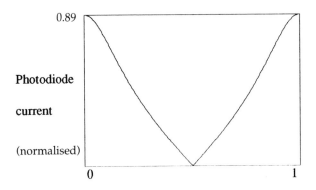

Figure 3.12 *Photodiode current vs path-length difference for Fabry – Perot interferometer.*

3.3.4 Dual-wavelength interferometers

The principle of operation of a typical dual-wavelength interferometer is similar to that of a single-wavelength type. One beam travels along a path of fixed distance whilst the path of the other beam involves a reflection from a movable mirror. The two beams are combined by a beam splitter and the resultant beam fed to a sensor. In a dual-wavelength interferometer, however, the two beams have slightly different wavelengths. On combination these produce a beat frequency. (This is because the sensor responds only to the intensity of the light.) The position of the movable mirror produces phase modulation of one of the beams, thus producing frequency modulation of the beat frequency dependent on the velocity of the movable mirror.

Figure 3.13 shows a typical dual-wavelength interferometer. The two output beams of frequencies f_1 and f_2 are polarized in the same plane. The polarized beam splitter (PBS1) allows the first beam of frequency f_1 to pass through and on to the quarter-wave plate. The subsequent reflection and second pass back through the quarter-wave plate rotate the plane of polarization by 90°. The reflected beam is thus reflected by the beam splitter and passed to the second beam splitter (PBS2) where it is again reflected. The second output beam passes from the source through the second beam splitter, where it is combined with the first beam. The polarizer (P) selects components of each beam in the same plane and interference takes place. The resultant beam is passed on to the sensor.

Possible dual wavelength sources are shown in Figures 3.14 and 3.15. In Figure 3.14 the source is a stabilized transverse Zeeman laser (STZL) the output of which consists of two components whose frequencies differ by several hundred kilohertz and whose planes of polarization differ. The two components are separated by the polarized beam splitter (PBS). Figure 3.15 shows how a single-wavelength source such as a laser diode may be used to produce two beams of differing wavelength. The output from the laser diode source (LD) is split by the polarized beam splitter (PBS). The two beams so produced are each fed to acousto-optic modulators (AOM1 and AOM2) which shift the frequency by frequencies f_a and f_j. These frequencies are typically 40–50 MHz and the difference between them $(f_a - f_b)$ is typically a few hundred kilohertz. (See also Takahashi *et al.*[9])

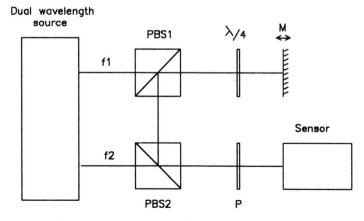

Figure 3.13 *Typical dual-wavelength interferometer.*

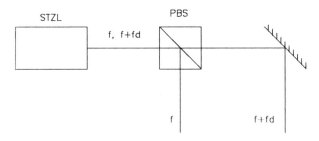

Figure 3.14   *Transverse Zeeman laser dual wavelength source.*

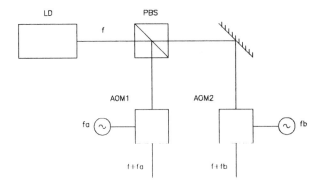

Figure 3.15   *Acousto-optic modulator dual wavelength source.*

If the optical output field from the movable mirror M has a component given by

$$Em = Am \exp j\,(2\pi f_1 t + (4\pi/\lambda)x(t)) \tag{3.12}$$

where $x(t)$ describes the motion of the mirror and λ is the wavelength of the incident beam (of frequency f_1) and the fixed source has a component given by

$$E_f = A_f \exp j\,(2\pi f_2 t) \tag{3.13}$$

then the photodiode current will contain a component given by

$$I = kSk_d\,(1 + \cos\,(2\pi ft - (4\pi/\lambda)x(t)) \tag{3.14}$$

where $f = f_1 - f_2$, k_d is the responsivity of the photodiode, S is the source power and k depends on the optical losses, which may be quite high due to the acousto-optic modulators.

Taking the same conditions as the single-wavelength examples, the resultant dynamic range is good at low frequencies due to the bandwidth redundancy of the FM technique, but less good at high frequencies because of the increased diaphragm velocities. A typical noise level would be 1 dB SPL in 20 kHz if the maximum frequency deviation was 240 kHz and the optical losses 20 dB. The maximum sound pressure would be 150 dB at 1 kHz and 125 dB at 20 kHz.

A disadvantage of the optical heterodyne or dual wavelength interferometer is that the technique requires either a specialized laser source or acousto-optic modulators, both of which are expensive and neither of which is integrable. The noise in the source cannot easily be reduced as in the single-wavelength interferometers, as this would require a stabilizing loop which worked up to frequencies of several hundred kilohertz. The acousto-optic modulators also each require an RF drive which must have good short- and long-term stability if excessive noise is not to be introduced. Another disadvantage of this technique is that the output is an FM signal which must be demodulated to produce an audio output. The demodulator must be of very high quality if the excellent performance of the microphone is not to be compromised. The output from the demodulator must also be fed to an integrator if the frequency response of the whole system is to be flat, as the system is velocity rather than displacement driven. The alternative to this is to design the microphone capsule to have the velocity of the diaphragm rather than the displacement independent of frequency. A major advantage of the technique is that the absolute path length difference does not matter, resulting in much simpler alignment. All single wavelength interferometers must have the path length difference accurately set, so that the system works in the linear region. An error of only 200 nm would result in the movement of the diaphragm being effectively full wave rectified in the electrical output!

3.4 Force feedback

It was stated previously that the main problems with the Michelson and Fabry–Perot interferometers were that the path length difference required accurate setting, the linearity was poor and the dynamic range was somewhat restricted. A single technique that can alleviate all of these problems, and gives other advantages besides, is that of negative force feedback. Broadly speaking a microphone with negative force feedback applied around it has an output that is a measure of the force required to hold the diaphragm still, rather than having an output proportional to the displacement of the diaphragm. If the feedback extends to DC, then the absolute position of the diaphragm may be controlled, and operation within the linear region is ensured. One effect of the negative feedback is to reduce the displacement of the diaphragm by a factor of one plus the loop gain. (The diaphragm would only actually be held still if the loop gain were infinite.) The maximum signal to which the microphone can accurately respond is likely to be determined by the maximum amount of force available from the feedback rather than the range of the measuring system. Another effect of the feedback is to reduce the distortion, again by a factor of one plus the loop gain, thus improving the linearity. The feedback has no effect, however, on the noise, and so the self-noise remains unchanged. This is not a significant disadvantage, as optical techniques can offer very low self-noise.

Perhaps the most powerful effect of the negative feedback is that the frequency response of the microphone is no longer dictated by the behaviour of the diaphragm, but is instead controlled by the electronics in the feedback path. This means that the position and shape of the electrode are no longer critical, and in fact were it not for the fact that we must somehow apply a restoring force to the diaphragm, the electrode could be dispensed with.

Feedback method

There is only one practical way to apply a feedback force to the diaphragm, and that is by use of electrostatics. The force (f) on a parallel plate capacitor is given by

$$f = \epsilon_0 V^2 A / 2d^2 \qquad (3.15)$$

where V is the applied voltage, A is the area of the plates and d is the separation of the plates.

The relationship between voltage and force can be seen to be a square law and thus highly non-linear. This relationship appears in the feedback path of the system and therefore dictates the overall system response. The relationship must therefore be very linear, and for this reason a single electrode and diaphragm forming a parallel-plate capacitor is not a suitable voltage-to-force transducer. If the diaphragm is flanked by two electrodes, however, and these are polarized to a high voltage as shown in Figure 3.16, and the two diaphragm-to-electrode spacings are the same, then the voltage-to-force relationship is given by

$$f = 2V_p V_s \epsilon_0 A / d^2 \qquad (3.16)$$

where V_p is the polarizing voltage, Vs is the signal voltage and d is the diaphragm-to-electrode spacing.

If the two spacings are not the same, then there will be a constant offset force on the diaphragm and a component of force proportional to V_s^2. This will introduce even-order harmonic distortion. An alternative configuration, which allows for this to be compensated, is shown in Figure 3.17. The relationship between voltage and force is the same as that given in equation (3.16) if the two spacings are again matched. Any small difference in spacing may be compensated by increasing or decreasing the drive to one of the electrodes and adding a DC offset. It is important to note that the relationship is linear only so long as the diaphragm does not move, as this will of course change the spacings. In a force feedback microphone the whole intention is to hold the diaphragm still and so this is indeed the case.

Maximum available force

The maximum force that the feedback can provide is limited by the available polarization and feedback voltages. If the polarization has a value of 200 V, the peak signal 200 V, the area is $8.8 \times 10^{-5} \, \mathrm{m}^2$ (this assumes a 15 mm diameter

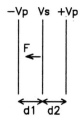

Figure 3.16 *Force feedback achieved by flanking diaphragm with two electrodes.*

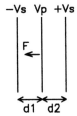

Figure 3.17 *Alternative arrangement of signals allowing compensation for spacing differences.*

capsule with 50% electrode coverage) and the diaphragm to electrode spacing is 50 μm, then the peak force is 2.5×10^{-2} N. This is equal to a pressure of 142 Pa over the diaphragm area, which is in turn equivalent to a sound-pressure level of 136 dB. The polarization and signal voltages represent the maximum that could be applied to a capsule of the given diaphragm to electrode spacing before ionization of the air in the gap occurs. The dynamic range of an optical force-feedback microphone is thus limited at one extreme by the self-noise due to the opto-electronics and at the other by the maximum feedback force available before ionization. Increases in the power of the laser source would result in a lowering of the self-noise, but acoustic noise would probably then dominate. This could be reduced by increasing the diaphragm to electrode spacing, but this would then require the use of greater feedback and polarization voltages. It is doubtful whether an exceptionally low self-noise is of any practical advantage as mixer or recorder noise would dominate in virtually all instances.

The combination of optical techniques and force feedback results in a microphone with very wide dynamic range and very predictable frequency response. Unless the components of the system are integrated, however, the high cost restricts the number of possible applications.

References

1. Keating, D.A. (1989) The optical microphone – fact or fiction. *Proc. I.O.A.*, **11**, (7) 277–287.
2. Ohashi, M., Umeda, N. and Suzuki, H. (1987). Study of diaphragm vibration in condenser microphones using optical heterodyne interference. *Jap. J. Acoustics*, **43** (12) 953–959.
3. Weardon, T. (1990). Integrated optics for interferometric sensors. *Transducer Technology*, June 1990, 10.
4. Senior, J.M. (1985) *Optical Fiber Communications* Prentice-Hall, London. pp. 231–295.
5. Dyson, J. (1970) *Interferometry as a Measuring Tool*, Machinery Publishing Co Ltd, London.
6. Mada, H, and Koide, K. (1983) Optical direct analog to digital conversion for microphones. *Applied Optics*, **22** (21), 3411–3413.
7. Mada, H. and Muramatsu, Y. (1986) Direct optical digital detection of diaphragm deflection: maximum resolution. *Applied Optics*, **25** 761–763.
8. Park, S.M. and Diebold, G.J. (1987) Interferometric microphone for optoacoustic spectroscopy. *Rev. Sci. Instrum*, **58** 772–775.
9. Takahashi, H., Masuda, C., Gotoh, Y. and Koyama, J. (1989) Laser diode interferometer for vibration and sound pressure measurements. *IEEE Trans. Instrumentation and Measurement*, **38** 584–587.

4 High-quality RF Condenser Microphones

Manfred Hibbing

4.1 Introduction

The transducer of a condenser microphone operates as a variable capacitor formed by the diaphragm and the back electrode. The distance between both electrodes varies as the diaphragm is excited by sound signals. The conversion of acoustic information via mechanical vibrations into electrical signals requires additional electrical biasing. A signal has to be applied to the transducer, acting as a carrier that is modulated by variations in the capacitance of the transducer.

Two alternative operating principles are used to achieve this. The low-frequency (LF) method uses DC biasing, whereas the radio-frequency (RF) method applies a high-frequency signal. In the first case a carrier with frequency zero is used and only its amplitude can be varied, whereas in the latter case the frequency or the phase may also be altered. In both cases the carrier can be rejected by simple R–C filters. The LF technique needs a high-pass filter and the RF technique a lowpass one. In the latter case the usable frequency range may be extended down to zero, thus enabling the measurement of static pressure. The LF method generates the audio-frequency signal directly, whereas the RF method needs appropriate demodulation to recreate the audio signal.

The RF technique may seem indirect. However, its benefits become evident when the impedance relations are considered. The electrical impedance of the transducer is inversely proportional to the signal frequency. If the LF technique is used, the signal frequencies range over three decades from 20 Hz to 20 kHz. The transducer impedance thus varies by the same ratio, typically from 200 kΩ to 200 MΩ for a transducer of 40 pF capacitance. Obviously, impedances of this order must be matched to the signal lines by additional means. The RF principle, however, decreases the transducer impedance to only 400 Ω if a 10 MHz carrier is used, and further reduction is feasible by means of resonance transformation. As an additional benefit, and contrary to the LF method, the resulting impedance is independent of the frequency of the sound signal.

The extremely high source impedance inherent to the LF operating principle requires amplifiers with input resistances around 1 GΩ to prevent low-frequency roll-off and degradation of the noise performance. Thus carefully selected insulation materials are demanded for the critical sections of the transducer and the amplifier input to reduce detrimental leakage effects caused by humidity. These problems have been virtually solved since RF condenser microphones became available, as the low-impedance characteristics of the RF design combines high-quality sound pick-up and reliability, even under extreme climatic conditions.

4.2 Historical review

The historical development of RF condenser microphones is reviewed by Beranek[1]. The RF principle was introduced by Riegger[2] in 1924 to enable sound pressure measurements down to 0.1 Hz, and was improved by Hull[3] in 1946. The capacitive transducer was incorporated into a resonant circuit excited by an RF generator at the slope of its resonance curve. The capacitance variations of the transducer altered the resonant frequency and caused amplitude variations of the applied signal. After amplification, rectification and low-pass filtering, an electrical signal corresponding to the sound signal was obtained. By an additional circuit the carrier could be partially compensated to increase the modulation depth. The main advantage was derived from the extended low-frequency range provided by the RF design. Noise performance was also stated to be superior to the LF design.

In 1947 Zaalberg van Zelst[4] improved the RF condenser microphone technique in two aspects. Firstly, the inherent noise of the RF generator was suppressed by a bridge design and, secondly, a cable was used between the microphone head and a separate RF unit. The microphone head comprised a series resonant circuit formed by the transducer, an inductor and a resistor. The resulting impedance was low enough to connect a cable directly. A second resonant circuit with identical parameters but with a fixed capacitor was connected to another conductor of the cable, and served as a reference. Both circuits were equally supplied from a high-impedance signal source, thus completing the bridge arrangement. The difference signal at the bridge output arising from capacitance variations of the transducer was amplified and, after adding back the carrier, demodulated by a Riegger discriminator. The complete circuit incorporated three pentodes and two diodes.

Until the advent of improved transistors in the late 1950s, RF technology was used only to improve the properties of measurement microphones, due to its expense, and was not commercially applied to condenser microphones for recording and broadcasting use. In studio microphones, a tube combined with an output transformer provided the necessary impedance matching. Accordingly, these microphones were fairly large. This was no problem in broadcasting and recording studios but became troublesome for television applications. It was obvious that the use of transistors in condenser microphones would yield many advantages. Besides the reduction in size of the microphone, the power consumption would be decreased as heating was omitted, and the feeding cables and power supply units could be simplified. Due to the reduced operating voltages, battery powering and transformer-less outputs would also become feasible.

Thus efforts where made to develop transistorized condenser microphones. However, direct replacement of the tubes by bipolar transistors based on the LF principle failed, as the serious impedance mismatching caused increased noise and low-frequency roll-off. Thus the only practical way to overcome the inherent problems was to adopt the RF principle, as it could provide the necessary impedance matching to bipolar transistors.

Various RF circuit designs were elaborated and patented in the late 1950s and early 1960s. Most of them favoured bridge designs to compensate for the generator noise.[5] Modifications mainly concerned the composition of the bridge elements, but the results were not completely satisfactory. Problems were caused by temperature and long-term drift of the transducer capacitance, and special precautions were needed, such as automatic bridge adjustment. Another essential disadvantage resulted from the fact that one side of the bridge was controlled by the variable impedance of the transducer while the other was fixed. Thus conversion linearity was affected at higher signal levels. The push–pull transducer was discussed as a solution to this problem but was not commercially utilized at that time.[6]

Since field effect transistors (FET) with characteristics similar to tubes became available, all microphone manufacturers but one returned to the more familiar LF principle. Sennheiser has been the only company working continuously on RF condenser microphones from then until the present. A newcomer at the beginning, the company was not bound by traditional conceptions and seized the opportunity to create unconventional and innovative solutions. The technical details of high-quality RF condenser microphones will be discussed on the basis of the solutions presented by this company, as this best describes the inherent continuity of the development and improvements of this technology.

A short summary may be given as follows. After several attempts with the bridge design, a breakthrough was achieved utilizing a crystal-controlled oscillator for reduced frequency noise and applying phase modulation (PM) instead of AM.[5,7] The transducer was incorporated into a phase-sensitive discriminator circuit tuned to the oscillator frequency. Variations of the transducer capacitance altered the resonant frequency of this circuit, and thus the phase of the applied oscillator signal. The limiting characteristics of the discriminator circuit rejected most of the amplitude noise of the oscillator. By this means the noise performance and dynamic range were considerably improved. Thus the PM design was the operating principle of RF condenser microphones for more than two decades.

At the onset of digital recording in the early 1980s investigations were made on the distortion characteristics of studio condenser microphones, which revealed the detrimental effects emerging from the conventional single-sided transducer design.[8] It turned out that a balanced push–pull design could basically improve the transducer linearity. Consequently, recent RF condenser microphones are based on this technique. By this means not only the linearity and the signal-handling capability but also the noise performance was considerably improved.[9]

4.3 General design aspects

In this chapter the main design features which are essential for high-quality condenser microphones will be presented. Though the considerations are valid for all types of condenser microphones, the resulting concepts are consequently applied here only to RF condenser microphones at the present. Thus this chapter also presents the design philosophy of the MKH series of RF microphones from Sennheiser.

4.3.1 Noise

Since digital audio has significantly expanded the dynamic range, low-noise performance has become an even more stringent demand on high-quality microphones. The improved signal definition at low levels may reveal inherent noise from the microphones which affects the 'filigree' of complex sound signals. Noise from condenser microphones is generated from the transducer and the circuitry. Though the capacitive transducer incorporates no electrical noise source, there are inherent acoustical noise sources that have to be minimized by careful design of the transducer.

The microphone receives acoustic information from air particles impinging on the diaphragm. The impacts are averaged with a certain statistical error which introduces noise. The uncertainty of this process depends on the number of particles involved. From statistical considerations, every doubling of the diaphragm surface area increases the signal-to-noise ratio by 3 dB. This suggests the optimal diaphragm should be as large as possible. However, practical restrictions arise, since directional characteristics suffer with large diaphragm diameters.

The noise performance of the microphone may be further improved for a given diaphragm size by decreasing the resistive acoustic damping of the transducer. This not only reduces the associated noise but also the physiological sensation of noise, since this is highly frequency dependent. The CCIR 468 weighting curve refers to this phenomenon. Decreased damping will improve sensitivity, especially around the resonance frequency of the transducer. By appropriate design the resonance frequency can be located next to the peak of the weighting curve. Thus the signal-to-noise ratio will be increased in the frequency range where the human hearing is most sensitive to noise and this will yield a considerable subjective improvement. As the associated frequency response of the transducer is no longer flat, electrical equalization has to be applied by the microphone circuit.

The noise generated by the microphone circuit can be kept low if the RF technique is applied, as it provides a low-impedance design and, as an additional benefit, contributes mainly pure white noise. This allows a high level of electrical bass emphasis without degrading the noise performance. This is an advantage over the LF technique, where the inherent flicker noise of the FET will raise problems.

All these aspects are considered in the design of RF condenser microphones from Sennheiser. Though the diaphragm diameter of the MKH.6 series is only 11 mm, the equivalent sound pressure levels are around 16 dB A-weighted and 26 dB CCIR-weighted. The MKH 40 push–pull series shows further improvements. The diaphragm diameter is increased to 15.5 mm, thus doubling the diaphragm area, and the circuit is designed even more diligently for further reduction of noise. Thus the equivalent sound-pressure levels are improved to around 12 dB A-weighted and 21 dB CCIR-weighted.

4.3.2 Sensitivity

Besides internally generated noise, there are also external sources that may contribute disturbing signals, For instance, among potential problems, electrical fields may induce signals into the microphone line, or hum may be superimposed on the supply voltage. Though the balanced technique rejects most of these interferences, additional protection arises from high-level output signals, due to high microphone sensitivity. This also reduces the noise contribution of the subsequent microphone amplifier.

Another reason for high transmission factors results from the improved noise performance of recent microphones and the accordingly reduced output noise signal. As the inherent output noise of transformer-less microphones cannot fall below a certain limit, the microphone sensitivity must be increased to prevent degradation of the noise performance.

As a consequence, the subsequent microphone amplifier may be overloaded by the maximum signal levels, which are above the studio reference level. To prevent this, a switchable pre-attenuation should be provided, but this facility should be used only under critical conditions, as it increases the equivalent noise level.

The nominal sensitivity of the RF condenser microphones from Sennheiser ranges from 20 to 50 mV/Pa (–34 to –26 dB re 1 V/Pa). High sensitivities are provided for interference line microphones as they are conveniently used at large recording distances and with faint sound signals. The output noise voltage of the microphones is above 5 µV, thus providing improved signal handling even with amplifiers which do not feature extreme low noise figures. Output voltages up to 3 V may arise from recent microphones which can be attenuated to 1 V with a –10 dB switch.

4.3.3 Directional characteristics

Various directional characteristics are required during recording. In most cases independence from frequency is demanded to avoid off-axis sound colorations.

Pure pressure microphones expose only one side of the diaphragm to the sound field. The resulting omnidirectional characteristic is degraded at high frequencies where the sound waves are no longer bent around, but reflected from, the microphone body. This causes a pressure increase up to 10 dB for on-axis sound. Since the onset of this effect depends on the transducer diameter, small diameters would be a preferable choice. However, this will deteriorate the noise performance.

A bidirectional or figure-of-eight characteristic results if the diaphragm is symmetrically exposed on both sides to the sound field. Diligently designed bidirectional microphones exhibit nearly frequency-independent and symmetrical directional patterns. Push–pull transducers are predisposed to exhibit ideal front-to-back symmetry due to their symmetrical design.

The majority of directional microphones respond partly to the pressure and partly to the pressure gradient of the sound field. Careful acoustical design is demanded to integrate the characteristics at low and medium frequencies with the natural pressure increase at high frequencies. As the pressure increase causes 10 dB treble emphasis for on-axis sound, the supercardioid and the wide-angle cardioid are ideal companions. Diligently designed supercardioids exhibit few disturbances in the cross-over range. The cardioid, which responds to equal parts of the pressure and the pressure gradient, has to be designed with special regard to satisfactory backward rejection over a wide frequency range.

All cardioid transducers need a second sound inlet which provides differing time delays for frontal and backward sound incidence. Condenser microphones use an acoustic R–C lowpass for this purpose. The compliance C is realized by a volume behind the diaphragm. The acoustic resistor R is often also used for damping the transducer in order to flatten the frequency response. High damping demands a high resistance R and a low compliance C and thus a small volume behind the diaphragm to achieve the same delay time. As a consequence, the associated resistor must be mounted in the vicinity of the small volume and thus separate from the rear inlet which is conveniently situated in the microphone housing behind the transducer. This design may cause resonances in the resulting cavity which deteriorate the directional characteristics at high frequencies.

A better solution is provided by the low acoustic impedance design which has been discussed with regard to improved noise performance. A low R needs a high C and thus a large volume behind the diaphragm which may ideally be extended to the periphery of the microphone housing. Then the resistance R may be located directly at the rear sound inlet, and cavity resonances will be avoided. Thus the low noise design yields an additional benefit.

From these considerations it follows that treble equalization with acoustic resonators should be avoided. Besides the fact that disturbing phase effects may be caused, this will only improve the on-axis response. However, the off-axis characteristics will be degraded. For the same reason the transducer front should be mounted flush with the microphone housing to avoid cavity resonances in front of the diaphragm.

4.3.4 Frequency response

Generally, a flat frequency response should be provided for all high-quality microphones. Switching facilities may provide additional features. The frequency range should be extended up to 20 kHz. The low-frequency limit depends on the directional type of the microphone and on the practical operating conditions.

Omnidirectional microphones should serve with an extended bass response down to 20 Hz or even less. A useful facility is an incorporated diffuse-field correction to compensate for the high-frequency losses under reverberant recording conditions which are caused by the increasing directivity at high frequencies, due to the pressure increase for on-axis sound. If electrical equalization is used, this may be easily provided by a switchable treble emphasis. Thus universal operation will be feasible for both direct and diffuse sound-field conditions. The normal flat frequency response features neutral pick-up at close recording distances. At larger distances, where reverberations become dominant, the alternative switch position is more suitable.

The frequency response of gradient microphones should range down to 40 Hz. Further extension to low frequencies mainly adds disturbing signals caused by air conditioning and other such sources of background noise interference. Steep cut-off slopes should be avoided under normal operating conditions as they degrade the phase response. A switchable soft roll-off may compensate for the proximity effect.

Interference microphones are mainly used under acoustically difficult conditions where the recording distance is fairly large and where wind may contribute additional low-frequency disturbances. Therefore, these microphones must exhibit an effective rejection of disturbing low-frequency signals by means of an increased lower limiting frequency and a sharper cut-off slope. The cut-off frequency may be switchable for even higher attenuation under worse conditions. In most cases treble emphasis is required, which may be a fixed or switchable feature.

4.3.5 Linearity

High signal-handling linearity is another stringent demand on high-quality microphones which has become increasingly significant since the introduction of digital audio. The linear quantization process of digital signal conversion improves the linearity, especially at high signal levels. This is contrary to the characteristics of analogous devices. Thus the inherent distortions due to microphones, which have previously been concealed by the deficiencies of the analogous recording technique, may now be exposed.

Sources of non-linearity can be found in the transducer and in the microphone circuit. The contribution of the circuit depends on the applied operating principle and will not be discussed here. The non-linearity of the transducer, however, is of more specific relevance. As the operation of the capacitive transducer is based on variations of the air-gap width, the trapped air is forced to move in the narrow slit against the inherent viscosity effects caused by molecular friction. The resulting acoustic resistance is inversely proportional to the third power of the actual thickness of the air layer. If the initial width of the air gap is d and the diaphragm displacement towards the back electrode is x, then

$$\delta = x/d \qquad (4.1)$$

defines the relative diaphragm displacement, and the resulting resistance can then be expressed by

$$R_1 = R_0/(1 - \delta)^3 = R_0(1 + 3\delta + 6\delta^2 + 10\delta^3 + \ldots) \qquad (4.2)$$

R_0 represents the initial resistance when the diaphragm is located at its rest position. R_1 is increased as the diaphragm is deflected towards the back electrode and is decreased for the opposite movement, but by differing amounts. Obviously, the resistive force which impedes the diaphragm motion is not a constant but rather a non-linear function of the diaphragm excursion. This is the basic source of the transducer non-linearity.

On a gross level, the transducer linearity can be improved in two ways. Firstly, the spacing between the diaphragm and the fixed electrode may be increased to reduce the inherent resistance and its non-linear effect. However, this is only a gradual improvement, which may additionally degrade the sensitivity and noise performance. A more basic modification takes advantage of the push–pull principle. An additional counter electrode identical to the back electrode is mounted symmetrically in front of the diaphragm. Thus a second air layer is formed which is represented by the complementary resistance

$$R_2 = R_0/(1 + \delta)^3 = R_0(1 - 3\delta + 6\delta^2 - 10\delta^3 + \ldots) \qquad (4.3)$$

The total resistance acting on the diaphragm results from the sum of both resistances:

$$R = R_1 + R_2 = 2R_0(1 + 6\delta^2 + \ldots) \qquad (4.4)$$

As all odd terms of the series development cancel, no even distortion components will occur. The remaining symmetrical components are less important at normal operating conditions where the diaphragm excursion is small. Practical investigations have shown that the push–pull design improves the linearity by a factor of more than 10.[8,9] The improvement is actually limited by the imbalance due to inevitable mechanical tolerances. As an additional benefit of the push–pull principle, the resultant electrical symmetry can be utilized in a bridge arrangement to improve the circuit linearity.

The linearity of microphones can be tested using the difference frequency method according to IEC 268–2. In Figure 4.1 the basic test set-up is shown. Two sounds of equal pressure are applied simultaneously and separately via two loudspeakers to the microphone under test. The main distortion component

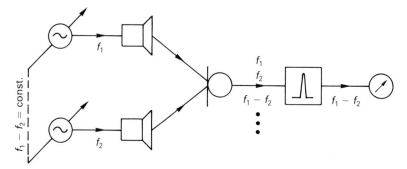

Figure 4.1   *Difference frequency test set-up.*

arising at the difference frequency is selected by a narrow band filter. The distortion due to the loudspeakers does not affect the test results. The lower limiting frequency for the test tones is approximately given by three times the frequency difference. Reliable results are obtained in the entire upper audio frequency range. This is important in as much as the distortions usually increase at high frequencies.

4.3.6 Transformerless output

Transformers in microphones are used for several reasons. The output signal can be floated, balanced and insulated from ground, and matching of the output to the internal operating conditions is also feasible. Due to the wide frequency range and the limited space available, transformers in microphones have to be very precise devices. The main problem is the distortion at low frequencies arising from the iron core. Consequently, RF condenser microphones are planned without transformers from the beginning. The A–B feeding standard provides this in a simple manner. The alternative phantom feeding is a more adequate means if output transformers are used, but the circuit design becomes more sophisticated, as will be shown in Section 4.5.

4.4 Phase-modulation design

This Section will deal with the analysis and technical details of the RF condenser microphone utilizing phase modulation. A comprehensive analysis regarding all mutual effects is very difficult and will not be presented here. Only the fundamental properties will be deduced under simplifying assumptions.

4.4.1 Analysis

Figure 4.2 shows the basic PM circuit. The transducer capacitance and the primary inductance L of the RF transformer L_1 form a resonant circuit. If C_0

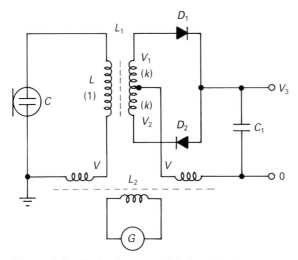

Figure 4.2 *Basic phase modulation circuit.*

represents the capacitance when the diaphragm is located at its rest position and δ is the relative excursion according to equation (4.1), then the actual capacitance of the transducer can be expressed as

$$C = C_0/(1 - δ) \tag{4.5}$$

provided that the active capacitance of the transducer is not shunted by dead or stray capacitances. δ is positive if the diaphragm is moved towards the back electrode due to a pressure increase in front of the transducer. δ is typically about 10^{-5} to 10^{-4} (0.001 to 0.01%) for sound pressures of 1 Pa and is thus very small under normal operating conditions.

The circuit is excited by the voltage V supplied from the RF generator G via the RF transformer L_2. The generator is tuned to the resonant frequency of the circuit when the diaphragm is not excursed ($δ = 0$):

$$f_0 = 1/2π\sqrt{LC_0} \tag{4.6}$$

The resulting voltage at the inductance L is given in complex notation by

$$V_1 = V[jQ/(1 + jδQ)] \tag{4.7}$$

where Q is the quality factor of the circuit. The phase between V_L and V is exactly 90° for $δ = 0$ and is increased by negative δ and decreased by positive δ.

The audio frequency signal V_0 is recreated by means of a discriminator circuit comprising the diodes D_1, D_2 and the capacitor C_1. It is assumed that both secondaries of L_2 supply identical voltages V to the resonant circuit and to the discriminator. Then the signals at the left-hand side of the diodes are given by

$$V_1 = V + kV_L = V[1 + j(δ + k)Q]/(1 + jδQ) \tag{4.8}$$

$$V_2 = V - kV_L = V[1 + j(δ - k)Q]/(1 + jδQ) \tag{4.9}$$

where k is the step-down ratio of L_1. From these equations the amplitudes A_1, A_2 of the signals V_1, V_2 can be derived:

$$A_1 = A\sqrt{[1 + (δ + k)^2Q^2]}/\sqrt{(1 + δ^2Q^2)} \tag{4.10}$$

$$A_2 = A\sqrt{[1 + (δ - k)^2Q^2]}/\sqrt{(1 + δ^2Q^2)} \tag{4.11}$$

where A represents the amplitude of the input signal V. The phase angles of V_1 and V_2 in reference to V are given by

$$Ø_1 = \arctan(kQ/[1 + δ(δ + k)Q^2]) \tag{4.12}$$

$$Ø_2 = -\arctan(kQ/[1 + δ(δ - k)Q^2]) \tag{4.13}$$

Figure 4.3 illustrates the phase and amplitude relations by phasor diagrams for $kQ = 1$ and $δ/k = 0.5$. The amplitudes of the signals V_1 and V_2 are equal if the diaphragm is not displaced, while their phases differ even then by 2 arctan $(1/kQ)$. In reference to V, V_1 always leads and V_2 always lags in phase. A_1 is greater than A_2 for positive δ and smaller for negative δ.

Figure 4.4 shows the signals V_1 and V_2 in the time domain for positive and negative diaphragm excursions. The dashed line represents the resulting instantaneous output voltage V_0. As D_1 conducts at the positive peaks of V_1,

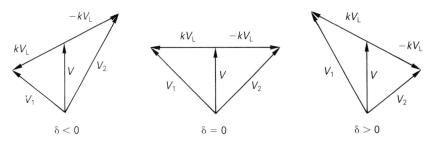

Figure 4.3 *Phasor diagrams of phase modulated signals for kQ = 1 and δ/k = 0.5.*

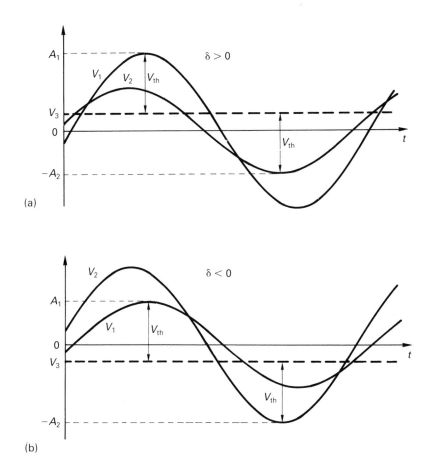

Figure 4.4 *Signals of the phase modulated circuit for positive and negative diaphragm excursion.*

where $V_1 = A_1$, while D_2 conducts at the negative peaks of V_2, where $V_2 = -A_2$, the following relations must be fulfilled:

$$A_1 - A_2 = 2V_0 \qquad (4.14)$$

$$A_1 + A_2 = 2V_{th} \qquad (4.15)$$

It is assumed that D_1 and D_2 are ideal diodes featuring a constant threshold voltage V_{th} independent of the forward current. From equations (4.10), (4.11), (4.14) and (4.15) the following basic relationships can be derived:

$$A_1 = 2V_{th}p/(p + n) \qquad (4.16)$$

$$A_2 = 2V_{th}n/(p + n) \qquad (4.17)$$

$$A = 2V_{th}a/(p + n) \qquad (4.18)$$

$$V_0 = V_{th}(p - n)/(p + n) \qquad (4.19)$$

where the following abbreviations have been used:

$$a = \sqrt{(1 + \delta^2Q^2)}$$

$$p = \sqrt{[1 + (k + \delta)^2Q^2]}$$

$$n = \sqrt{[1 + (k - \delta)^2Q^2]}$$

Figure 4.5 shows graphs of these relationships for various values of kQ. In order to generalize the results, all signal amplitudes are given in reference to V_{th}, and δ in reference to k. All curves are symmetrical with respect to $\delta = 0$. As the sum of A_1 and A_2 is always equal to $2V_{th}$, the amplitudes vary oppositely. If δ is zero, then A_1 and A_2 are both equal to V_{th} and the output voltage V_0 vanishes. The minimum value of the input signal amplitude A results for $\delta = 0$ and is given by

$$A = V_{th}/\sqrt{(1 + k^2Q^2)} \qquad (4.20)$$

The dip around $\delta = 0$ gets deeper if kQ is increased.

The operational range is restricted to the area between the peaks of the *s*-shaped curves. Thus the absolute value of δ must be smaller than the step-down ratio k, and V_0 cannot exceed the threshold voltage V_{th} of the diodes. The peak value of V_0 and the usable excursion range are further reduced if kQ is small, and this will also degrade the sensitivity and the conversion linearity. However, due to the symmetry of the curves, only uneven distortion components will arise.

The bending of the curves indicates that the conversion sensitivity T defined by

$$T = V_0/\delta \qquad (4.21)$$

is not a constant but depends on δ. At normal operating conditions δ is very small and T may be approximated to

$$T = (V_{th}/k)/(1 + 1/k^2Q^2) \qquad \text{for} \quad \delta << k \qquad (4.22)$$

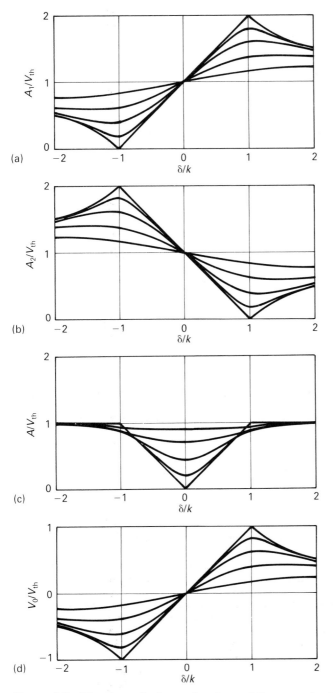

Figure 4.5   *Signal amplitudes as functions of the excursion δ/k for kQ = 0.5, 1, 2, 5 and 100. The shape of the curves gets sharper with increasing values of kQ.*

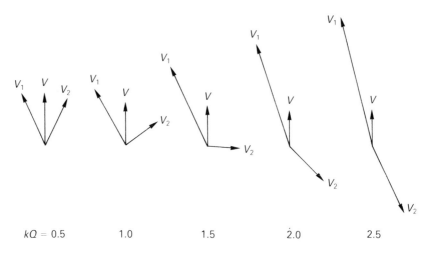

Figure 4.6   *Phasor diagrams of phase modulated signals for δ/k = 0.5 and various values of kQ.*

If the transducer is shunted by dead and stray capacitances then equation (4.5) must be modified and the resulting curves will become asymmetrical and additional even distortion components will occur.

Figure 4.6 shows some phasor diagrams which illustrate the dynamics of the amplitude and phase relations even at small values of kQ. $δ/k$ is fixed in this case to 0.5. Evidently, the amplitudes and phases of V_1 and V_2 vary drastically if kQ exceeds unity.

4.4.2 Circuit design

Figure 4.7 shows the simplified circuit diagram of the MKH. .6 T RF condenser microphones from Sennheiser. The various parts of the circuit will be described in detail.

(a) Discriminator

Compared to the basic structure from Figure 4.2, the discriminator circuit is extended by two additional diodes and a capacitor to form a push–pull arrangement. In this instance both halves of the RF signal period are utilized for the demodulation process. This doubles the conversion sensitivity and maximum output voltage.

The circuit parameters are set to about $k = 0.1$ and $Q = 10$. From these values $kQ = 1$, which is a recommendable choice as revealed by practical experience. The threshold voltage of the silicon diodes is around 0.75 V. Thus for small diaphragm excursions, the conversion sensitivity $T = 8$ results, being twice the value computed from equation (4.22). The maximum output voltage of the discriminator is around 0.6 V, according to the curve shown in Figure (4.5d). From this a maximum r.m.s. voltage around 0.45 V results for sinusoidal signals.

The DC voltage at the output terminals of the discriminator marked with A and B is adjusted to zero with the core of L_1. Consequently, the resonant circuit will be exactly tuned to the oscillator frequency.

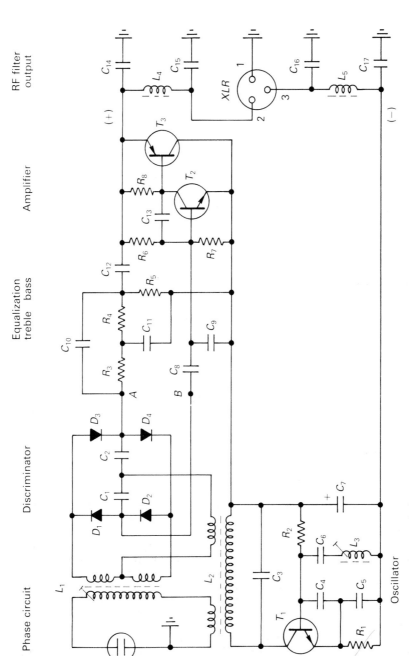

Figure 4.7 *Simplified circuit diagram of RF condenser microphone using phase modulation.*

(b) Oscillator

During the theoretical discussion it was assumed that the high-frequency signal generator accepts the voltage swing according to equation (4.18) and Figure 4.5 c. Thus, the generator has to act mainly as a current source. Other demands include high-efficiency and low-noise characteristics to provide maximum signal-to-noise ratio of the microphone.

Though the discriminator rejects the amplitude noise of the applied high-frequency signal considerably, the inherent phase noise must be minimized by careful design of the oscillator circuit. Especially, the Q-factor of the frequency-determining resonant circuit must be high. Satisfactory results have been previously achieved only with a quartz crystal, as its extremely high Q-factor kept the noise spectrum low. This was caused by imperfections of the oscillator transistor. A drawback of this technique was, however, that if the crystal was exposed to extreme mechanical shock it could be damaged. Therefore, as the noise characteristics of transistors became improved the crystal was replaced by an L–C circuit without degradation of the noise performance. Thus the durability and reliability of the microphones were improved.

The frequency determining circuit is composed of C_4, C_5, C_6 and L_3 and is tuned to 8 MHz by final adjustment with the core of L_3. C_4 and C_5 provide the positive feedback and are much larger than C_6, thus preventing degradation of the Q-factor. The output current from the collector of T_1 is supplied via the circuit L_2, C_3 to the PM circuitry. The DC input of the oscillator is bypassed for AC signals by the capacitor C_7. The resistors R_1 and R_2 set the DC current to 6 mA at a supply voltage of 7.5 V.

(c) Amplifier

Generally, the discriminator output may be connected directly to the signal line as the output impedance is only some hundreds of ohms. But the linearity would suffer from inevitable loading especially at high signal levels. Thus the linearity will be improved if the discriminator is decoupled from the output by an additional amplifier.

The amplifier consists of the transistor pair T_2 and T_3. The resistor R_8 sets the current of T_2 to around 0.5 mA to provide noise matching to the low source impedance. C_{13} provides appropriate high-frequency roll-off to ensure stable operation. The voltage drop at the amplifier stage is determined by R_6 and R_7 and the base–emitter voltage of T_2 is about 2.5 V.

The discriminator output signal is coupled via C_8 to the amplifier input. C_8 and R_6 operate as a high-pass filter attenuating subsonic signals. C_9 rejects the residual carrier signal. The second terminal of the discriminator output is connected to the output of an equalizing network and closes the feedback loop of the amplifier. The equalizing network consists of two parts. C_{12} and R_5 provide a low-frequency emphasis which is limited by R_6 at low frequencies. The bridged T-filter composed of R_3, R_4, C_{10} and C_{11} provides treble emphasis and cuts off above 20 kHz. Due to inevitable losses in the network circuit the minimum gain at medium frequencies is around 2 to 3 dB.

The amplifier output signal is connected via the RF filter C_{14}, C_{15}, L_4 to the positive output pin and via C_7 and a second RF filter C_{16}, C_{17}, L_5 to the negative pin. The RF filters prevent interference from external RF signals. The low output impedance of about 20 Ω provides voltage matching, thus making the output voltage and the sensitivity nearly independent of the output load.

Table 4.1 *Technical data of PM-type MKH microphones*

Type	MKH 106	MKH 406	MKH 416	MKH 816	MKH 110	MKH 110-1
Acoustic operating principle	Pressure	Pressure gradient	Interference + pressure gradient		Pressure with extended low frequency response	
Directional pattern	Omni	Cardioid	Lobe/super-cardioid		Omni	
Frequency range (Hz)	20–20 000	40–20 000	40–20 000	50–20 000	1–20 000	0.1–20 000
Sensitivity	20 mV/Pa (–34 dBV) MKH 416 P48: 25 mV/Pa (–32 dBV)			40 mV/Pa (–28 dBV)	20 mV/Pa (–34 dBV)	2 mV/Pa (–54 dBV)
Maximum SPL		124 dB		118 dB	120 dB	140 dB
Equivalent noise SPL	26 dB 13 dBA	25 dB 14 dBA	25 dB 14 dBA	25 dB 14 dBA	26 dB 14 dBA	44 dB 32 dBA
Output	10 Ω/balanced and floating				100 Ω/unbalanced	
Connector	XLR-3				DIN	
Feeding	T types : 12 V/6 mA A-B feeding P12 types : 12 V/6 mA phantom feeding P48 types : 48 V/2 mA phantom feeding				8 V/8 mA	
Diameter (mm)	19/20		19			19/20
Length (mm)	125	165	250	555		125

(d) Feeding

The circuit shown in Figure 4.7 is provided for A – B powering according to IEC 268–15, which enables the microphone to be fed and return its signal via a two-wire cable. This technique enables transformerless floating and balanced output. An ungrounded microphone circuitry is necessary in this system. The RF principle is predisposed to fulfil this requirement without extra expense. As can be seen from Figures 4.2 and 4.7, the transducer is grounded while all other parts of the circuit are insulated from the ground by separate windings of the RF coils. On the contrary, with the LF design the transducer must be connected to the amplifier circuitry and thus demands insulated mounting.

Both amplifier and oscillator are connected in series and are thus supplied by the same DC current of about 6 mA. The A – B feeding standard provides a 12 V DC source with 180 Ω resistors connected in series with either terminal. The DC voltage at the microphone output is 10 V. The resulting power consumption is 60 mW.

4.4.3 Microphone types

The phase-modulated RF condenser microphones of the MKH series feature several directional characteristics and feeding arrangements. Besides types for A–B feeding (MKH. .5, MKH. .6 T) also phantom powered types for 12 V and 48 V according to IEC 268–15 have been developed with transformerless outputs (MKH. .6 P12, MKH. .6 P48). The circuit design of the phantom powered versions corresponds to that of the balanced bridge-type microphones presented in Section 4.5 and will be described there.

The directional characteristics vary for differing applications from omnidirectional types (MKH 105/106) through cardioids (MKH 405/406) to short and long interference types (MKH 415/416, MKH 805/815/816). The acoustical interference effect is used to increase the directivity above the limits set for gradient microphones. As this technique is lacking at low frequencies the interference microphones use a transducer with a cardioid or supercardioid characteristic for extended directivity at low frequencies. Figure 4.8 shows as examples from this microphone series the cardioid MKH 406 T and the short interference microphone MKH 416 T.

For measurements in the subsonic range, microphones with omnidirectional characteristics and extended frequency responses down to 1 Hz (MKH 110) and 0.1 Hz (MKH 110–1) have been developed. These microphones comprise a DC amplifier and need special 8 V feeding.

Another benchmark in its time was the clip-on microphone MK 12 which was launched in the middle of the 1960s. It could be directly connected to a pocket transmitter which also contained the oscillator for the microphone. The complete phase discriminator circuit was incorporated into the MK 12. The RF signal was fed to the microphone via the same coaxial cable which also returned the audio signal.

4.5 AM push–pull bridge design

Though the phase-modulation technique works sufficiently well in most cases, there are some inherent problems. Firstly, the mutual tuning of the oscillator and the phase circuit incorporating the transducer must match for years without maintenance. This demands an extremely stable transducer system built as a unit on a rigid ceramic support. The manufacturing process is difficult and required

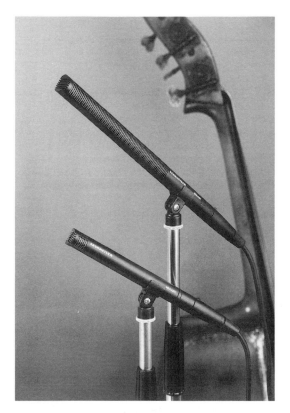

Figure 4.8 *RF condenser microphones MKH 416 T and MKH 406 T (Sennheiser).*

many years of experience to master. Secondly the linearity of the PM technique is degraded at high signal levels. Since digital audio made stringent demands on linearity, alternative solutions have had to be found. As the non-linearity of the transducer could essentially be reduced utilizing the push–pull design, it was reasonable to resurrect the AM bridge technique, since now, contrary to former times, symmetrical operation could be provided.

4.5.1 Analysis of bridge design

Though some details are similar, the theory of the balanced AM bridge design is much simpler and more comprehensive than the PM design. The basic circuit is shown in Figure 4.9. The signal from the high-frequency generator G is supplied to a RF transformer. The outer pair of secondaries applies symmetrical voltages V and $-V$ to the push–pull transducer. The bridge output is bypassed for low-frequency signals by the choke L_1.

It is assumed that the transducer features ideal symmetry, so that the capacitance of each part is C_0 when the diaphragm is at its rest position.

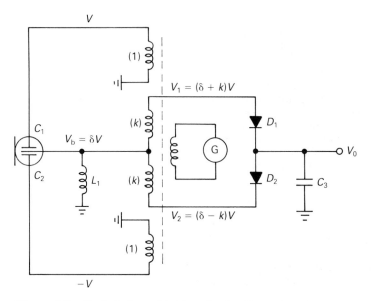

Figure 4.9 *Basic balanced bridge circuit with push–pull transducer.*

Corresponding to equation (4.5), the actual transducer capacitances C_1 and C_2 are related to the relative diaphragm excursion δ by

$$C_1 = C_0/(1 - \delta) \tag{4.23}$$

$$C_2 = C_0/(1 + \delta) \tag{4.24}$$

provided that no dead and stray capacitances are present. The total capacitance of the transducer is constantly $C_0/2$, independent of δ. The voltage across the transducer is divided according to the impedances of C_1 and C_2 as the signal current through both capacitances is identical. The voltages across C_1 and C_2 are given by

$$V_{C1} = VC_0/C_1 = (1 - \delta)\ V \tag{4.25}$$

$$V_{C2} = VC_0/C_2 = (1 + \delta)\ V \tag{4.26}$$

Thus the bridge output signal arising at the diaphragm terminal is then simply given by

$$V_b = V - V_{C1} = -V + V_{C2} = \delta V \tag{4.27}$$

The push–pull transducer acts as a capacitive voltage divider which is controlled by the sound signal. Naturally, this relation is based on the assumption that the output is not significantly loaded.

Equation (4.27) represents the fundamental relation from which the benefits of the push–pull bridge design are derived. Firstly, the bridge output signal is strictly proportional to the diaphragm excursion, thus providing ultimate linearity. Secondly, while sound is absent the bridge output signal is zero, independent of the instantaneous properties of the generator signal. Thus any noise contribution from the generator is completely rejected. If the diaphragm is moved by sound signals, then the bridge becomes unbalanced and noise is contributed from the generator. But as this noise is proportional to the excursion caused by the sound signal, the signal-to-noise ratio will remain constant. Additionally, the noise will be masked by the sound signal.

Another advantage of the AM bridge design is the simple phase relationship. Evidently, the bridge output signal δV is in phase with V for positive excursions δ and out of phase for negative excursions. As the phase contains the information about the direction of the diaphragm displacement, a phase-sensitive discriminator is demanded for correct demodulation.

The discriminator circuit shown in Figure 4.9 comprises the diodes D_1 and D_2, the capacitance C_3 and a symmetrical pair of secondaries supplying voltages kV and $-kV$. The signals at the left-hand side of the diodes are

$$V_1 = V_b + kV = (k + \delta)V \tag{4.28}$$

$$V_2 = V_b - kV = -(k - \delta)V \tag{4.29}$$

If the absolute value of δ is below k, then the signals V_1 and V_2 are out of phase, otherwise they are in phase. Normal operation will be represented by the first case and overloaded by the latter.

The amplitudes of V_1 and V_2 can be expressed by

$$A_1 = (k + \delta)A \tag{4.30}$$

$$A_2 = \pm(k - \delta)A \tag{4.31}$$

where A represents the amplitude of the signal V. The plus sign stands for normal operation and the minus sign for overload conditions. Figure 4.10 shows the signals for both cases in the time domain. Again, equations (4.14) and (4.15) have to be fulfilled. This leads to the following results for normal operation, where the absolute value of δ is below k:

$$A_1 = (1 + \delta/k)V_{th} \tag{4.32a}$$

$$A_2 = (1 - \delta/k)V_{th} \tag{4.33a}$$

$$A = V_{th}/k \tag{4.34a}$$

$$V_0 = (\delta/k)V_{th} = \delta A \tag{4.35a}$$

$$T = V_0/\delta = V_{th}/k = A \tag{4.36a}$$

Accordingly, it follows for overload operation, where the absolute value of δ exceeds k:

$$A_1 = (1 + k/\delta)V_{th} \tag{4.32b}$$

$$A_2 = (1 - k/\delta)V_{th} \tag{4.33b}$$

$$A = V_{th}/\delta \tag{4.34b}$$

$$V_0 = (k/\delta)V_{th} = kA \tag{4.35b}$$

$$T = V_0/\delta = kV_{th}/\delta^2 \tag{4.36b}$$

Figure 4.10 shows graphs of these relationships.

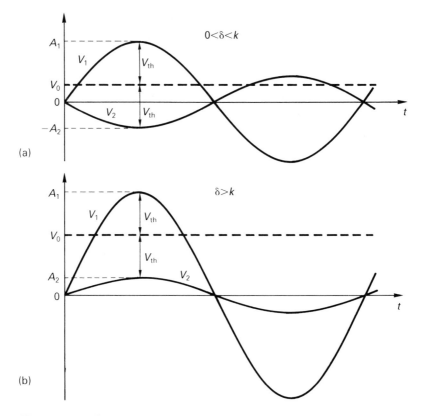

Figure 4.10   *Signals of the bridge circuit for positive diaphragm excursion under normal (a) and overloaded (b) conditions.*

At normal operating conditions, T is constant as expected and simply related to V_{th} and k, or, more simply, to the amplitude A of the signal voltages at the transducer. At overload conditions T decreases rapidly, thus indicating the onset of clipping. Contrary to the PM design, A is constant within the normal operating range. A_1 and A_2 change oppositely and are both equal to V_{th} for $\delta = 0$.

It is convenient to operate the bridge at its resonant frequency determined by the total inductance of the primary windings and the capacitance of the transducer. In this case the output impedance of the bridge circuit is extremely low, as it results from the parallel connection of the upper and lower parts of the bridge which are in series resonance. The output impedance is purely resistive and in the order of Z_0/Q if δ is small. Z_0 represents the impedance of C_0 at the resonant frequency and Q defines the quality factor of the bridge circuit.

4.5.2 Circuit design

The simplified circuit diagram from Figure 4.12 represents the MKH 40 P48 push–pull rf condenser microphone line from Sennheiser. The various parts of the circuit will be described in detail in this section.

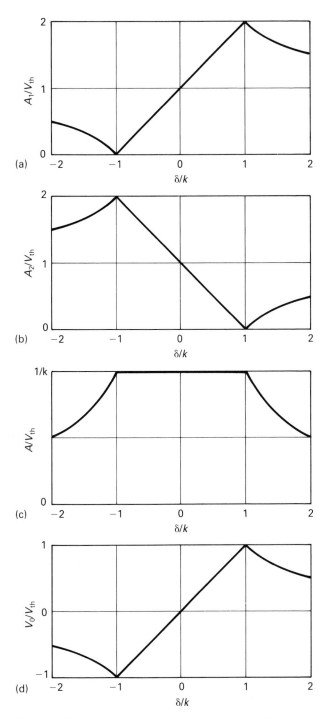

Figure 4.11 *Signal amplitudes as functions of the excursion δ/k.*

(a) Bridge and discriminator

Compared with Figure 4.9, the bridge and discriminator circuits exhibit several modifications. The discriminator comprises additional Zener diodes D_3, D_4 and capacitors C_2, C_3 for the following reason. The simple circuit limits the output signal to the threshold voltage V_{th} of the diodes, which is in the range of some hundreds of millivolts, depending on the diode type. This is prevented by additional DC biasing. The Zener diodes virtually increase the threshold voltage V_{th} by the Zener voltage V_z. In the previous analysis, V_{th} then has to be replaced by $V_{th} + V_z$. The turns ratio k must also be modified if the voltage amplitude A and the accompanying conversion sensitivity T are not to be altered.

With these considerations all design parameters can be set individually. V_{th} determines the maximum amplitude of the output voltage V_o, and k determines the maximum amplitude of the relative diaphragm displacement δ. The sensitivity T and the amplitude A of the voltage at the transducer are given by V_{th}/k.

The second modification concerns the additional RF transformer L_3, which is inserted between the bridge output and the discriminator input to provide a switchable 10 dB attenuation according to the turns ratios. This introduces no additional noise as the impedance of the transformer is very low. As an additional benefit, the transformer provides a floating signal output due to insulation of the discriminator circuit from the ground. The primary inductance of L_3 is tuned to the operating frequency with C_1.

Thirdly, the diaphragm of the transducer is grounded instead of the centre of the windings. This is reasonable as the diaphragm is usually mounted and contacted to the conductive housing of the transducer. Otherwise the transducer must be insulated from ground.

Additionally, a possibility is provided to adjust the bridge circuit for slightly unbalanced transducers. Therefore, the windings connected to the transducer are no longer tightly coupled to each other. This is provided by mounting the windings fairly well apart from each other. The associated inductances may be varied oppositely by core adjustment. Zero DC voltage at the output terminals of the discriminator marked with A and B indicates correct adjustment.

From this measure a mutual inductance M results which causes an additional inductance $-M$ in series with the bridge output and thus increases the output impedance. This effect can be compensated by an external inductance M connected in series with the bridge output. This inductance is represented by L_2 in Figure 4.12.

Some calculations will reveal several inherent characteristics from the bridge circuit. All values used in the calculations are approximated. The main circuit parameters are $k = 0.15$ and $V_{th} = 5\,V$. Thus, both the voltage amplitude A and the conversion sensitivity T are 34 V. The r.m.s. voltage applied to the transducer is 24 V. As the operating frequency is 6.5 MHz and the transducer capacitance is 40 pF, the impedance of each side of the transducer is $Z_0 = 600\ \Omega$ and the associated current is 40 mA. The total reactive power trapped in the resonant circuit is thus 2 W. The active power results from division by the quality factor $Q = 70$. Thus the oscillator has to supply an RF power of 30 mW to the bridge circuit. This power is increased by losses from the discriminator to about 35 mW. The output impedance of the bridge output is $Z_0/Q = 10\ \Omega$. The maximum relative diaphragm deflection is 16% without attenuation and 50% with 10 dB attenuation. The maximum r.m.s. output voltage is above 3 V.

The dynamic range of the discriminator circuit is limited at the lower end by the inherent noise of the diodes. Very low values can be achieved with Schottky diodes as they combine low threshold voltages with short switching times.

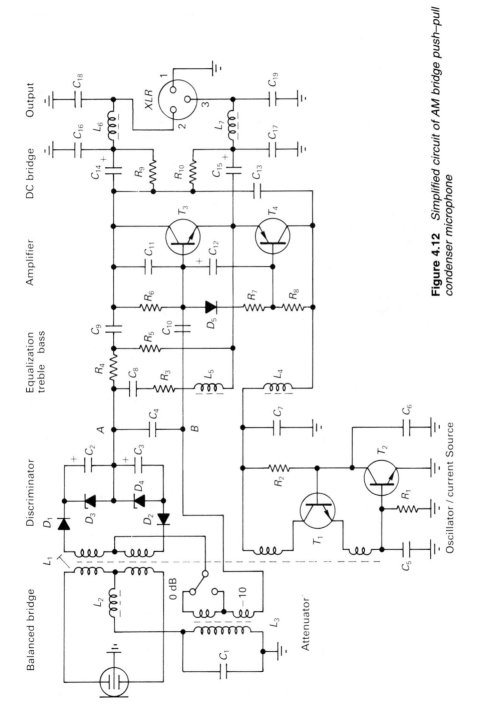

Figure 4.12 *Simplified circuit of AM bridge push–pull condenser microphone*

(b) Oscillator

The bridge circuit is excited at its resonance frequency by the oscillator circuit. The necessary positive feedback is provided to the oscillator transistor T_1 by an additional winding of L_1. Thus the mutual tuning of bridge and oscillator is guaranteed without maintenance. This is significant because it eliminates a problem which was inherent to the PM circuitry.

Transistor T_2 controls the DC current of the oscillator and thus the current of the microphone due to the base–emitter voltage drop across R_1. R_1 is composed of three resistors including an NTC resistor to provide temperature stability of the current setting. The oscillator functions additionally as an electronic choke which decouples the amplifier section for AC signals from the ground. This is essential to enable a transformer-less balanced output. For the same reason the DC input of the oscillator is bypassed by C_7 only for RF signals. The choke L_4 assists to reject RF signals.

The microphone current is 2 mA and the minimum voltage is 25 V. The minimum DC power input of the oscillator is thus 50 mW. As the efficiency is about 70%, the output power of 35 mW is sufficient to drive the bridge and discriminator circuit.

(c) Amplifier

The amplifier is composed of the complementary transistors T_3 and T_4, which are connected in parallel for AC signals by C_{12} and C_{13}. Thus push–pull operation is provided featuring low distortion. The DC voltage across the stage is determined by R_6, R_7, R_8 and D_5. The diode reduces the temperature drift.

The equalizing network provides bass emphasis by C_9 and R_5, which is limited by R_6 and R_8. Treble emphasis is provided by R_4 and the series resonance circuit composed of L_5, C_8, R_3. This design keeps the output impedance of the network low and thus reduces the associated noise contribution in the emphasized range.

The output signal of the amplifier is coupled via C_{14} and C_{15} to the microphone output. R_9 and R_{10} are resistors of identical value which collect the DC supply current from both conductors of the signal line. This completes the DC bridge, the other part of which is localized in the power supply unit. The DC current flows through the transistors T_3 and T_4, which are connected in series for DC, and then through the oscillator to the ground.

The RF filters prevent interference due to external RF signals. The output impedance of the microphone is set to 150 Ω with additional resistors connected in series with L_6 and L_7, which are not shown in the simplified circuit diagram.

The voltage drop across the amplifier circuit is 9.5 V, which enables an RMS output voltage up to 3 V for sinusoidal signals. The maximum r.m.s. output current is twice the supply current of 2 mA divided by the r.m.s. factor 1.4. From this a minimum output load of about 1 kΩ results. This value is actually increased due to internal loading from the equalizing network.

(d) Feeding

The circuit is developed for 48 V phantom feeding according to IEC 268–15, since this feeding is the most widely used in recording studios. The supply voltage is 48 V with a tolerance of ± 4 V. The supply current is split by two 6.8 kΩ resistors incorporated in the power supply unit into equal parts which are fed via the two

Table 4.2 *Technical data of push–pull MKH microphones*

Type	MKH 20	MKH 30	MKH 40	MKH 50	MKH 60	MKH 70
Acoustic operating principle	Pressure	Pressure gradient			Interference + pressure gradient	
Directional pattern	Omni	Figure-of-eight	Cardioid	Super-cardioid	Lobe/supercardioid	
Frequency range (Hz)	20–20 000		40–20 000		50–20 000	
Sensitivity		25 mV/Pa (−32 dBV)			40 mV/Pa (−28 dBV)	50 mV/Pa (−26 dBV)
Maximum SPL		134 dB			125 dB	123 dB
Equivalent noise SPL	20 dB 10 dBA	22 dB 13 dBA	21 dB 12 dBA	21 dB 12 dBA	17 dB 7 dBA	17 dB 6 dBA
Output	150 Ω/balanced and floating					
Connector	XLR-3					
Switchable sensitivity attenuation	10 dB					
Switchable bass attenuation	No	−5 dB at 50 Hz (compensation of proximity effect at approximately 0.5 m)			−3 dB at 140 Hz 18 dB/oct below 50 Hz	
Switchable treble emphasis	+6 dB above 8 kHz	No			+5 dB above 8 kHz	
Feeding	Phantom feeding 48 V/2 mA					
Diameter (mm)	25					
Length (mm)	155	175	155	155	285	415

wires of the microphone cable to the microphone and returned via the cable shield. According to the circuit under consideration, the current through each wire is 1 mA and the current drop at the 6.8 kΩ resistors is 7 V. If the supply voltage is down to 44 V, then the DC voltage at the microphone is 37 V. As the DC current is collected inside the microphone by two resistors which cause a loss of 2 V, only 35 V remain at the DC input of the amplifier circuit. Due to the voltage drop of 10 V from the amplifier, the oscillator is supplied by 25 V. If the phantom supply voltage is up to 52 V, then the oscillator is fed by 33 V. The voltage swing does not alter the microphone properties as the current is stabilized and the RF voltage of the oscillator is held constant by the Zener diodes in the discriminator. The DC power consumption of the microphone is 75 to 90 mW, depending on the supply voltage.

4.5.3 Microphone types

The RF push–pull condenser microphone series has directional characteristics ranging from omnidirectional (MKH 20) and figure-of-eight (MKH 30) through cardioid (MKH 40) and supercardioid (MKH 50) to short (MKH 60) and long (MKH 70) inferference types for differing applications. Figure 4.13 shows the cardioid microphone MKH 40 P48, which was released as the first type of this series.

As previously mentioned, all microphones feature transformerless outputs and a switchable 10 dB attenuator. Furthermore, there is at least one additional switchable feature incorporated to the microphones which was not shown in the simplified schematic.

The omnidirectional microphone MKH 20 incorporates a diffuse-field correction facility by means of electrical treble emphasis to guarantee optimum results

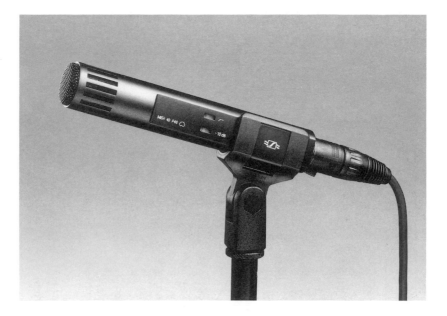

Figure 4.13   *RF push–pull condenser microphone MKH 40 P48 (Sennheiser).*

for both direct and diffuse sound-field conditions. A supplementary ring may be put upon the front end of the microphone to acoustically provide an additional 3 dB treble emphasis without degrading the directional pattern. This allows fine tuning of the sound balance at medium recording distances.

Gradient-type as well as interference-type microphones incorporate a switchable bass attenuator. It compensates for the proximity effect at about 0.5 m for gradient types and cuts off subsonic signals for interference types. Interference microphones additionally feature a treble emphasis switch to compensate for losses from longer recording distances.

As all microphones use electrical equalization, it was no problem to extend the low-frequency response of gradient-type microphones to 40 Hz. Without equalization much larger diaphragm diameters would be required. This merits mention especially for the figure-of-eight type as it eliminates the bass lack inherent to this type of microphone. All RF push–pull condenser microphones exhibit almost flat frequency responses up to 20 kHz which, once more, is not self-evident for bidirectional microphones.

The medium-sized transducer diameter combines low-noise performance with satisfactory directional characteristics. The directional patterns of all microphone types have been carefully engineered to provide a high degree of frequency independence.

The nominal sensitivity of the types MKH 20 to MKH 50 is uniformly set to 25 mV/Pa (– 32 dB referred to 1 V/Pa) and allows problem-free combination and interchange of these microphones. This simplifies, for instance, stereophonic recording using the M–S technique. The interference types feature a higher sensitivity up to 50 mV/Pa, as they are conveniently used at larger recording distances.

4.6 Summary

The development of RF condenser microphones led to extraordinary features, a part of which is bound to this technology. A few examples will be given as follows. The low noise design which has been introduced to the transducers of RF microphones may be also transferred to LF microphones. Problems may arise from the noise characteristics of FETs if low-frequency emphasis is applied. As no uniform electrical equalization is feasible for the various directional types, additional problems arise from a modular microphone concept. Problems may also arise if the push–pull transducer design is applied to LF condenser microphones because the front half of the transducer system is directly exposed to humidity. This is contrary to the conventional single-sided transducer where the sensitive electrical system is sufficiently protected by the diaphragm against instantaneous influence of moisture. Thus the push–pull design needs essentially the RF technology for satisfactory practical application.

References

1. Beranek, L.L. (1950) *Acoustic Measurements* (1st edn), Wiley, pp. 215–217.
2. Riegger, H. (1924) On high-fidelity sound pickup, amplification and reproduction. *Z. für tech. Phys.*, **5**, 577–580.
3. Hull, G.F. (1946). Resonant circuit modulator for broad band acoustic measurements. *J. App. Phys.*, **17**, 1066–1075.
4. Zaalberg van Zelst, J.J. (1947–1948). Circuit for condenser microphones with low noise level. *Philips Tech. Rev.*, **9**, 357–363.
5. Griese, H.J. (1965). Circuits of transistorized RF condenser microphones. *J. AES.* **13**, 47–52.

6. Griese, H.J. (1969). Recent developments in the field of condenser microphones. *37th AES convention*, unnumbered preprint, October.
7. Griese, H.J. (1965). Self-calibrating condenser microphones with integral RF circuitry for acoustical measurements. *J. AES*, **14,** 17–19.
8. Hibbing, M. and Griese, H.J. (1981). New investigations on linearity problems of capacitive transducers. *68th AES Convention*, preprint No. 1752, March.
9. Hibbing, M. (1989). Design of a low noise studio condenser microphone. *77th AES Convention*, preprint No. 2215, March.

5 Radio Microphones and Infra-red Systems

Erhard Werner

5.1 Introduction

5.1.1 The necessity for short-range wireless transmission

The natural propagation of audible signals normally allows people with normal hearing capabilities to pick up a wide dynamic range of sounds generated at distances from the direct neighbourhood to some 1000 m, depending on the level of the source and on the level of masking by other sound sources. The use of wireless links for distances mostly less than 100 m therefore needs an explanation.

One reason for wireless operation is that listening in the sense of getting a message needs acoustical selectivity, which under usual environmental conditions very seldom exists if the source is farther away than some metres. In daily life, too, many sources try to reach the receiver simultaneously, and amount to a mixture which even the well-known cocktail party effect capabilities can only split into useful parts at close distances. So, if no 'public address' but individual connection is wanted, special means to install a selective channel have to be taken. If at least one of the communication partners, source or receiver, is moving around, it is obvious that wired connections will not fufil the requirements sufficiently.

Another reason can be that either the source or the receiver are not able to communicate normally. This is usually either because the source has a disease of the articulatory tract, or the receiver's hearing is impaired. Both disadvantages can be reduced a certain amount by aids to the handicapped partner (voice amplifier at one end or hearing aids at the other), but wireless links with technical equipment at both ends can improve the situation far more.

A further reason is the need to process the signal on its way from the source to the listener. The most obvious use is for stage performances with special dramatic intent. Here the wireless link adds possibilities to already existing solutions, realized by the use of a multitude of microphones in fixed positions, sophisticated mixing consoles with processing of amplitude, frequency and time, and reproduction with loudspeakers (often driven with computer-distributed signals).

These reasons, very often in combination, led engineers to work on short-distance wireless links, which started nearly 50 years ago.

5.1.2 Some aspects of radio wave applications

Speech or even music are fast-varying signals needing high resolution in amplitude and time. When engineers began to think of suitable solutions for short-distance wireless transmission of audio signals, their first choice was the use of radio frequencies.

Broadcasting techniques had already been established for more than 30 years and the necessary components had been improved remarkably. It became possible to transmit wide band audio signals without cables by the methods used for generation and modulation of radio waves. Broadcast methods were adopted and features specially needed for short-distance operation were developed.

Every system working with radio frequencies has to respect the conditions for their use laid down by the various authorities. The main reason for the existing regulations is to guarantee undistorted operation of approved services like broadcast, radio navigation, wireless telecommunication, etc. Many wide-range services work under international regulations because of their operating distance, or their use under mobile conditions crossing borders. Wireless microphones, however, are still mainly regulated under national auspices. Due to the increased mobility of their users and their importance for more than private use at home, authorities have been forced to improve on a former CEPT recommendation,[1] to a state where equipment for short-distance communication can also be used under the same conditions in different countries.

The regulations set limits for the technical performance of the units. This means that the optimum technical solution may not always be realized because of other priorities.

One of the most interesting topics is the allocation of the usable frequency bands. Because of the extensive use of the radio bands, the later wireless microphones in most countries were allowed only to share frequencies with other services, mainly below 50 MHz. As a consequence of this, disadvantages have to be accepted with regard to aerial radiation efficiency, and the incidence of interfering sources like ignition systems of cars.[2] The situation has improved for users in the radio and television business, but many other applicants are looking for better possibilities and good results from the international standardization efforts mentioned above.

5.1.3 Some aspects of infra-red light applications

Due to the intensive use of radio bands by other services, only a limited number of channels could be provided for short-range sound transmission. Even sharing the frequencies for multiple use did not improve the situation, but increased the probability of interference. Furthermore, the propagation mechanism failed to guarantee privacy, and also identical systems could not be used in immediate proximity without problems of interference.

So the discussion on other possibilities intensified. Besides radio waves, inductive loops were used for several applications but their possibilities were found to be quite restricted.

If electromagnetic radiation is considered for even higher frequencies than those normally used for broadcast applications, it is obvious that light waves could be regarded as a possible carrier. Unlike radio waves and alternating magnetic fields, light waves can be absolutely retained in a room, so that wireless systems using light as carrier will not interfere with each other in adjacent rooms. For different reasons invisible light sources are the best technical choice.

Due to the fact that indoors the level of interfering sunlight is low, economic solutions have been designed with infrared emitting diodes (IREDs) light sources based on GaAs. The intensity of the IR light can be kept far below dangerous values. Extra suppression of long-wave infrared light and visible light can be achieved by the natural spectral properties of the detector material, combined with that of the plastic material for the lenses[3] and in conjunction with a filter material with low-pass characteristics.[4]

Silicon diodes are used as receiver elements, in spite of technical limitations due to sensitivity, capacitance and chip area. Optical enlarging of the area is realized mainly by sticking the detector chips on to half-spherical lenses.[5]

The use of infra-red wireless transmission has increased since 1975 to such an extent that standards and regulations have become necessary, not only from the point of compatibility, but also because of interference problems. The IEC and other organizations are presently working on these topics. A condensed survey of the present situation is given in an IEC report.[6]

5.2 Wireless links on radio frequencies

5.2.1 General aspects of radio-frequency use

(a) Propagation problems

The amplitude and phase of electromagnetic waves vary during their propagation in time and space. Electric and magnetic field strengths have fixed phase relations. At greater distances from the transmitter, and under free field conditions, the field strength will decrease reciprocally with distance. Obviously at extreme distances the receiver will get such a low input signal that satisfactory use will be impossible.

Free field propagation is an idealization. Indoors many reflections occur, and the overlay of different reflections results in reinforcement as well as reduction, sometimes even in total signal cancelling. These signal variations are a big problem in radio microphone techniques because the transmitters are normally mobile and thus the field conditions are continuously varying. Total extinction is relatively rare compared with levels low enough to result in noisy and often distorted audio output from the receiver.

The problem of partial or total cancellation is reduced by picking up the electromagnetic field at more than one point. This type of diversity operation can be realized with various different signal processing methods.

As good audio quality is the real aim, an automatic selector circuit identifying the best audio signal available from the different paths would be the optimum solution. Unfortunately, laboratory models showed that a system based on an audio signal was far more complicated, more space and power consuming, and also more expensive, than solutions based on radio-frequency signals.

The RF solution is based on the fact that except for some very rare cases, the RF signal strength and the audio quality are well related to each other. Under idealized conditions with only one receiver input this relates to the signal-to-noise (S/N) ratio. At low input levels the noise will be unacceptable but when the signal reaches some microvolts the S/N ratio quickly ascends to its maximum. Normally every other audio property of interest goes along with this. Possible exceptions like distortion from multi-path propagation with long time delay, distortion by non-linear interaction with other RF signals, or concentrated radio noise sources near to only one of the aerials, can be neglected.

The simplest method (combining two or more aerials) can be highly advantageous when used in the right manner, (Figure 5.1). With a passive RF combiner in this case, both aerial signals are added, but may still result in lower quality than the best of all the inputs. The combined signal can be split again to feed several receivers, in order not to install too many aerials.

A better method is the use of a switched diversity mode (Figure 5.2) where the best of the aerial signals is always selected. Early receivers used a separate

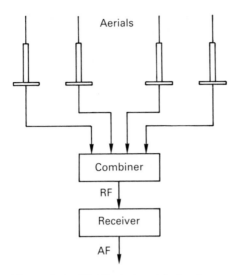

Figure 5.1   *Distributed aerials feeding one receiver.*

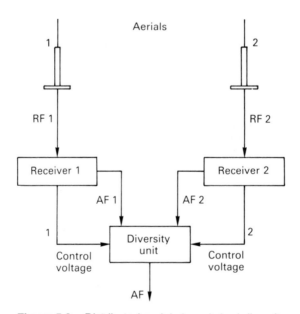

Figure 5.2   *Distributed aerials in switched diversity mode.*

diversity unit for this purpose. Nowadays receivers are ready made to include this facility.

The improvement using this diversity mode can be seen from Figure 5.3. In a small corridor of 40 m length the RF signal from one aerial is found to vary by approximately 50 dB. By the use of diversity, the variations were reduced to about

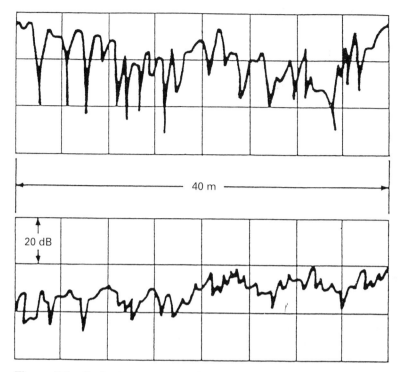

Figure 5.3 *Reduction of level changes by switched diversity.*

20 dB. This not only improves the audio S/N ratio and channel selectivity but also reduces the probability of interference distortion by non-linear effects in multi-channel operation.

These variations cannot be suppressed totally but the probability of cancellation is reduced to the product of the various single probabilities.

Besides the practical results, a computer simulation of the field conditions with reflections was made to get some figures for comparison of the different diversity methods.[7] As a simulation of natural surroundings would have been very complicated and the results even more confusing because of the influence of too many details, a very simple model was used with only two highly conductive perpendicular planes, modelling the floor and one wall of an infinitely extended room.

Two vertically oriented receiving aerials were taken for the calculation at fixed positions from wall and floor. The transmitter aerial was also regarded as vertical and moved at a constant distance from the floor. The range of distance variations from the transmitter to the nearest aerial was always used for the calculation (1 to 32 m with a radio wavelength of 8 m).

The electrical field strength at every point could be derived from the vector potential of the infinitesimal dipole and substituting the influence of the conducting planes by applying the image theory and replacing the planes by equivalent virtual sources.

The receiving aerial signals were determined for 1558 transmitter positions. The computed levels were referred to the level produced by a single dipole at a

Table 5.1 *Classification of relative antenna levels*

Class	Relative level (dB)
9	+ 5 and more
8	0 to + 5
7	− 5 to 0
6	− 10 to − 5
5	− 15 to − 10
4	− 20 to − 15
3	− 25 to − 20
2	− 30 to − 25
1	− 35 to − 30
0	− 35 and less

quarter wavelength distance under free-field conditions (0 dB) and classified after Table 5.1.

The reception conditions were calculated for three cases: single receiver (one antenna signal), vectorial sum of the signals of two fixed antennas, and true diversity by selecting the better signal of both antennas. The results in Figure 5.4 confirm practical experience.

The use of only one RF path (dotted bars in Figure 5.4) leads to a widespread distribution of levels with nearly 20% in classes below 5, expressing levels less

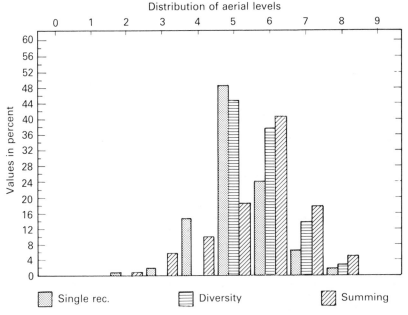

Figure 5.4 *Distribution of aerial level classes under different operating conditions.*

than or equal to – 15 dB to the reference. Even only 1% in class 2 with – 25 dB may lead to critical performance of the channel.

In the case of summing the two antenna signals (diagonally-hatched bars in Figure 5.4) the majority of all calculated samples shifts from class 5 to class 6, meaning an improvement of 5 dB. However, drop-outs are not avoided. Low-level signals still occur, because the vectorial addition of both aerial signals does not always produce signals of greater value, due to significant differences in the phases of the superposed single signals. As the noise is superposed too, summing the signals also affects the audio S/N ratio.

The results for switched diversity (horizontally hatched bars in Figure 5.4) show no samples below class 5. As the single aerial signals are not vectorially added but selected, the increase for high signal levels is not as great as for summation, but this kind of operation, in this example, totally avoids all drop-outs and low-level signals less than 15 dB below reference.

Besides the computations, practical measurements were made with similar antenna arrangements but in non-ideal surroundings and with all three operation modes, recording both aerial signals and the vectorial sum at the same time. For indoors and outdoors measurements, the theoretical results were confirmed: with the difference in reliable operation between switched diversity and two-way summation, respectively, single antenna operation results even increased.

(b) Limitations in radiated power

Radio programmes can be received over distances of hundreds of kilometres. Wireless microphones are known to be limited to a range of up to a maximum of about 100 m. This difference requires an explanation. The far operational distance of radio stations results from fixed installations with high power and optimized aerial conditions. Transmitters for wireless microphones have to be worn by the operator, and this restricts size and weight. It follows that the necessary batteries should not be too big. For this reason most radio microphone transmitters can only produce low power and an operating time of a few hours.

Besides this reason for low power output there is also the inefficiency of the emission because of the aerial design. For practical use the aerial can only be a relatively short wire hanging down from the transmitter, leading to undefined radiation conditions. Many applications have to be run at frequencies near 40 MHz so that, in addition, the aerials are far shorter than the optimum length. Also absorption of electromagnetic energy by the human body has to be taken into account, influencing the radiated power over a far wider range of frequencies.[8]

In addition to the technical limits of emission, there also exist limitations set by the authorities. These result from the guarantee of interference-free use of radio-frequency equipment for every user with an operating licence. In most countries the regulations are made by the postal authorities, who also supervise the transmission activities. Type approval and operating licence procedures differ widely from nation to nation with respect to the kind of operation, the user group, technical specifications and so do the fees for approval or operating licence.

(c) Modulation and RF bandwidth

From the very many types of modulation, the optimum had to be chosen for the transmission of high-quality audio signals, taking into account the requirements for economic use of available RF bandwidth, high audio bandwidth, low distortion, high audio S/N ratio, good suppression of interfering signals and also low power consumption, at least for small portable transmitters.

Some of the conditions have also been valid for the introduction of high-fidelity sound broadcasting. So, even without the need for compatibility, the same frequency modulation was used for short-range systems, using nearly all the parameters from the broadcast specification.

Frequency modulation is a kind of angle modulation where the instantaneous frequency Ω varies linearly with the actual value of the audio signal $x(t)$:

$$\Omega(t) = \Omega_0 + \frac{M}{a} x(t) \tag{5.1}$$

For sinusoidal audio signal

$$x(t) = a \sin \omega t \tag{5.2}$$

the resulting modulated carrier can be described by

$$A(t) = A_0 \sin (\Omega_0 t + M \sin \omega t) \tag{5.3}$$

The spectral description of this complex equation can be given by expanding it into a series of Bessel functions showing infinite extension under ideal conditions. Usually the real spectrum can be cut off without audible disadvantage, so that a real spectrum of a modulated transmitter looks like Figure 5.5.

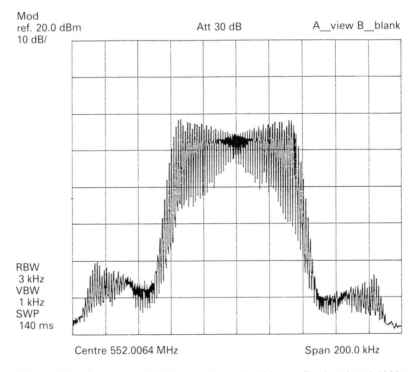

Mod
ref. 20.0 dBm   Att 30 dB   A__view B__blank
10 dB/

RBW
3 kHz
VBW
1 kHz
SWP
140 ms

Centre 552.0064 MHz   Span 200.0 kHz

Figure 5.5   *Spectrum of FM transmitter with 1 khz audio signal and 40 khz deviation.*

By this means, even high-fidelity transmission with FM needs only a bandwidth marginally wider than twice the sum of maximum deviation and maximum audio frequency. The sound broadcast channel grid is 300 kHz, allowing the transmission of up to 15 kHz audio frequency at a peak deviation of 75 kHz and extra services like multiplex stereo components, traffic news, pilot signal and radio data. Short range RF wide-band systems transmit only a monophonic sound with upper frequency limits to 20 kHz and nominal deviation near 40 kHz, which fits them into a 200 kHz channel grid.

Even if less RF channel bandwidth is available, the audio bandwidth will not necessarily be reduced to AM broadcast quality, where the RF channel offers only 9 kHz for 4.5 kHz audio bandwidth. Medium bandwidth FM systems are applied for 40 or 50 and 100 kHz carrier distances with reduction in deviation to 7 to 15 kHz and audio bandwidths between 7 and 12.5 kHz to cover the high-fidelity audio range.

The audio S/N ratio of wireless transmission is determined by several parameters. Besides microphone and amplifier noise already existing for wired links, the contribution of the RF part including the electromagnetic field plays an important role. With low-noise high-level audio inputs the RF link determines the audio floor noise at the receiver output. Over a wide range the value is proportional to the deviation as long as it does not reach the audio input S/N figure. Therefore one manufacturer even offers systems with 150 kHz peak deviation to gain another 6 dB against the above-mentioned 75 kHz; this, however, wastes valuable RF bandwidth.

Low audio distortion in FM-modulated links is achievable with good engineering design. Wide-band modulation and demodulation circuits offer values below 1% total harmonic distortion (t.h.d.) over practically the full audio range. Mainly due to the characteristics of available filter components, medium bandwidth set-ups show slightly increased t.h.d., which for most applications is accepted, instead of having to use bigger units with optimized discrete filter and demodulator stages.

Also for the suppression of interfering signals, frequency modulation is of advantage compared with other possibilities. Amplitude variations of the modulated carrier by fading effects or in-band signals are successfully suppressed by the limiter amplifier in the receiver. Furthermore, the use of a high modulation index leads to increased suppression as the effective S/N ratio S_{FM} is proportional to the carrier amplitude V_{RF}, the modulation index M, and the reciprocal of the amplitude V_{IF} of the interfering signal, and the square root of the receiver's upper audio frequency limit f_0, after the relation[2]:

$$S_{FM} \sim \frac{V_{RF}}{V_{IF}\sqrt{(2f_0)}} M \tag{5.4}$$

Low power consumption for portable transmitters is a consequence of the excellent immunity of wide-band frequency modulation against interference. Low field strength at radiated powers of some mW is sufficient for reliable operation at most-used distances on stages, sportsgrounds or e.g. in music halls. Harmonics produced by power-saving RF amplifiers in B or C mode do not cause the audio quality to deteriorate and can be prevented from being radiated into the air by effective filtering.

(d) RF non-linearity

Nearly all signal-processing systems suffer from non-linear distortion. For analogue signals mainly two characteristics are of interest: harmonic distortion and intermodulation effects. Harmonic distortion hardly ever occurs alone, because generally the spectrum of the input signals contains more than one frequency component. So non-linear characteristics lead to a mixture of harmonics and intermodulation products. In audio signals the intermodulation components have a more annoying effect than the harmonics, which mainly lead to a coloration of the sound.

A mix of radio frequencies is influenced in the same way by non-linearities. Additional spectral components in the radio-frequency range may result in spectral changes of the demodulated audio signal. To ensure a high audio quality for wireless microphones, such effects have to be reduced to a minimum. The non-linear behaviour has even to be considered for single channel applications, because in most cases other radio services will be present and will react with the transmitter or the receiver. If several groups are using radio applications at the same time on stage the mobility and technical requirements lead to the need for separate radio-frequency channels for each actor.

The characteristic of a non-linear element may be described by a Taylor series:

$$y = a_1 x + a_2 x^2 + a_3 x^3 + \dots \tag{5.5}$$

with x as input signal, y as output and a_i as coefficients of the powers of x. With real elements like diodes or magnetic material driven outside the linear range, the coefficients become smaller with increasing order. Therefore it is normally allowable to truncate the series after the fifth term.

The generating mechanism for harmonics and intermodulation components can be seen by introducing the composite input signal x as:

$$x = \sum_{i=1}^{n} x_i \sin \omega_i t \tag{5.6}$$

with x_i as amplitude of the single component and ω related to the frequency f_i of the single component by:

$$\omega_i = 2\pi f_i \tag{5.7}$$

with i as index. By using (5.6) as input for (5.5) and rewriting the power terms of the trigonometric functions by the combination frequency components the output signal will be

$$y = \sum_{k} A_k \sin (\omega_k t + \phi_k) \tag{5.8}$$

where the amplitudes A_k of the components as well as the phases ϕ depend on the coefficients a_i, the input amplitudes x_i and the highest power of the series (5.5) included in the calculation. The same holds for the number of components indicated by k.

The frequencies of the output signal are linear combinations of the input frequencies described by

$$\omega_k = \sum_{i,j} b_j \omega_i, \qquad b_j = 0, \pm 1, \pm 2, \dots \tag{5.9}$$

with the coefficients b_j limited in number only by the order of truncation of (5.5).

Table 5.2 *Non-linear frequency components up to 5th order (2 input frequ.)*

Order / Additional frequency lines (input f_1 and f_2)

2	o (d.c.)	$f_1 \pm f_2$	$2f_1$	$2f_2$		
3	$2f_1 \pm f_2$	$f_1 \pm 2f_2$	$3f_1$	$3f_2$		
4	$3f_1 \pm f_2$	$2f_1 \pm 2f_2$	$f_1 \pm 3f_2$	$4f_1$	$4f_2$	
5	$4f_1 \pm f_2$	$3f_1 \pm 2f_2$	$2f_1 \pm 3f_2$	$f_1 \pm 4f_2$	$5f_1$	$5f_2$

Table 5.2 shows the simplest case with two entry frequencies f_1 and f_2 and the resulting frequencies up to nonlinearity of fifth order. Besides a DC value every power of the Taylor series from 2 onwards supplies spectral components with double the number of its order to the output signal. It is obvious that a greater number of input frequencies will greatly increase the number of non-linear spectral lines. Which components have to be regarded for channel planning is described in Section (f) on interference effects.

(e) Spurious emissions

Every signal which is not really necessary for the operation of an RF unit may be regarded as spurious. Because of the interference risk, only the radiated RF components will be discussed. Spurious signals may not only interfere with other radio services but even cause unwanted effects in a wireless system itself. For that reason the officially allowed values are as low as the engineering state of the art can realize economically. Many national regulations state that transmitters with low power output are not allowed to produce more than 4 nW of radiated power for spurious signals. Receivers are mostly restricted to half this value.

There are several reasons for spurious signal generation. Many transmitter circuits use oscillators on low frequencies and multipliers for the final radio frequency. Without careful filtering of the unwanted oscillator harmonics, these will be radiated as spurious signals. Furthermore, the final RF stage itself may generate harmonics, if for power efficiency reasons class B or C amplification is used. Well-designed filters between this stage and the antenna can attenuate these signals down to the allowed limits. Besides these components, other signals outside the used channel may occur due to overmodulation, mainly spoiling the neighbour channels.

Also heterodyne receivers use internal oscillators to feed the mixer for the intermediate frequency. The oscillator frequency includes possible subharmonics in the case of multiplying, and also the IF must be kept within the system by proper means. Some products on the market show remarkable radiation of these signals.

Unexpected spurious signals may occur from the interaction of a unit with external signals. This effect is directly related to the immunity against strong RF signals, the reduction often being limited by design parameters such as low power consumption. This is still a field for engineering progress.

Measurement of the spurious signals should always be done under radiation conditions, as not only the antenna but also the box and other cabling may contribute to the total radiation.

(f) Interference

Interference in radio systems has two different aspects, the generation of interfering signals and the immunity against these.

Already operating radio services are filling the ether with a 24-hour mixture of frequencies at widely different powers. Normally every RF user will regard the rest as interfering sources. Short-range operation may neglect the majority of weak signals and may avoid the neighbourhood of strong sources.

However, short-range systems may interfere with themselves. In many installations every actor uses his separate RF channel. A greater number of microphone channels consequently leads to many spectral lines, not only from the operating frequencies but also from the limitations in filtering spurious components. Many of these remain outside the operating bandwidth of the receivers, but all odd-ordered harmonics have 'subsets' in the neighbourhood of, or directly on, the carrier frequencies.

In the field, high harmonics and intermodulation products of high order decrease rapidly with distance. Figure 5.6 shows input frequencies and non-linear components of second and third order.

Of the third-order products, only difference frequencies $2f_i - f_j$ may fall into the operating range if the system frequencies are as near to each other as usual. It is obvious that the worst case is when intermodulation frequencies and frequencies of interest become equal, which is for example the result of equidistant carriers (Figure 5.7).

The use of non-equidistant carriers does not guarantee freedom from interference. Figure 5.8 shows that difference frequencies may fall into the receiver bandwidth and will then be processed like signals of an adequate transmitter. In FM systems this leads to strange audible effects often called 'birdies'. Experience has shown that satisfactory transmission quality can be achieved if third-order or at most fifth-order distortion products are kept out of the channels.

Not only the carriers themselves, but also the modulated radio output, have to be examined for interfering signals. With audio modulation, the intermodulation frequencies do not remain constant but shift, depending on the kind of modulation. In FM systems the instantaneous value of the carrier frequency depends on audio amplitude and frequency and covers with its fundamental modulation content, a width of twice the sum of maximum deviation and the highest audio frequency. For the difference frequencies of third order, one of the two contributing frequencies is doubled. As a result, the bandwidth of this

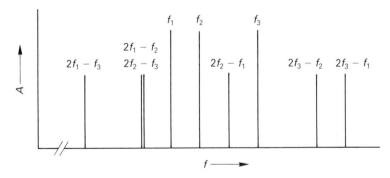

Figure 5.6   *Non-linear components for third-order intermodulation (three input frequencies f_1, f_2, f_3, non-equidistant).*

component is twice the bandwidth of the wanted signal (Figure 5.9). Tests should therefore also be made with modulated transmitted frequencies.

The property of immunity against interference is covered by several effects. Transmitter properties are mostly unaffected by external signals when standard technical precautions are made. The situation for the receiver is far more difficult, as the design has to be optimized, even to detect very weak signals from the related source.

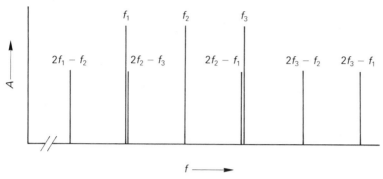

Figure 5.7   *Non-linear components for third-order intermodulation (three input frequencies f₁, f₂, f₃, equidistant).*

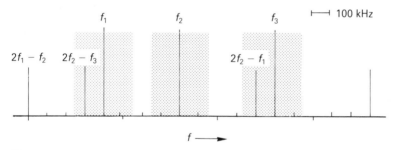

Figure 5.8   *Non-linear components for third-order intermodulation and receiver bandwidth (dotted).*

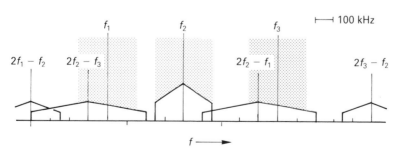

Figure 5.9   *Non-linear components for third-order intermodulation with carrier f₂ modulated.*

The worst case for a receiver is when the interfering signal falls in the channel bandwidth. The interference may not only result in audible distortion but can lead to a complete extinction of the original signal by the interfering source if the amplitude is high enough. The effect can be easily reproduced by adjusting home broadcast receivers to radio channels used by several programs (capture effect).

Signals out of band will have serious effects on the receiver if their level exceeds the receiver's immunity threshold. Non-linear effects as described in Section (d) can contribute to the in-band signal as distortion. Usually a non-linear characteristic means that the lower the input levels are, the greater the signal-to-noise ratio will be. However, the input level cannot be reduced drastically because of field-level variations (see Section (a)).

Another effect is *desensitisation* or even blocking of receivers due to strong out-of-band signals leading to changed bias values of RF amplifier or other stages. Here, instead of audible distortion, the audio signal is attenuated or even lost.

Also worth mentioning is spurious response, which means that a receiver not only detects signals in the defined channel but also on other frequencies due to, for example, intermodulation signals in the mixer oscillator, or low i.f. shielding.

All of these effects show that the use of a good receiver is the best way of increasing the number of channels suitable for use (for details see Section (g) below).

5.2.2 Radio microphones

(a) General remarks and short review

Radio microphones are the solution to the original problem, which is to avoid the cable between microphone and amplifier. The introduction of wireless microphones on radio frequencies was only possible when the limitations of wired installations could be answered by the development of suitable components, mainly for the radio microphone transmitter. Even when consumer magazines in the 1940s offered something like this they were little more than a pleasing novelty for home entertainment.[9] At that time signal amplification and generation of radio frequencies could only be done by vacuum tubes requiring high anode voltages and extra heater energy.

Nevertheless, rising demand with increasing 'show activities' and TV 'live performances' led to the first 'miniature' transmitters in the 1950s.[10] Such a transmitter (Figure 5.10) was from today's point of view extremely basic. The reason for this was to keep size and power at acceptable values. Most of the volume of 156 mm × 105 mm × 27 mm was filled with batteries.

Figure 5.10   *Pocket transmitter SK 1001, 1958. (Courtesy of Sennheiser).*

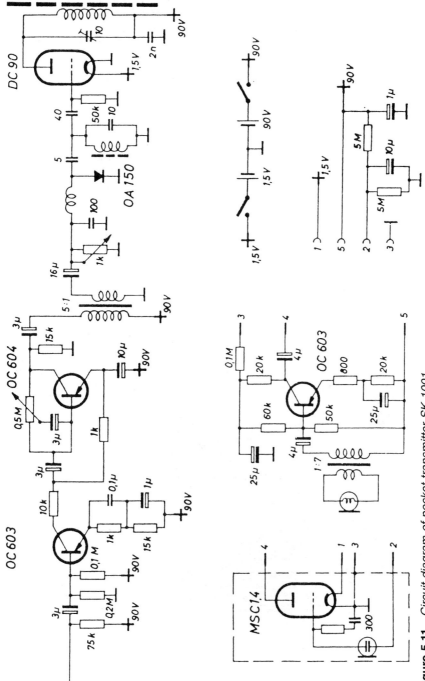

Figure 5.11 *Circuit diagram of pocket transmitter SK 1001.*

The circuit diagram (Figure 5.11) shows that the RF part of the transmitter used only one low-power valve and the AF part already benefitted from the newly introduced AF transistors. Two plug-in microphones were available. The dynamic type could operate with another transistor stage, whereas the condenser type needed another vacuum tube because of its high impedance.

Technical improvements in semiconductor technology resulted in size reduction of the transmitters. The possibilities were only partially exploited, because at the same time higher transmitter power was asked for. So after a size reduction at low power with two 9 V batteries IEC 6 F 22 (Figure 5.12) to about 115 mm $\times$ 75 mm $\times$ 24 mm, an increased size was necessary for a version with 100 mW output power (Fig. 5.13), using three 9 V batteries for sufficient operation time.

Realization of higher legally accepted output power and improved technical qualities due to highly sophisticated audio processing characterize the present situation. Modern transmitters with 100 mW power output (Figure 5.14) and many AF features need only one sixth of the volume of the SK 1001 with its far lower power.

Besides the reduction in size and increase in power, the frequency range was remarkably extended during more than 30 years of radio microphone history. The first permissions were granted for about 36 MHz and far less power than 1 mW. Already at the beginning of the 1970s, versions up to more than 200 MHz and 100 mW radio output were available. Today, wireless microphones cover nearly the complete frequency range between 30 and 1000 MHz.

Figure 5.12   *Small transmitter SK 1004. (Courtesy of Sennheiser).*

Figure 5.13 *Increased power with SK 1007. (Courtesy of Sennheiser).*

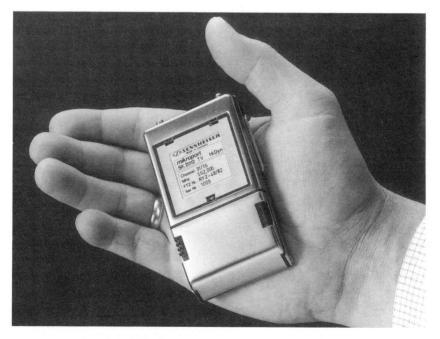

Figure 5.14 *Miniature pocket transmitter SK 2012. (Courtesy of Sennheiser).*

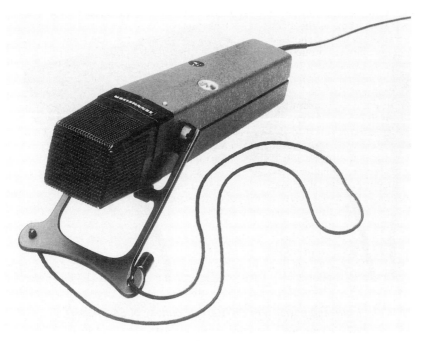

Figure 5.15　*Hand-held transmitter SK 1008. (Courtesy of Sennheiser).*

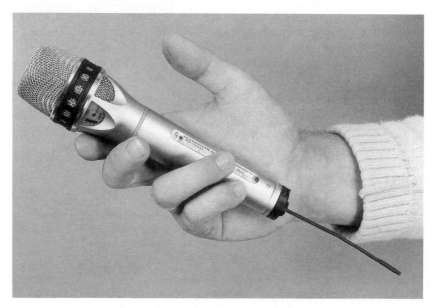

Figure 5.16　*Hand-held transmitter SKM 4031 in microphone size and design. (Courtesy of Sennheiser).*

Pocket transmitters were the first realization but did not quite replace the one-piece wired solution, as the microphone itself was worn separately and connected to the transmitter by a short cable. Therefore as soon as the size reduction of components allowed, a one-piece model was introduced (Figure 5.15).

To offer a certain flexibility for different acoustic environments, either an omnidirectional or a unidirectional dynamic microphone head could be plugged onto the transmitter body. Some years later condenser capsules were also manufactured. This transmitter used the 9 V battery, thus leading to a slightly bigger size than that of conventional hand-held microphones.

Hand-held transmitters for many years did not reach the same usage as their wired counterparts because, similar to the pocket version, technological progress was first used to bring in more refined AF and RF circuits, and to increase the power. Only when smaller semiconductors and better primary alkaline batteries were developed did it become possible to design transmitters with the dimensions of the usual stage microphones (Figure 5.16).

Today the use of surface-mounted elements and DC/DC converters allows manufacturers considerable freedom in reducing the size, so far as the RF radiation is not affected by the interaction between the electromagnetic wave and the hand holding the transmitter.

The transmitter aspect has been described above. A suitable receiver completes the 'wireless' microphone. Receivers are comparable in design to the home receivers used for VHF radio. As a consequence the earliest receivers were modified tunable home receivers using vacuum tubes because no RF semiconductor elements were available. When official operational licences were granted on fixed frequencies continuous tuning had to be replaced by fixed frequencies.

With the development of transistors for higher frequencies the receiver could be fully transistorized. Some features were taken over from the older home techniques. The EM 1008 of 1966 (Figure 5.17) offered two switchable channels and an AFC circuit which was uncritical, because mostly only one radio microphone was in operation.

When more audio channels came into use, mainly for professional applications, crystal-controlled oscillators had to be introduced because of the AFC always

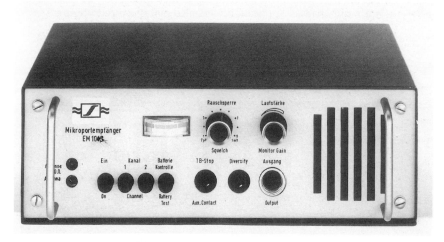

Figure 5.17 *Receiver EM 1008, 1966. (Courtesy of Sennheiser).*

Figure 5.18 *Receiver EM 1010, 1972. (Courtesy of Sennheiser).*

trying to draw the receiver to the strongest signal. Such a five-channel receiver is shown in Figure 5.18.

The advantageous diversity operation was then introduced. The easiest way was to drive two receivers in parallel, with the squelch threshold adjusted to a reasonable level. Both receivers were switched on and their AF output signals added, as long as both input signals passed this threshold. When the input signal was lower than this value, the corresponding receiver was muted. With careful adjustment of the squelch this procedure worked quite well, with only a small risk that both inputs would be lower than the threshold. For easier operation and less risk, the next generation used separate diversity units. They derived a control voltage proportional to the RF input level and also took over the AF receiver output, from extra connections to the receiver, and used the stronger control voltage to switch the AF path to that particular output.

The extreme operating conditions in multi-channel application showed that, for a long time, the vacuum tube had advantages in linearity compared with transistors. Broadcasts used mains operated receivers with vacuum tubes for many years. Only with the development of superior high-current transistors did the vacuum technology disappear. At that time 19 inch rack-mounted designs for space-efficient multi-channel application were also manufactured (Figure 5.19).

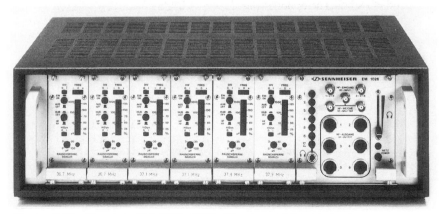

Figure 5.19 *Receiver EM 1026, 1979. (Courtesy of Sennheiser).*

From that time on, diversity operation was offered with far more advanced techniques. Either cassettes for one reception path had all the necessary connections to operate with a second one automatically, in a diversity mode, or in compact receivers all wiring needed for this was already included. Today also smooth operation with selection at the RF entry is possible, thus allowing a considerable reduction in the amount of receiver circuitry.

Besides mains-operated fixed installations, portable receivers were needed. They came at a time when transistors could be used for all stages. Real miniaturization, however, could not be achieved until integrated circuits and passive miniature components replaced their predecessors. Often the mechanical design of the transmitter was also used to form the corresponding receiver. This approach made sense for technical and economic reasons and is still in use, e.g. the pocket transmitter of Figure 5.14 has a receiver pendant in the same housing.

(b) The interaction of applications and technical solutions

The restricted technical possibilities at the time when the first radio microphones were constructed made it necessary to meet every application with the same model or merely to stay with the wired microphone. Because of the different microphone applications, users however needed several variations of the wireless pendant unit.

Obviously the pocket and the hand-held version of the transmitters are examples of a customized solution. Pocket versions are mainly meant to be used where the microphone operation should not be visible. Hand-held wireless microphones, however, are often a visible part of a stage performance and sometimes an 'anchor' for the artist.

Besides its influence on the mechanical design, the electrical performance has to be designed to suit the needs of the particular application. Single-channel versus multi-channel operation or even the transmission of special signals are examples of this trend, sometimes changing the wireless microphone far more than was originally planned. (Sections 5.2.3 to 5.2.6).

Limitations of wired installations

The early limitations of wired microphones were mainly attributable to general restrictions attributable to the state-of-the-art components at that time. Thus suspended carbon microphones mounted in marble housings used at first were soon followed by microphones working on the dynamic and the condenser principle. Today the latter types can be found in many situations where optimum quality is demanded.

Live music, lectures, interviews and sound recordings are examples of the use of normal cabled microphones. This older technique is used with a high degree of reliability whenever the performer's mobility is of little importance and sometimes when size or appearance are of little interest.

The limitations of cable-bound microphones are evident. Real mobility of a speaker or singer is not achievable with microphones connected by cables to a fixed installation. As the distance between microphone and sound source should be short to get best acoustical results, the use of distant microphones on floorstands is not a satisfactory solution. The use of long cables, on the other hand, is also inconvenient. Additional problems arise if several microphone users try to move around at the same time.

Besides the aim of unlimited mobility, there are cases where the microphone has to be invisible, for example at theatre performances or film productions. Often in these cases both demands are combined, so that even the very small and nearly invisible miniature tie-clip microphones are of little help. Apart from wireless techniques, one could only use special shot-gun microphones. Because of their good directional properties, these microphones may be used some metres away from the speaker, but have to be carefully aimed towards the sound source. Highly experienced technical staff, sophisticated technical equipment and situations that change only slowly are necessities for successful sound techniques under these circumstances.

If one is transmitting sound from vehicles on land, water or in the air, strictly cable-bound techniques will be of no use. Wireless techniques using radio frequencies solve the problem. Of course it is obvious that, according to the level of difficulty, experienced operators will bring forth the best results.

Technology requirements for radio microphones

The importance of the progress made by component suppliers has been pointed out. This holds true not only for size but also for electrical performance.

The increasing demand for higher audio and RF quality needs elements with low noise. In the audio chain, for example, metal film resistors are used instead of carbon types with higher noise. Also semiconductor elements have to be chosen carefully with regard to their noise contribution. Noise in the RF stages may be produced by either active or passive elements, e.g. crystals with unstable operation.

Linearity over the complete range of signals is needed to keep down signals which could be a source of interference within the system and outside. When non-linear operation is needed, as in mixers or multipliers, the distribution of by-products has to be considered carefully and should be optimized by the best choice of components.

Stability of the electrical performance is needed not only because professional use does not allow failure, but also to ensure non-interference with other radio users. Temperature and humidity may be important. Even if the temperature range during operation is modest (approx. 15 to 45°C), the storage conditions sometimes are extreme, and very often there is only a short time for the equipment to change from storage to operational conditions. Humidity has become a major problem, since modern musical performances often use pocket transmitters for some singing dancers. Sometimes their transmitters are even hidden under warm isolating costumes so that they get bathed in moisture.

The power supply should offer high efficiency. For mains-operated receivers it may only be a matter of heat energy and ventilation. The batteries for transmitters, however, contribute largely to their size. High energy in a small volume, low source impedance for high loads and wide temperature and humidity tolerance are main requirements.

The mechanical design should meet the demands not only of size but also of interaction with the electrical circuits. The housing and inner cases are used for shielding, so that no unwanted interaction between stages or radiation takes place. Furthermore, mechanical rigidity is needed to prevent the electrical function from being influenced by changing capacitances and, of course, from mechanical damage.

(c) Single-channel applications

Even if impressive numbers of wireless microphones operating at the same time on stage are an engineering challenge, the majority of the sets produced worldwide are for serving single users.

Many announcers and orators wish to address their audience without being bound to a fixed lectern. In announcers' equipment (Section 5.2.6) or one-man shows, radio microphones are often combined with battery-driven loudspeaker systems, leading to lightweight mobile solutions. Services in churches often use wireless microphones instead of switching microphones between the priest's different positions. Film amateurs used wireless microphones during the 1970s when the 'Super 8' film format with sound track came on to the market.[11] Distant sound could then be transferred to the camera. With the increasing use of mobile video recorders, this is still of great value in situations where long microphone cables are neither practical nor desirable.

Some special applications are related to certain professional uses or to physical diseases (Sections 5.2.3 to 5.2.6).

The necessary technical performance of single-channel radio microphones is different from the specification for multi-channel operation as the precautions against interference from other channels can be less. On the other hand, diversity operation can be impossible for single operation if, for example, the receiver also is mobile, as for video application.

Besides optimization for portable receivers, other aspects depending on the usage can differ from multi-channel specification (Sections 5.2.3. to 5.2.6).

(d) Multi-channel operation

In many applications very many wireless microphones are needed at the same time and in the same place. Theatre performances are not usually possible with only one actor wearing a wireless microphone. Due to the continually varying acoustical conditions, as between different actors, acoustic levels and spectral differences cannot easily be set under these circumstances.

Sharing the frequencies with other contributors is also impossible. The frequency modulation used does not allow an RF summation without producing extreme audio distortion and noise. Even the use of synchronizable carriers and, for example, amplitude modulation, would not allow one to treat the audio channels separately, as required for equalization and dramatic effects. As a consequence of this, every actor of importance has to be given a separate RF channel.

The situation with moving transmitters leads to extreme variations in the RF level at the receiver antenna (Section 5.2.1(a). This means that the relevant signal at a receiver input may be far lower than that of other channels in the installed system. Possible negative effects on the receiver operation have already been mentioned (Sections 5.2.1d and f).

Several means of diminishing the effects of intermodulation are available. Apart from suitable hardware, one can attempt to choose frequencies with computer programs and tables to ensure that intermodulation products up to the third or fifth order remain outside the channels in use.[12] These considerations originally came from narrow-band communication systems. Their application to wireless microphones is restricted because of the wide-band modulation, covering 200 kHz and more per channel, so that only a few channels can be located ideally within the authorized frequency bands.

The generating algorithm for intermodulation products of third order (Section 5.2.1(d)) can be rewritten in a way that allows conditions for interference-free operation with regard to the carrier situation to be derived. The result is that third-order carrier interference is avoided when no two differences between any possible pair of channel frequencies are equal.[12]

More complicated, but with a similar result, is the situation when carrier intermodulation up to the fifth order is considered. Rewriting the algorithm, the conditions for interference-free operation are that not only the third order requirements have to be fulfilled, but also the difference between any two carrier frequencies should not be equal to the sum of any other two such differences.[12]

In the calculation of frequency sets, it is easier to ignore the real frequencies and to introduce channel numbers based on a fixed grid, by this dealing only with integer numbers. It is obvious that the conditions for interference-free operation do not depend on the real location of a multi-channel system but on the relative position of the carriers from each other. The conditions for third order mean that for n channels at least a switching range equal to the sum of the numbers 1 to n is necessary. Following the complete conditions, slightly increased values are calculated. With fifth-order arrays the switching range increases remarkably. Table 5.3 shows the results under the assumption that for selectivity reasons the neighbouring channel is left out.

Table 5.3   *Minimum switching ranges for third- and fifth-order compatibility*

Number of channels	Minimum switching range 3rd order	5th order
3	6	–
4	10	13
5	15	26
6	21	>40
7	29	
8	40	

The calculations show that the choice of frequencies for a greater number of channels compatible for third- or even fifth-order intermodulation leads to extreme loss of available RF bandwidth, by leaving the biggest part of the range unused. Therefore this procedure can be used only for a restricted number of channels and with enough bandwidth to be left unused. This situation exists where TV channels may be used for wireless microphones.

Fortunately, the TV broadcast stations do not cover the complete TV bands in one location but the TV channels are distributed over the regions with regard to the number of programmes and the limited selectivity of home receivers and special services. That means that at fixed locations, there are always some TV channels not in use for programme broadcasting. So the VHF TV band as well as the UHF band are used for wireless microphones, mainly by the broadcast companies themselves.

Depending on the TV band and the standard used, TV channels may be, say, 7 or 8 MHz wide. As an example, the European UHF band is divided into channels 8 MHz wide. The TV signal does not use the upper 1 MHz, which serves for communication links and other uses by the broadcasters. The remaining 7 MHz of locally unused channels may be used for wireless microphones. By introducing a raster of 125 kHz, preferred channel allocations could be computed. They take into account a minimum carrier distance of 375 kHz, and thus avoid the direct neighbourhood of TV sound and video carrier allocation, even taking into account possible long-distance interference to operating TV channels. Under these assumptions it is possible to place in one TV channel six wireless microphone channels for fifth-order compatibility or 9 channels for third order.

The use of six high-quality channels in one TV channel is very often sufficient; however, more and more performances need higher numbers, either for quick changes with consecutive groups or many independent actors at the same time. Therefore additional means had to be found for an increase of the number of channels.

The potential exists to use more than one TV channel. The UHF range offers at least a band between 470 and 790 MHz divided into 40 TV channels. For fixed installations it is likely that more than one TV channel is not used or is of such low activity that it can be used for radio microphones. If these TV channels are not direct neighbours, the RF front-end selectivity of radio microphone receivers splits the total system into non-interfering subsystems. They can be regarded as separate and optimized individually. By this means, multi-channel operation of wireless microphones has been installed with up to 48 channels.

The equipment for simultaneous use of so many wireless microphones already needs special extra efforts to supply the sound engineers with the necessary information on the actual channel states. Remote supervision is sometimes needed when the receivers are located separately from the control room. With the relevant interface at the receiver, information on RF levels and AF signals can be displayed on a colour monitor. An impression of the effect for a 27-channel installation is given in Figure 5.20.

Figure 5.20 *Equipment for 27 channels with remote display. (Courtesy of Sennheiser).*

Trouble-free operation relies not only on an optimum choice of frequencies but may also depend on attention to every other possible means of improving the situation for different amplitudes, multiple-path reception and non-linear operation.

In spite of the mostly unpredictable level variations of the electromagnetic field, some operating conditions may be optimized by regarding the propagation under free-field aspects. Because in this case the different distances between transmitter and receiver mainly influence the aerial level, the variations of distance should be kept as small as possible. This can be achieved by proper location of the receiver aerials with regard to the area of transmitter operation. Many installations following this rule have operated quite successfully.

The aspect of adequate distance can also be used advantageously for the simple summing of aerial signals. Mainly where larger areas are covered by the actors, as in open-air theatres, the probability of decreasing the sum level will be less than the increase, by always approaching the nearest aerial by the necessary amount.

The positive effect of distributed receiver aerials within adequate distances is in any case increased by always switching to the best input. Besides the classical solution of selecting the AF output, RF input selectors have also been realized (Section 2.1.6).

Multiple-path reception with wireless microphones produces mainly the effect of amplitude variations. Remarkable delay times of the different beams can be neglected because of the extremely weak amplitude of far reflections. But even with near reflections, the use of directive receiver aerials can improve the overall quality. For open-air performances unidirectional receiver aerials were installed in the back of the auditory and facing the stage. By this means the aerial gain supplied enough signal strength also at greater distances, and furthermore, radio sources from other directions were suppressed.

(e) Audio performance characteristics

As already mentioned, the first aim with radio microphones was to avoid the cable without changing the audio performance. As far as audio bandwidth and non-linear distortion are concerned, the FM radio operation very quickly reached the necessary quality level.

A long-lasting problem was the necessary dynamic range. There were quite audible differences between real sound, audio signal transmission via high-quality cabled microphones and wireless links. For better understanding, it is worth looking carefully at the different influences (Figure 5.21).

All sound levels are normally referred to the capability of the human ear, the threshold of 0 dB being 20 μPa in absolute terms. Natural sound fields may extend the dynamic range of the ear, with its upper limit of 130 dB. Normally the lowest levels will not be heard because of the level of environmental background noise. Even in very quiet rooms this level is 20–30 dB(A). Any processing equipment placed between sound source and listener will further reduce the dynamic range. This equipment may be necessary either for storage/recording purposes or for transmission to a distant point. Transmission may be via cable or radio. Each system has an upper limit for its S/N ratio, so it should preferably operate up to this level. Lower levels result in a decreased dynamic range; higher levels may produce distortion.

The microphone as the first technical element in the chain produces noise equivalent to sound-pressure levels of 13 dB(A), or 20 dB according to CCIR-weighting and peak measurement.

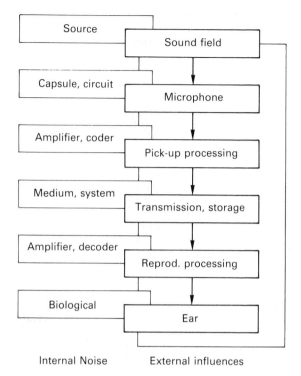

Internal Noise   External influences

Figure 5.21   *Noise contributions between sound source and ear.*

These low noise values are only slightly increased by the first processing stage if it is properly designed and works at a high supply voltage. The highest sound pressure a microphone can pick up with full quality determines its dynamic range. The best studio microphones of today can handle about 120 dB with one gain setting. Even far higher sound pressures than the ear can withstand may be picked up at reduced gain, so the dynamic range remains practically constant.

Wired transmission to a distant point may result in a very small reduction of the original S/N ratio. With wireless transmission, however, additional loss in dynamic range occurs because of the limited available bandwidth. Linear FM modulation leads to values less than 70 dB. To allow the usual safe headroom, another 6 dB are also lost. All transmitter stages, including the audio amplifier, oscillator, modulating units, and radio-frequency stages contribute to the total transmitter noise. Each stage contributes a different amount. In many radio microphones the audio-frequency input stage and oscillator are major noise sources. Standard transmitters usually accept dynamic microphones with a sensitivity of about 1 mV/Pa.

At extreme low sound levels the audio amplifier will determine the noise of the complete wireless microphone, but in most cases the RF limitations will be responsible. With a dynamic range of 70 dB the system noise during pauses can always be heard if the maximum audio level at the output is higher than this level, plus an extra margin for masking effects by background noise from other sources in the reproduction room. Taking into account that frequently peak audio levels during reproduction reach 110 dB, there remains a gap of 40 dB down to the hearing threshold.

Even if the receiver RF input is short-circuited, the AF output will deliver a measurable noise floor at the rated AF gain. The amount is smaller than in the preceding stages, but it also requires the use of low-noise circuits in the signal train.

The noise from processing the signal for reproduction is negligible if the transmission path carries a high-level output signal. Background noise in the listening room may further reduce the perceived dynamic range. Wireless microphones are intended for sound reproduction in relatively quiet rooms like theatres and living rooms. Thus this last noise contribution can be ignored.

The evaluation of the different noise sources showed that audible progress would require remarkable improvement of the stages from transmitter modulator to receiver demodulator. The possibilities of pre-emphasis and de-emphasis were already used. Higher peak deviation did not greatly improve the value, and reduced the number of channels in a given switching range. The only remedy was to compress the audio signal in the transmitter to the dynamic range of the RF system and to expand it at the receiver to the original values.

Compressor and expander circuits had already been used for tape recording, because the dynamic range of the tapes was also less than that of the original signal. For wireless microphones the circuits realized for this purpose were either far too big or at least too power consuming. But in 1979 a suitable integrated circuit came on to the market and could be adapted to the requirements of wireless microphones.

This circuit was used by many manufacturers and performs logarithmic compression and expansion with a factor of 2. The operating characteristics are shown in Figure 5.22 in comparison with linear transmission.

Only at the reference point do compression and expansion have unity gain. The advantage of purely logarithmic compression and expansion is that differences in the reference levels of transmitter and receiver do not change the level relations between spectral components. From the compandor curves it can be seen that

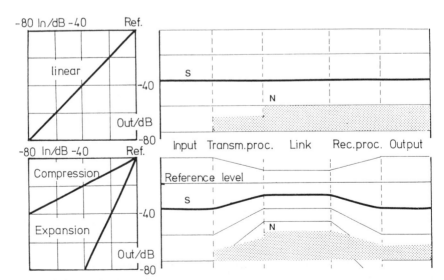

Figure 5.22   *Level diagram for linear (above) and 2:1 compandor (below) operation.*

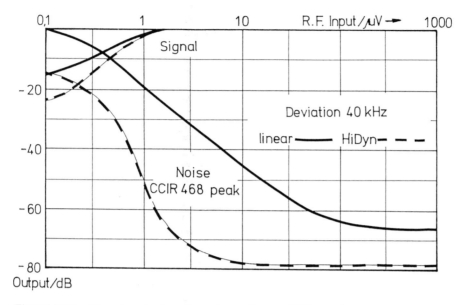

Figure 5.23 *Signal and noise at receiver output vs RF input (solid line = linear, dashed line = 'HiDyn' compandor).*

additional noise introduced between the compressor and expander is reduced at the audio frequency output as long as the signal input stays below reference level. The greatest effect is obtained if the modulated signal works at the highest possible deviation with marginal headroom.

At reference level, of course, there is no noise suppression effect by the compandor circuit, but the available S/N ratio is high enough for the masking effect to make the noise floor inaudible.

The technical realization of compressor and expander has to be done very carefully. Attack and release times have to be set so that neither overload nor 'pumping' effects can be heard and also AF distortion by the time-varying gain is minimized.

The optimized systems show remarkable advantages and are also fully accepted by professional users. At low RF levels the increase in S/N ratio went up by nearly 40 dB (Figure 5.23). The maximum S/N ratio of a 1981 model shifted from 65 dB to nearly 80 dB, even with the most critical noise-measuring procedure (CCIR 468–3 peak). With the A-weighted r.m.s. measurement usually used for consumer electronic equipment, the value would be more than 10 dB better, due to the noise spectrum. Meanwhile further development efforts have improved the values by more than 20 dB.

(f) Characteristics of radio microphone transmitters

Many important characteristics of audio and radio performance have already been discussed in the preceding sections. Some additional characteristics, however, are strongly related to the electrotechnical design, and will therefore be presented together with the block diagrams of transmitter and receiver.

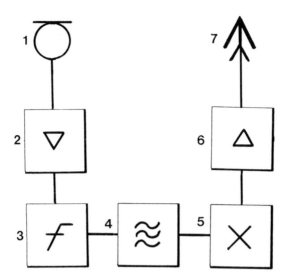

Figure 5.24   *Simplified block diagram of radio microphone transmitter.*

A simplified block diagram of the radio microphone transmitter is given in Figure 5.24. The microphone (1) in this Figure has to be chosen after considering the requirements of the sound field and perhaps some needed corrections. An example for special treatment is the construction of microphones with a suitable filter to suppress the chest resonances of the speaker, which may be picked up by lapel microphones.

Even with the use of condenser microphones or dynamic microphones with higher impedances, the AF voltage may reach peak values of some millivolts, the following audio amplifier (2) will also have to handle signals from low-level sound sources and should operate linearly and with marginal noise contribution.

As already mentioned, the full dynamic range of the acoustical input cannot be transmitted linearly by the radio channel. The necessary gain reduction in the limiter amplifier (3) can be done by several means, resulting in quite different acoustical quality. Clipping diodes, limiters, special compressors, compandors or a combination of these can be used.

Clipping diodes are the simplest means. They can be used to meet the regulations for out-of-band signals, but lead to poor audio quality at high levels. Limiter amplifiers should also react only during peaks of the signal. With proper design of attack and release time, a quite acceptable audio quality can be achieved. The attenuation of high audio levels will become annoying when the average level is set too near to the peak values.

An even 'softer' characteristic nearing compression distributes the effect over a wider level range, but is not always accepted under musical aspects. Only when the compressor function is compensated in the receiver can the dynamic compandor already mentioned bring back the desired substantial quality.

Also included in the processing step (3) is the raising of higher audio frequencies by use of pre-emphasis circuits. This also leads to better use of the available bandwidth with regard to the spectral content of normal sound programmes.

The resulting audio signal is used to modulate the radio frequency. In most realizations this is done by direct influence on the oscillator (4). Special circuits are necessary to optimize the contradiction between a highly stable radio frequency and easy modulation.

Most transmitters on higher frequencies do not run the oscillator at the final frequency but use multiplier stages (5). In several stages the modulated oscillator frequency is amplified and brought to, for example, 32 times its original value. By the use of different frequencies in each stage, feedback risks and other possible interactions are reduced. The multiplication generates many harmonics so that the desired multiple has to be selected by adequate resonant circuits. New solutions have been offered by use of synthesizers. With the phase-locked-loop technology (PLL) it is possible to lock the average frequency of an easily modulated voltage-controlled oscillator (VCO) to a very stable reference crystal oscillator. Quite different from the RF generation in the block diagram, at not too high frequencies, the VCO may operate directly on the output frequency, and comparison with the reference oscillator is done after dividing the RF frequency by a suitable value.

The RF power stage (6) is normally not designed as a multiplier but the need for low-current operation mostly leads to class B or class C operation mode. These modes again produce harmonics which have to be suppressed by passive filters. The RF signal has not only to be checked for harmonics and other spurious frequencies, but also for the power delivered to the immediate neighbouring frequencies. Regulations exist for this 'adjacent channel power', which, due to the proximity of the carrier, can not be reduced as easily as the spurious signals. Where 4 nW is achieved for spurious frequencies a value of some microwatts for the adjacent channel is regarded as a good result.

The design of the transmitter aerial (7) is of extreme importance for the radiation efficiency. The worst conditions arise if the length is short compared to a quarter wavelength. This is the case for many private-user frequencies. However, even at higher frequencies where the length can be sufficient, absorption by the user's body has to be regarded. For hand-held transmitters the aerial should not disappear into the palm; body-worn transmitters should be carried with the aerial as distant from the skin as possible. Experienced operators know a lot of possible ways to fix transmitter and aerial and even take into account the RF interaction between microphone cable and aerial.

(g) Characteristics of radio microphone receivers

For the receiver also, a discussion of some remaining characteristics of importance will be given with reference to the block diagram (Figure 5.25).

Several possibilities and operational recommendations for the aerial (1) have already been described. It should also be noted that the cabling between aerial and receiver input may lead to signal attenuation when inadequate material or extreme length are used. Certain studio applications with extreme distances therefore need front amplifiers for the aerial signal and also extreme RF linearity.

The RF input amplifier (2) should be as selective as possible. It would be ideal if every receiver had to process only one channel, because then non-linearities in the course of signal processing would result only in harmonics of the input signal. This selection cannot be achieved, for technical and economical reasons. The receiver passband characteristics have to be flat for the whole channel. Even the steepest possible cut-off slopes will extend to the neighbouring channels. So even with optimum alignment for the receiving channel, adjacent signals may still enter the first amplifier stage with enough strength to produce extra signals.

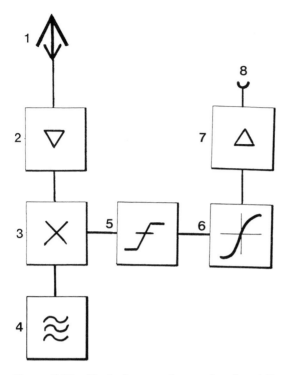

Figure 5.25   *Block diagram of a receiver for wireless microphones.*

Linearity for a wide input signal range is therefore the only means to avoid intermodulation products within the receiving band, and this requires the best available components on the market. This holds for passive elements, of course, as much as for active ones. The degree of linearity of a receiver can be described, for example, by its intercept point.

The multiplier sign $\times$ stands for the mixer (3) which converts the amplified RF input to the intermediate frequency (IF) by non-linear processing with the radio signal of the local oscillator (4). This necessity can be the reason for a multitude of unwanted signals. Different kinds of mixer circuits are available. For high performance only the best are used to avoid any unnecessary and even risky by-products.

Even if the mixer circuit is carefully designed, a poor oscillator signal can also result in a bad performance. Optimum oscillator level and low harmonics or subharmonics are necessary conditions for proper operation. Crystal oscillators in receivers mainly run below 100 MHz, so that for VHF or UHF frequencies, again multiplier circuits are needed. Their content on spurious spectrum must be suppressed efficiently before the mixing process. The already mentioned PLL technology is meanwhile also used for the generation of the oscillator frequency. By this means the discrete crystals that would otherwise be required are replaced by a single one, combined with programmable dividers and memories (PROMs) to set the receiver only to the allowed frequencies.

Wide band radio microphones mainly operate on 10.7 MHz IF in the receiver because well-designed components from the home receiver industry can be used.

The IF amplifier (5) has not only to raise the signal selectively to an appropriate level but also to maintain a constant amplitude-limited level that guarantees optimum suppression of amplitude-modulated interference effects.

In case of medium- or narrow-band operation, the necessary higher selectivity can be achieved by double conversion with a second IF, in frequency ranges where narrow band filters are available at reasonable cost, e.g. 455 kHz from the AM broadcast. The blocks (3) to (5) would then be repeated for the second IF. Double conversion may, however, introduce extra non-linearity problems. Therefore, for the price of higher expenditure, high-performance crystal filters on 10.7 MHz are used to keep the receiver as a single conversion unit.

Once the IF signal has attained the right amplitude it is ready for demodulation in stage 6. Demodulation circuits for FM are virtually an established branch of engineering, but careful examination of the performance of the different solutions is necessary. Development engineers now mainly use sophisticated integrated circuits with application rules often given by the laboratories of the big semiconductor companies.

The demodulated signal needs further processing. The audio amplifier (7) has not only to provide the right matching values for the next stage but also to invert every procedure applied to the signal at the receiver, mainly to expand compressed signals and to equalize pre-emphasis by the corresponding de-emphasizing network. The necessary matching values depend on whether the audio signal (8) is fed to a line, a loudspeaker or a headphone. They are laid down in several international standards, e.g. IEC Standard 268.[13] These are highly recommended. Even simple things can cause trouble, as may be seen from a well-known example. If the audio polarity on the way from source to receiver output is changed, a mixed signal at the mixing console from two transmitters picking up the same source will show extreme coloration by partial cancellation of low frequencies.

Not shown in the block diagram are the circuits for diversity operation. As has been described in the historical survey (Section (a)), very different approaches to this operation have been realized. The classical solution with two complete receiver tracks and automatic AF output selection still applies to the majority of all diversity receivers, but other possibilities like RF input selection will have a good chance to take over in the future because of reduced circuit complexity or less expense.

5.2.3 High-power microphone links

Further increase of power results in an increase in size which goes far beyond that of conventional microphones. These high-power wireless microphones are used for outside broadcasting under special conditions. A very early model (Figure 5.26) was developed in 1963 only five years after the pocket transmitter SK 1001. The low power output from RF transistors made it necessary to use up to eight of them for an output power of 1 W! This made it possible, for example, to give traffic reports from a helicopter over longer distances. The high-quality audio signal is often combined with normal radio programmes. In this sense these O.B. transmitters are a logical extension of wireless microphones.

When tests were made for the transmission of Olympic games events in 1972, even 1 W proved not to be enough for sailing reports from accompanying boats. A booster for 10 W and a larger accumulator pack were the solution. Meanwhile several intermediate steps led to a transmitter offered with several frequency ranges, including a talk-back receiver, working either in wide-band or in narrow-band modes on communication frequencies (Figure 5.27).

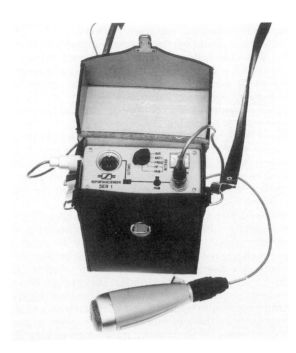

Figure 5.26   *Early OB transmitter SER 1 with 1 W RF power. (Courtesy of Sennheiser).*

Figure 5.27   *Outside broadcasting transmitter SER 20. (Courtesy of Sennheiser).*

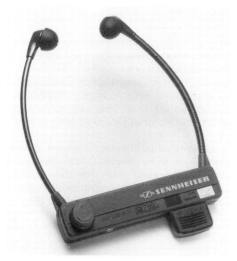

Figure 5.28 *Tour guide receiver HDE 300. (Courtesy of Sennheiser).*

5.2.4 Tour guide systems

An application where a multitude of portable receivers is combined with one transmitter is for guided groups. In many cases the acoustical conditions for this situation are very bad due to background noise, the number of participants, or the low voice level of guides. Wireless short-range links bring the information directly to the ear of the visitor. Besides pocket receivers with earphones connected by a cable, a compact and extremely lightweight version in the ear mould uses the plastic tubes from the central circuit to the earpieces as support for a coiled antenna. (Figure 5.28).

5.2.5 Radio aids for the handicapped

Hearing-impaired people can profit from the development of radio microphones. Conventional hearing aids amplify the acoustical signal picked up by a microphone worn by the user. Unlike normal hearing individuals, people with impaired hearing find that the intelligibility of distant sources is reduced, even if the level is adjusted by the gain of the hearing-aid amplifier. However, the use of a transmitter by the speaker and the corresponding receiver by the handicapped person can effectively shorten the acoustical distance to that between the orator's mouth and his microphone.

This remarkable increase in intelligibility plays an important role in the education and speech training of children with hearing impairment. Children born nearly deaf succeed far better in speech training; highly hearing-impaired children may follow the lectures even in schools not especially acoustically designed for their handicap.

Several manufacturers offer small-sized transmitters and receivers which sometimes combine standard hearing-aid facilities with the wireless option.

Not only hearing problems may be overcome by wireless aids, but also loss of speech volume. Even 'whispering' speakers (after larynx surgery) will be understood by a distant audience, a requirement necessary for teachers, lawyers and people of similar professions.

5.2.6 Other applications

Wireless microphones are very often combined with other audio equipment in complex units when ease of handling or reduction in overall size is of importance. Examples for this are loudspeakers with built-in RF receivers which may even be battery-operated, so that cabling for temporary installations is not necessary. By this means events in larger areas can be supplied with audio signals. Besides mobile RF public address equipment, smaller units of low weight serve as portable address units in rooms of medium size or for outdoors address of small groups.

Film production teams or broadcast teams need lightweight equipment with even more facilities. Multi-track sound may have to be provided. Sometimes side channels are necessary to communicate with remote staff. For this type of operation a wireless mixer has been developed which offers multi-channel operation in two directions combined with the necessary audio settings (Figure 5.29).

5.2.7 Future aspects

The technical progress in more than 30 years of radio microphones has made operation extremely easy and safe. Miniaturization has brought products down to the size limit for handling. The technical specification of actual equipment is superior to that in the past in nearly every characteristic. Only the type of modulation and channel bandwidth is still the same.

Further development of the classical FM link will be mainly based on higher integration of electrical functions, replacing some old analogue functions by digital means and the use of improved component and production technologies. Many possibilities can already be derived from the consumer electronics market. The limited market for radio microphones always forced the manufacturers to look for components made for bigger markets and to adapt them to their needs. Wherever features of other applications were useful for RF microphones, people tried to transfer them. On the other hand, some ideas had to be postponed because of non-available components. Special functions needing highly integrated circuits can only be realized when complicated integrated circuits for

Figure 5.29 *Wireless mixer WM 1. (Courtesy of Sennheiser).*

analogue and digital operation (ASICs) are designed and produced at reasonable cost, e.g. for TV receivers or consumer video equipment. Changes in this direction have to be expected, because of the changing attitude of the semiconductor industry.

The evolution of analogue processing is a separate question from that of digitizing the audio signal. The introduction of the compact disc and the start of digital audio broadcasts (DAB) intensifies the discussion. However, given the current state of technology, it is still not yet possible to meet the major demands such as the size and power consumption of the transmitter, the RF bandwidth occupied, the complexity of transmitter and receiver and, last but not least, an acceptable price. The situation may change as progress is made in the field of digital signal processing.

5.3 Wireless links on infra-red

For good reasons electromagnetic waves in the broadcast range were chosen for wireless microphones. Partially these reasons are still valid, even if many applications of infra-red (IR) now exist. A serious examination of these applications shows that most of them are not really replacing a wired microphone but a wired reproduction unit, mainly a headphone. The following sections will give a condensed survey in order to keep within the scope of this book.

5.3.1 Historical survey of infra-red links

Optical signals for the transmission of information have been used by our ancestors for long-distance communication. Of course only simple messages at low speed could be handled with the available sources and codes. However, even today improved versions of the old 'smoke or light bits' are still in use.

Infrared light as a possible medium for wireless microphones was mentioned as early as 1958,[10] but the technical obstacles then were too high. Apart from some scientific experiments or very expensive equipment, the widespread use of light as a carrier of sound information was very limited until the production of modern semiconductor elements. With IR-emitting semiconductor diodes invisible light could be used also for inexpensive equipment. The application quickly spread from speech and music into other regions, e.g. two channels for hard-of-hearing pupils or 'fold-back' signals in studios.[14]

The first products used in larger quantities were small IR transmitters and stethoscope receivers for the transmission of monophonic TV sound for home use (Figure 5.30).

This combination, introduced in 1975, used a 95 kHz carrier frequency, which was standardized world-wide later.[15] This wide-band system was also used by hearing-impaired people.

Only one year after the mono system had been introduced a two-channel system was developed which realized stereo operation in special classes for hard-of-hearing children, where formerly inductive loop systems were in operation. This system worked with narrow-band frequency modulation on carrier frequencies of 200 and 280 kHz.[16]

Beside this two-channel narrow-band version, the development of a two channel wide-band version succeeded in a system presented in 1977 for high-fidelity use.[16] This system not only proved to be useful in home applications but also in professional use, e.g. for foldback systems during live music events and also post production.

Wireless multi-channel conference systems have used inductive loops with amplitude modulation on carrier frequencies near 100 kHz. Large installations

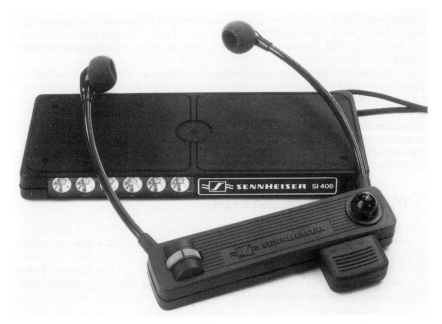

Figure 5.30   *Infra-red transmission set for TV sound. (Courtesy of Sennheiser).*

with multiple use of the same channels in one conference centre were affected by spilling over. Therefore for the same application in 1977, a nine-channel IR system was presented.[17] Here also the FM technique showed good reliable results. The system was originally meant only for conference systems but later was used also for addressing the floor people in TV studios and theatres. The immunity of the receiver against interfering light was increased, so that even under TV studio conditions, trouble-free operation was achieved. Thus, the then existing problems of availability of channels for RF systems could be reduced.

The year 1981 brought a further step towards even better performance in high-fidelity two-channel stereo systems by the introduction of dynamic range compandors.

The improvement in already used applications continued. Multi-channel frequency multiplex systems ran up to 16 channels. Besides this solution, a pulse-distance system was introduced which offers a multitude of channels and also some back-channels. These realizations were possible because faster-emitting diodes came on to the market.

Besides the majority of units used with FM techniques, further solutions with different pulse modulations were developed. They will be described later in the special applications sections.

5.3.2 General aspects of the use of infra-red light

(a) Generation and propagation of infra-red light

Infrared light emitted by semiconductor diodes is used as a carrier. Up to now, nearly all systems use GaAs-based diodes with a light efficiency of up to 30%. Low-cost silicon-doped diodes achieve approximately 10%, combined with long lifetime at specified current limits, but cannot be operated at very high speed. Aluminium-doped diodes allow higher operating frequencies, but in many cases show higher degradation. The highest efficiency today is by diodes using several layers. Meanwhile, high-speed diodes with good efficiency are available which can be operated up to more than 10 MHz. So the choice of the right type for different applications is wide. Because the emitted light of all of these sources is not monochromatic, all existing systems use as a first step intensity modulation which carries the information by a second modulation step.

The spectrum of these diodes is a very narrow band near to the visible red. By different doping materials, the peak emission wavelength can be fixed currently at any position between less than 800 nm and 1050 nm. By this means the sources offer different light wavelength channels which, however, under the common receiving conditions in free propagating light and because of receiver component selectivity, unfortunately cannot economically be used today for widespread applications.

Though the radiation is invisible, one has to deal with limitations due to possibly harmful power density. Limits from laser regulations are not applicable, but most applications use power densities below the IR content of bright sunshine and thus have to be regarded as safe.

Excluding ideally bundled parallel beams of light, normally the light density drops inversely as the square of the distance between light source and object. This free propagation is changed by object shadows and by light reflection and absorption. Total absorption is the worst condition. The best case would be a total reflection, since all light energy would then remain in the room. The extreme case involving an infinite number of reflections would then lead to light energy being homogeneously distributed in the room. The normal case will be somewhere between direct beam and homogeneous distribution, except for shadow zones.

Infrared light behaves in the same way as visible light. Due to the close proximity of the IR light frequency to visible light, reflection coefficients will also not be very different from those for visible light. This propagation is advantageous, because by using walls or curtains, the signals may be restricted to predetermined areas.

The distribution of the IR radiation can be influenced by the position of the radiator units and also by the choice of diodes with different directional characteristics, or extra directional elements in the radiator.

(b) Limitations in radiated power

It has already been mentioned that IR transmission needs only low power density; however, compared to RF systems the power is still far higher. For wideband FM systems in normal illumination six diodes of the early mono system supplied rooms of up to approximately 25 m^2. The total infra-red power of these six diodes was about 50 mW, so as a rule of thumb 2 mW/m^2 might be recommended. But because of the low electro-optical efficiency and, for the remaining circuits, the corresponding transmitter needs more than 1 W DC power (40 mW/m^2), which is more than 10 times the amount needed for a RF wireless microphone covering far wider areas. As a consequence, the majority of IR applications work with mains-operated transmitters.

A true wireless microphone on IR with low power consumption could be realized by another modulation technique. Pulse-modulated systems offer power reduction at the price of increased bandwidth. An FM pulse system for speech works with pulses of 1 μs duration. With a peak current of 800 mA and 9 transmitter diodes, an area of about 40 m^2 is supplied. Both transmitter and receiver are portable. During speech transmission an average pulse rate of 40 kHz leads to a duty cycle of 1:25. Thus, the DC power goes down to $10\,\text{mW}/\text{m}^2$, and during pauses even to half of this value, thus allowing some hours of operation with reasonable battery weight.

(c) Modulation and channel bandwidth

The necessary intensity modulation may be done with any suitable design. Main characteristics are similar to the RF modulation possibilities, besides the fact that the value of the radiation intensity will never be negative, unlike monochromatic sine waves.

A classification of modulation methods refers to time and spectral location of the message. The majority of applications use continuous modulation and frequency-modulated subcarriers. Pulse modulation techniques operate preferably with rectangular waveforms, whereas systems with frequency sharing need sinusoidal forms. The most important parameters for the choice of modulation are the available light intensity, the necessary data rate, the signal-to-noise ratio, immunity against other sources and sometimes, as a major factor, the area to be supplied with the signal.

Many properties can be directly derived from the knowledge of the modulation techniques for radio waves. Especially with regard to telemetry systems, a comparison between RF and IR was described in 1981.[18] The channel capacity is given by the IRED modulation bandwidth, now some MHz, and can be split up into different channels using continuous or discontinuous modulation techniques. Continuous FM still offers an optimum between channel capacity and signal quality. For rooms of normal size only mains operation of the transmitters is suitable because of their high power consumption.

Similar to the channel raster in RF applications, the available bandwidth for IR may be divided into channels of different width. Wide-band application offers the best features for the signal. The available modulation bandwidth (not light frequency bandwidth) allows at least some stereo channels. Due to the availability of IF components the channel grid for narrow-band installations was chosen with 40 kHz spacing.

Interpreter systems make use of a multitude of channels. Because of its many advantages, the IR solution has replaced the former inductive loops. The IEC standard from 1983[15] specified nine channels with 40 kHz bandwidth.

Pulse modulation has been mentioned for a special wireless microphone (Section 5.3.1). Besides the use for sound transmission, there are many control systems, and nowadays data links working with baseband- or carrier-modulated pulses. These systems are likely to interfere with other IR systems and always have to be taken into consideration if multiple IR applications are used within a single room.

(d) Spurious emission

The optical bandwidth of the emitting diodes used is only some 10 nm. Out-of-band power decreases rapidly, so the amount of emission can be neglected for interference purposes.

Different from spectral components, the spurious part of modulation power in the light bandwidth used has to be considered. Depending on the modulation used, the radiated signal may cover a wide range in the band available for subcarrier operation. So besides the choice of the most suitable modulation system, technical measures for the reduction of out-of-band spectral content have to be taken. Similar to the situation in radio wave use where the antenna has to be fed with an especially processed input, the IRED driving current also has to be preconditioned for minimized subcarrier bandwidth. This processing may include the use of opto-electrical feedback.

Unfortunately many other light sources emit energy in the near IR range. This spurious signal has to be investigated very thoroughly for interference, especially when it contains modulation-like components, mainly due to HF-operated ballasts of fluorescent lamps where the hydrogenium emission spectrum, in addition to the wanted ultra-violet part, also contains a strong infra-red line at 1014 nm.

(e) Immunity against interference

Interference may occur as well with other IR communication systems, e.g. sources emitting IR for other purposes or even as a by-product. As for the content of the message, neither audio signals nor data should be distorted by other sources in an unacceptable manner.

As described in Section (c) the modulation bandwidth can be extremely different, depending on the system used. Parallel use of frequency-modulated systems is very easy with separate channels. Pulse systems normally are not synchronized, and will interfere with each other and also disturb FM systems. A mixed operation, e.g. in conference halls, is therefore not possible, but even in the use of low rate pulse systems for home application, remote control has to be taken seriously in order to prevent malfunctioning.[19]

The situation with other signal sources is similar to RF use, with the advantage that only sources of interference in the room where the IR is used have to be regarded.

Daylight has a wide spectrum, including IR. The intensity in bright sunshine is so high that normally no IR operation will work there (special systems, however, will still do their job in this situation).

Other light sources may also influence the transmission. The visible light can be attenuated by an optical filter, but many sources emit a large amount of infrared energy, e.g. tungsten light and even gas-discharge light sources.

Modern spotlights used in studios produce more than 10 000 lx at a distance of 10 m. The IR power of those sources is still so high that IR sound transmission is no longer economical under such conditions. The average illumination in a TV studio, however, is about 1.5 klx and is acceptable for the use of specially designed IR receivers. To get an impression of the capabilities of such receivers one should keep in mind that a light density of 1000 lx is produced by a 100 W tungsten source from a distance of 35 cm!

In FM systems daylight and artificial light can be suppressed efficiently. Even fluorescent lamps will not cause interference as long as they are driven at standard mains frequencies.

The fluorescent lamps already mentioned with higher operating frequencies of about 30 kHz may cause heavy distortion of IR sound and remote control systems.[20] The high operating frequency leads to modulated energy in the IR range, up to more than 200 kHz. Comparison with the spectrum of a nine-channel conference system under the same conditions shows that the lower channels are already overridden by the distortion. The only way to avoid interference from the HF-driven lamps is to work only at signal carrier frequencies above approximately 250 kHz, or better still to use other forms of illumination, because an increase in carrier power is not recommended, for reasons of reduced safety and increased cost.

Besides the necessary precautions against interference from other light sources, the electronic circuitry of transmitters and receivers has to be well designed to avoid interference with the electromagnetic energy from other equipment. Standard EMC techniques normally meet this demand.

(f) Audio signal properties

Every relationship between modulation characteristics and audio quality which is known for RF applications can be directly transferred to the IR subcarrier signal. Baseband systems will also have effects in other technical realizations, even in wired communication links. Thus no difference from an equivalent RF system exists. It should, however, be kept in mind that not every modulation technique used with radio frequencies is advisable for IR, as explained before.

5.3.3 Infra-red headphones

Because of radiation efficiency all systems with continuous frequency-modulated IR subcarriers are powered by the mains. Only the receiver part can be designed with low-power battery consumption and, in combination with a low-power audio amplifier, will be able to drive a headphone for many hours.

The mechanical design of such a headphone can be in one piece or with a separate miniature receiver. Only the first is a true IR headphone, whereas the miniature receiver of the second solution could also be combined with other equipment such as active loudspeakers, tape recorders, telephone lines or similar systems.

Already the TV mono headset of 1975 from Figure 5.30 was a 'real' IR headphone; however, in these days, high-fidelity fans enjoy their stereophonic 'open air' headphones. Compared with the first stereo IR headphone of 1977, today world-wide production of stereophonic IR headphones offers a variety of quality levels, mostly operating on the standardized frequencies 95 and 250 kHz, some with compandor circuits for higher dynamic range.

It is quite obvious not only that these headphones are used by single persons in the home but also that they offer 'personal' listening even in groups of people, e.g. in pensioners' hostels or hospitals. For this application the IR transmitter power had to be increased. This was done either by satellite radiatiors getting their modulation from the central transmitter or by power radiators with more than 100 IREDs.

Receivers with electrical output instead of acoustical output can be connected to headphones or other circuitry needing low input power. This combination is useful e.g. for handicapped people, (Section 5.3.6).

The channel coding for stereophonic operation is different from that in FM broadcast. There, for compatibility reasons a multiplexed signal is broadcast. Infra-red headphones receive instead two separate channels and, with the related high channel separation, are optionally able to receive two different sound tracks, e.g. as offered with some TV programmes. Operation on the standardized subcarrier frequencies allows the use of filters for both channels without conversion to an intermediate frequency.

5.3.4 Infra-red loudspeakers

The combination of IR receivers with active loudspeakers allows flexible positioning of the speakers. Self-contained versions with high-energy batteries may even help in assemblies for easier distribution of sound from a fixed point over a wide area. The available power at both ends of the link made it possible to develop IR systems that transmit the digital code of CD players to the corresponding receiver circuits in the loudspeakers. The high bandwidth of the digital signal made it necessary to use the fastest available emitting diodes for this purpose.

5.3.5 Infra-red information and conference systems

As discussed in the historical survey, multi-channel IR systems have replaced the former inductive loops used for multilingual conferences. Besides the early standardized frequency-multiplex modulation, which exists in hundreds of installations, also other realizations for multi-channel operation were made to fulfil special situations and requirements. None of them replace others completely but offer special features, mostly in exchange for other attributes.

A pulse-position multi-channel system is also in use for conferences. It needs less IR power than the FM solution but must be installed very carefully and covers a far wider spectrum of the present scarce IR modulation bands. An advanced version also offers back-channels that transmit their IR signal to nearby receivers.

The realization of a six-channel pulse AM system is going to replace the former inductive loop techniques for information systems, e.g. in museums. With synchronized time slots, it is possible even to supply overlapping regions in bigger halls with different information on the same channel[21] and without severe distortion, e.g. that generated by two overlapping FM carriers. By holding the upper end of the receiver (Figure 5.31) near to the ear the visitor can listen to one of the six selectable channels, which can be used either for different languages or for different levels of information.

The dependence of the audio level with amplitude modulation on the distance by $1/r^2$ helps one to install the next information source relatively near to the first.

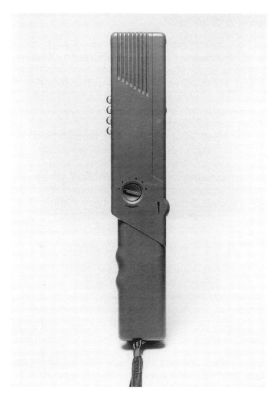

Figure 5.31 *Pulse AM receiver EMI 1060. (Courtesy of Sennheiser).*

By this means it improves the situation as compared to a pure acoustical field, which not only drops with $1/r$ but very soon reaches the diffuse field level of no further decrease.

A special information system combined with background music has been installed by putting IR receivers into the shells of ear protectors. In big manufacturing areas with high noise levels the workers have to wear ear protectors. Important messages cannot then be given acoustically to them. This problem has been solved by using a combination of ear protector and IR receiver. Even workers who previously had often refused to wear the protectors because of discomfort, were quite enjoying the programme supplied.

Another IR information system was developed for a huge entertainment park in 1982. Two different technical versions are used. The first one is a 'theatre' application which is similar to the multi-channel conference system.

Far more sophisticated is the 'vehicle application'. In this case the information has to move parallel with vehicles passing different scenes along a track of some 100 m length. The switching of the information is done by special multiplexers which supply a large number of radiators distributed along the track. The link between trackside emitters and the vehicle receiver is the first infrared transmission in the system. The vehicle receiver contains a demodulator and a booster amplifier. The demodulator supplies the audio signal for the main channel (English) to power amplifiers feeding the vehicle loudspeakers. The booster amplifier takes the composite signal and drives a small number of infrared emitting diodes in the front of the vehicle passengers, who carry receivers for special mother languages.

5.3.6 Infra-red aids for handicapped people

The pulse frequency modulated system described in Sections 5.3.2(b) was developed for use in special schools for hearing-impaired children.[22] Different from earlier teaching situations with a fixed stereo microphone position and cabling to the pupils' desks, or later IR receivers with high-power headphones, it allowed the teacher to move around without increasing the acoustical distance to the pupils' ears, because the transmitter microphone on the teacher's chest always remained at the same short distance from his/her mouth.

Quite surprisingly, it may be that blind people also may benefit from IR transmission. Some preliminary information dealt with the possibility of informing blind people via IR light, on special objects around them. The idea was taken up by a German group giving necessary information to blind people on railway stations, e.g. the different destinations for the tracks, announcement for the next trains, etc. An installation in a multi-track open-air railway station, now in use for nearly one year, shows encouraging results.[23]

5.3.7 Other applications

The limited and always crowded RF bands were the reason for many different applications of IR light. Besides those already mentioned, use was made for teaching young physicians by direct transmission of cardiac or other body-generated sound. Numerous hospitals use a system which allows heart sounds to be heard at the same time by the examining physician and the whole auditory. Comments may be inserted into the same reproduction chain via a radio microphone.

In the meantime an oesophageal stethoscope has also been developed by an American company as an aid in surgery. For safety reasons battery operation is preferred, but here the short distance between transmitter and receiver allows frequency modulation.

Further clinical attempts are known to have been made to transmit the cardiac sound of a foetus to an IR receiver for diagnostic purposes.

There are numerous applications for IR transmission of signals other than sound. Millions of remote control units for home entertainment systems are in use. The lamp and installation manufacturers have added control for switches and dimmers; even data links for computer communication will not complete the list.

5.3.8 Future aspects

The extreme increase in IR application has given rise to a situation that needs coordination. International standards organizations and evaluation groups are endeavouring to find recommendations for preferred values and to improve the interference situation. [6,24]

The electronics industry is looking for further possibilities from improved semiconductor material. If, for example, a separation of several light wavelengths could be realized independent of the incident angle at reasonable expense, the available bands for operation would multiply.

Transmission of digitized signals like CD music will also need less channel capacity by further progress in data-reducing algorithms. Similar to the situation with radio microphones, design engineers for IR systems are also looking for possibilities to adapt solutions used in mass production to their needs.

Sometimes the possibilities of using laser diodes are brought into discussion. It should be pointed out that direct modulation of monochromatic (coherent) light is nearer to the RF application. The advantage of the multifrequency IR signal from IREDs with no overall cancellation effects is lost, as is known from laser interference technologies. The problem of possible danger to the human eye would also have to be considered.

Today transmission of sound and data with IR light already plays an important role. This will continue with extended applications of the existing technical possibilities and will increase with the introduction of improved components and processing tools.

References

1. CEPT (1977). *Recommendation T/R 23 relative aux émetteurs et récepteurs pour les systèmes de microphones sans fil de faible puissance.* 1 December.
2. Pohl, W. (1972). Zweckmäßige Frequenzen für drahtlose Mikrofone. *Fernseh- und Kinotechnik* **5**; 160–162.
3. Greif, H. (1972). *Lichtelektrische Empfänger.* Geest & Portig, Leipzig.
4. Griese, H.-J. (1976). Audio auf Infrarot, Möglichkeiten und Standards, *radio-mentor-elektronik*, **11**, 440–442
5. Marhic, M.E., Kotzin M.D. and van den Heuvel A.P. (1982). Reflectors and immersion lenses for detectors of diffuse radiation, *J.Opt. Soc. Am.*, **72** (3), 352–355.
6. IEC 1147 (1993). Uses of infra-red transmission and the prevention or control of interference between systems. Technical Report. First edition, 1993–06.
7. Werner, E. and Beckmann B. (1985). Improving the reliability of radio microphone links by multiple input. *77th Convention Aud. Eng. Soc.*, Hamburg. Preprint 2218, March 5–8.
8. Neukomm, P.A. (1979). *Body-Mounted Antennas.* Diss. ETH No. 6413, Zürich.
9. Merrell, R. (1982). Wireless mikes put you on the scene . . . anywhere, *Broadcast Communications*, April, pp.35–47.
10. Griese, H.-J. and H. Koch (1958). Drahtlose Mikrofone, *Radio Mentor*, **2**, 87–89.
11. Werner, E. (1977). Zukunft Direktton – reizvoll, aber schwierig, *Film 8/16*, **2**, April/May.

12. Edwards, R. (1969). Selection of intermodulation-free frequencies for multiple-channel mobile radio systems. *Proc. IEE*, **116**(8), August, 1311–1318.
13. IEC Standard 268, part 15 (1987). *Preferred Matching Values for the Interconnection of Sound System Components* (2nd edn).
14. Werner, E. (1976). Wireless audio transmission with infrared light for studio applications. *53rd Convention of the Ac. Eng. Society*, preprint D-7, Zürich.
15. IEC Standard 764 (1983). *Sound Transmission Using Infra-red Radiation* (1st edn), Genéva.
16. Griese, H.-J. (1977). Mehrkanalanlagen mit Infrarot-Übertragung, *Radio-Mentor-Electronic*, **7**, 261–263.
17. Griese, H.-J. (1978). *Conference Systems Using Infrared Light Techniques*. AES preprint 1335 (D-3), Hamburg.
18. Weller, C. (1981). The use of infrared in medical and biological telemetry. *Proc. Telemetry Workshop*, Lyon, 29–30 October, p. 46 ff.
19. AEHA (1987). *Measures to Prevent Malfunctioning of IR Remote-controlled Electric Home Appliances*. Association of Electric Home Appliances, Tokyo, July.
20. Infrarotstörer (1982). *Funkschau*, No. 9, 14.
21. Werner, E. (1985). Sprechende Bilder, Drahtloser Museumsführer. *Funkschau* **2**, 44–46.
22. Griese, H.-J. (1982). Sound transmission with free propagating infra-red light, Pt. 2: Discontinuous modulation techniques. *71st Conv. Aud. Eng. Soc.*, Preprint 1880 Montreux, 2–5 March.
23. BILOS und seine Effektivität (1991). *Verkehrs- Nachrichten* **3/4**, Bundesministerium für Verkehr, Bonn, 27–30.
24. *Home Systems Specification*, Part 3: *Media Specifications Infra-Red* (1991). ESPRIT Project 2431, Esprit HS Consortium, January.

Bibliography

Alard, M. and Lassalle, R. (1987). Principles of modulation and channel coding for digital broadcasting for mobile receivers. *EBU Review* – Technical No. 224, August.
Bush, B. (1981). Radio microphones, *Studio Sound*. July, 58–60.
Hibbing, M. and Werner, E. (1985). Ton aus der Luft gegriffen, Lautsprecherzeilen drahtlos. *Funkschau* **2**, 55–56
Koch, H. und Werner, E. (1973). Eine neue Funkführungsanlage. *Funkschau*, **8**, 263–266.
Mäusl, R. (1976). *Modulationsverfahren in der Nachrichtentechnik*, Hüthig, Heidelberg.
Pohl, W. and Werner, E. (1982). Extension of the Dynamic Range of Wireless Microphones. *J. Ac. Eng. Society*, **30**, 318–323.
Rohde, N. and Burchard, D. (1991). Künstler an der Funkstrippe. *Funkschau*, **8**, 68–71.
Werner, E. (1978). Drahtlose Kommunikationstechnik im Theaterbetrieb. *Bühnentechnische Rundschau*, October.
Werner, E. (1979). Empfänger für Parallelbetrieb drahtloser Mikrofone. *Fernseh- und Kinotechnik*, **33** (Jg. Nr. 6), 210–212.
Werner, E. (1982). Sound transmission with free propagating infrared light, Pt. 1, Continuous modulation techniques. *71st Convention of the Ac. Eng. Society*, Preprint No. 1879, Montreux.
Werner, E. (1982). Improvement of multichannel radio microphone operation by use of advanced receiver techniques. *J. Ac. Eng. Society*, **30**, 203–207.
Werner, E. (1982). Zur Technik drahtloser Mikrophone unter Berücksichtigung neuer Frequenzbereiche, *Fernseh- und Kino-Technik*, **36**(Jg. Nr. 21), 57–61.
Werner, E. (1985). Modulation systems for transmission of biologic-medical data with infrared light. *Biotelemetry VIII*, Braunschweig, 155–158.

Annexe

The properties of high quality radio microphone systems can be seen from Table 5.4.

Table 5.4   *Typical characteristics of high quality radio microphone systems*

Characteristic	←——— Overall ———→	
	Transmitter	Receiver
Carrier frequency	←——— 30 MHz to 1 GHz ———→	
RF channel bandwidth	←——— 200 kHz ———→	
Switching range*	1.5 to 100 MHz	1.5 to 25 MHz
Number of simultaneous channels	←——— up to 9 in single subranges ———→	
RF radiated power*	10 to 500 mW	–
Adjacent channel power	<10 μW	n.a.
Spurious emission (broadcast bands)	2 to 20 nW	<2 nW
Rated AF input (rated deviat.)	4 mV to 800 mV	n.a.
AF bandwidth	←——— 40 to 20 000 Hz ———→	
Pre/de-emphasis	←——— 50 μs ———→	
Rated deviation	40 kHz	n.a.
Peak deviation	56 kHz	n.a.
T.h.d at rated dev.	←——— <1% (1 kHz) ———→	
AF compandor	2:1	1:2
S/N at peak dev. (comp. ON)	>110 dB (A, r.m.s.) equ.	>98 dB (CCIR peak)
RF sensitivity for 26 dB S/N	n.a.	<1.5 μV
Capture ratio	n.a.	<2 dB
Selectivity (>400 kHz re carrier)	n.a.	>80 dB
AM suppression	n.a.	>50 dB
IF rejection	n.a.	>80 dB
RF intercept point	n.a.	15 dBm (1.3 V)
Squelch threshold	n.a.	1 to 100 μV
AF output (opt. adjustable)	n.a.	6 dBm (1.55 V)
Power supply*	1 to 3, AAA or AA	Batt/Mains
Operating time*	3 to 10 h	n.a.
Size in mm*	92 × 52 × 7 (pocket)	pocket to 48 cm (19 in)
Weight* in g (incl. batt.)	200	
Options offered by several manufacturers	Remote transm. battery level display Video monitoring of channel status (RF, AF level, diversity, designation . . .) Diversity operation Remote control of receiver set-up (frequency, on-off, . . .) Transmitter power booster	

*These characteristics are highly dependent on the used technologies and design.

An example of data on a high-quality wireless microphone is given in Figure 5.32.

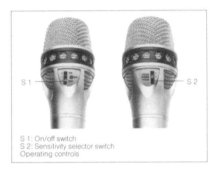

S 1: On/off switch
S 2: Sensitivity selector switch
Operating controls

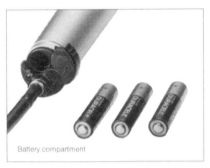

Battery compartment

Technical Data   Sennheiser SKM 4031

Carrier frequency	single frequency within the 30 to 45 MHz range
Max. frequency drift at temperatures between – 10° C and + 55° C,	
at battery voltages from 2 to 5 V	± 15 kHz
RF output power / radiated power	10 mW / approx. 1 mW
Type of modulation / pre-emphasis	FM / 50 μs
Compander system	"HiDyn", 2:1 compressor
Nominal sweep / peak sweep	± 40 kHz / ± 56 kHz
Maximum acoustic pressure for peak sweep	
a. Switch at "Norm" setting	approx. 136 dB
b. Switch at + 10 dB setting	approx. 126 dB
c. Switch at – 10 dB setting	approx. 145 dB
Dynamic range	96 dB (A)
S / N ratio (peak sweep) measured as per DIN 45 500, Curve A, eff.	typically 96 dB
S / N ratio (peak sweep) measured as per CCIR 468, peak	typically 82 dB
AF transmission range	70 to 20 000 Hz
Max. distortion (1 kHz, nominal sweep)	< 1 %
Power supply	Three 1.5 V AlMn batteries, style IEC LR 03 "Micro"
Battery life (continuous operation)	approx. 12 h
Dimensions (length x max. diam.)	205 x 49 mm
Weight with batteries	260 g

Figure 5.32 *Data on a high-quality wireless microphone.*

Modulation technology

Wide-band/narrow-band

To achieve the greatest resistance to interference signals, frequency modulation (FM) has been selected as the type of modulation to be used in wireless transmission systems.

One must differentiate between wide-band and narrow-band FM. As its name implies, in wide-band technology a broad band is assigned to each channel within a prescribed frequency range. Since the clearance between channels must be at least 300 kHz (necessitating assignment in steps of 300 kHz) in order to avoid cross-channel interference, it will be possible to assign only a few channels to a particular frequency range. When using narrow-band FM, on the other hand, the clearance is reduced to 40 kHz, which means that a far greater number of channels can be set up within a frequency range. The frequency response range which can be transmitted using narrow-band FM technology is from 20 to 12.000 Hz, whereas a range of 20 to 20.000 Hz can be attained using wide-band technology.

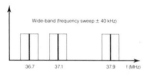

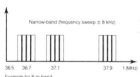

Example for 8 m band

Frequency sweep

In FM technology this term is used to designate the momentary deviation in the frequency from the carrier frequency. The frequency deviation is determined in relationship to audio frequency amplitude. The nominal sweep is ± 40 kHz for wide-band FM, while peak sweep is limited to ± 56 kHz. Narrow-band FM operates with ± 8 kHz nominal sweep and peak sweep of ± 10 kHz. Nominal sweep is achieved at full modulation of the transmitter or receiver. Peak sweep represents a maximum value which may not be exceeded, since interference in the adjacent channels could otherwise result. To avoid such interference the transmitters for wireless microphones are equipped with an automatic sweep limiter. use either high-quality limiter amplifiers which attenuate excessive audio signal levels or diode circuitry which limits the level by clipping.

Regulations and frequencies

Mandatory registration

High-frequency transmission systems must be licensed. Only equipment which has been approved by the telecommunications authority may be used.

Available frequencies

Sennheiser offers RF systems for frequency ranges 30 to 45 MHz, 138 to 260 MHz and 460 to 960 MHz. Since the frequency ranges reserved for operating wireless transmission systems will vary from one country to the next it is not possible to make any recommendations which are valid for all areas. Your authorized Sennheiser distributor can provide information as to the frequencies reserved locally for this purpose.

Frequency selection

Selecting the frequency is quite simple when planning a single-channel system. The situation becomes more complex when dealing with multiple channels, Since arbitrary can result in interference caused by the intermodulation products of the frequencies used.

Diversity-technique

As a result of multi-directional radiation and reflection of the electromagnetic waves from the walls of a room it is possible to encounter "holes" or "shadows" in field strength (where antenna voltage drops to a minimum); this can cause reception drop-outs. Often simply repositioning the receivers antenna can rectify such problemes. But these drop-outs can be eliminated completely by using the diversity procedure. This eceiver configuration provides for duplicated antennas and receivers (Fig. 3) with both receivers operating at the same frequency. The signal from the receiver with the higher antenna voltage will be switched through to the output.

RF-input voltage

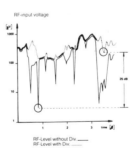

RF-Level without Div. _____
RF-Level with Div. _____

Effect of diversity-reception

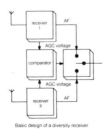

Basic design of a diversity receiver

Notes for projects and planning

● The following should be observed when positioning the antennas:

  – The receiver antenna must always be located in the same enclosure as the transmitter.

  – In order to avoid distortions caused by intermodulation within multi-channel systems, install antennas in a way that under no means the transmitters get closer than four metres to the antennas. This could be done by installing the antennas at four metres height.

  – Do not set up antennas in niches, alcoves or passageways.

  – When using two antennas the distance between them should not be half or three – quarter of a wavelength.

  – Close proximity to metal will change the reception characteristics of the antenna. It is for this reason that the antenna should be kept at least one metre away from metallic objects and reinforced concrete walls. Wooden partitions present no problems.

● One transmitter and one receiver will be necessary for each transmission channel, since the receiver can process only a single frequency, regardless of whether it can be tuned to different frequencies. Furthermore, simultaneous operation of two or more transmitters on the same frequency will cause intermodulation interference.

● Planning should always take the use into consideration. This means that where maximum transmission reliability is necessary – in theatres and auditoria for instance – diversity reception should always be used.

● The antenna inputs at the receivers are designed to accept 50 Ohm coaxial antenna cable (such as type RG 58). The cable should never be more than 100 metres long. Where the cable is more than 60 metres long, low-attenuation cable (such as RG 213/U) should be used if at all possible, in order to keep the voltage drop in the cable as low as possible.

Figure 5.33 *A synopsis of wireless microphone technology.*

6 Microphone testing

Harald Sander-Röttcher and Kersten Tams

6.1 Introduction

Most microphone testing is done to get some information about acoustic properties. Besides the on-axis frequency response, the directional characteristic plays an important role if the subjective sound impression is to be supported by objective measurements. Especially in evaluating the different stereo microphone set-ups this problem becomes more and more complex.

But there are some non-acoustic parameters that define the quality of a microphone. These are its sensitivity to environmental influences like mechanical vibrations and all kinds of air-current excitations. The results are expressed as an equivalent sound-pressure level. An acoustic excitation covering the given frequency spectrum produces the same microphone output as the specific non-acoustic excitation that is to be examined. In this way, microphones of different types can be compared.

In order to allow a world-wide comparision of test results, standard conditions like free or diffuse sound fields have to be used, although these conditions are usually not met in practical applications. Because real sound fields are very complex, standard conditions represent a convenient simplified approach which can be easily described by mathematical means. Nevertheless, the technical realization can be very costly in some cases. This is the reason that, in production testing, simplified set-ups are used. These do not give exactly the same results, but they show deviations from a reference microphone of the type which is to be tested. In series production this is sufficient to ensure the quality of the product.

6.2. Measurement of acoustic parameters

6.2.1 Principles of acoustic microphone testing

Measuring the acoustic transfer function of a microphone is always a comparision between a calibrated standard microphone and the one under test. This means that all results depend on the accuracy of calibration and the characteristics of the standard microphone with which the sound field in the specific measurement environment is to be analysed. To meet the requirements of accuracy, a pressure-type reference microphone has to be used. This type of microphone can be calibrated with a tolerance of ± 0.05 dB, using a reciprocity procedure which gives the best results[1]. (A directional microphone cannot be calibrated this way.) The accuracy obtained in a free-field calibration would be ± 0.3 dB.[2]

The comparision with the reference microphone can be carried out in two ways:

(a) substitution method
(b) simultaneous comparison method

236

(a) Substitution method

The two microphones are measured in turn at the same place in the sound field. With today's powerful digital equipment this method is no longer time-consuming. First the loudspeaker is equalized for a constant sound pressure at the reference point over the whole frequency range. This must be done by a regulation procedure because of amplitude non-linearities in the speaker driving unit.

If a swept sine excitation is used, the amplifier output amplitude for each frequency is simply stored in a data file and down-loaded each time a measurement is executed. Results with a repetition error less than ± 0.1 dB can be achieved. For pulse excitations from which the transfer function can be calculated by an FFT-algorithm there are two possibilities:

(a) A digital filter is adjusted during the regulation procedure to calculate the correct excitation spectrum for constant sound pressure. This requires advanced DSP hardware with a lot of storing capacity to give results with a tolerance of ± 0.1 dB.
(b) Much simpler, the two transfer functions are measured alternately and divided by each other. This means that the microphone is not tested at constant sound pressure level, but with a fixed analogue filter in the generator–amplifier chain the pressure level can be held within a range of 80 ± 5 dB which is sufficient, because amplitude non-linearities of microphones do not usually come into account at levels less than 120 dB SPL.

It is important that the loudspeaker should not have changed its characteristics since the equalization procedure has been carried out. Several effects can be responsible for this type of error:

(1) Warming-up of the voice coil alters the impedance and reduces the sound pressure. Passive electrical elements between amplifier and speaker terminals should be avoided because they increase this effect. Speakers with a large power-handling capacity and a good efficiency should be used.
(2) Ambient temperature and humidity affect the driver suspension. The tuning of the speaker's resonance is altered and the sound pressure changes, particularly at the resonance frequency.
(3) The radiation impedance and therefore the resulting sound pressure depends on the ambient air pressure.

So that ambient conditions are taken into account under stable weather conditions, the equalization procedure should be repeated daily to obtain a repeatability of ± 0.25 dB.

(b) Simultaneous comparision methods

The disadvantage of varying loudspeaker characteristics can be avoided if the two microphones are measured at the same time at two different places in the sound field.

If a swept sine excitation is used, the reference microphone with a flat frequency response is connected to a regulating amplifier in the generator signal path to keep the sound pressure at a constant level. This is the same basic measurement technique, which for practical reasons has been performed for many years with analogue equipment.

The use of a two-channel FFT analyser with pulse excitation is the fastest method of obtaining the microphone transfer function. Again the measurement may not actually be carried out at a constant sound pressure level, due to the errors explained above.

Some further considerations have to be taken into account when the simultaneous comparision method is used.

Firstly the sound pressure at the two places has to correspond within the wanted tolerance range. At frequencies below 1 kHz it is usually no problem to achieve an accuracy of ± 0.15 dB. At high frequencies the polar pattern of the sound source determines the error involved. Normally careful setting of the microphone positions has to be carried out before accurate measurements are possible.

Secondly the distance between the microphones has to be chosen so that interference between the two transducers is minimized. A larger distance is better than a small one. For normal microphones, 15 cm gives good results, but if artificial head microphones are to be measured, the substitution method is a better choice.

These peculiarities can represent severe limitations on the accuracy of the simultaneous comparison method. This is the fastest method but the substitution procedure will be more accurate if the transducers are measured directly one after another.

6.2.2 Measurement environments

Several different sound fields and methods of generating them are used:

(a) far-field measurements
 - free field
 - guided waves in a duct
 - diffuse field
(b) near-field measurements
 - artificial mouth (proximity effect, etc.).

(a) Free-field testing

Testing under free-field conditions means that the microphone is placed in a plane travelling wave. This is the type of sound field where most of the testing during microphone development is done. The conditions can be realized outdoors, in an anechoic room or in a duct. Outdoors we can have very good acoustic performance, but the results are affected by ambient noise, wind and weather. Anechoic rooms provide a good signal-to-noise ratio and sufficiently good acoustic performance down to the cut-off frequency for which the wedge length is a quarter of the wavelength. Below that frequency standing waves occur and sound pressure and sound velocity are no longer in phase. If we calibrate the sound field at the reference point with the reference microphone, testing of pressure transducers give correct results. But if we measure directional microphones which are sensitive to both pressure and velocity, we obtain incorrect data below the cut-off frequency. Therefore it is a good idea to generate the low frequency plane waves in a duct. This method will be discussed later.

Some considerations have to be met concerning the loud-speaker.

Firstly, it has to generate a sound pressure level of 80 dB (94 dB = 1 Pa, IEC 286 T4) over the whole frequency range of 20 Hz to 20 kHz. The influence of amplitude non-linearities on the transfer function should be less then 0.5 dB. If this cannot be achieved, a filter has to be used to measure only the fundamental frequency.[3]

Secondly, it has to produce plane waves. This is usually done by setting up an almost ideal point source which approximates a plane wave in a distance longer than half the wavelength. Again pressure-gradient microphones are affected by this frequency limit because pressure transducers show no difference worth mentioning if tested in a spherical or plane wave, as long as wave-bending effects can be neglected. This is another reason for using guided waves in a duct for low-frequency testing of directional microphones.

To meet the above requirements it is obvious that a two-way speaker system must be used. The question now is whether the tweeter should be placed in front or beside the woofer. In the first case, the measuring distance is not constant; in the second, we have different incident angles at low and high frequencies. It turns out that the latter set-up has a more severe effect on directional microphones than the first. This can be seen from the following considerations.

Assume a measuring distance of 1 m (this is the standard distance). If we place the tweeter in front of the woofer, the distance between woofer and reference point is, for example, 1.02 m, and that between the tweeter and reference point 0.96 m. This is a 6% change of curvature of the spherical wave front. If we now assume that a plane wave is generated in a distance longer than half the wavelength, this change of curvature is negligible for all frequencies above 170 Hz.

Figures 6.1 and 6.2 show a practical set-up. A 2.5 cm tweeter is mounted in front of a 25 cm woofer. The volume between the back of the high-frequency unit and the woofer membrane is filled with absorbing material to damp the resonance which builds up in the cavity behind the tweeter and its circumferential aperture. A further effect of the damping material, together with the spherical horn baffle of the tweeter, is to avoid wave-bending effects at the edges of the tweeter, the woofer and the woofer baffle. These provisions reduce peaks and dips in the polar pattern from about 12 dB down to 2 dB.

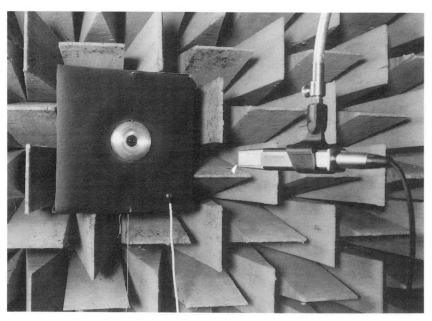

Figure 6.1 *Measuring loudspeaker built into the wall of an anechoic chamber.*

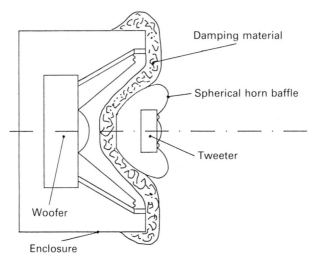

Figure 6.2   *Schematic cross-section of measuring loudspeaker.*

The next point concerns the cross-over. If a stepped sine-wave excitation is used, a frequency-dependent switch is the best alternative. Otherwise both speakers radiate at the same level at the cross-over frequency, which results in an interference field, due to the distance between both sources. Such a switch can be realized by triggering a counter on the positive-going edges of the generator's synchronization output. A decoder decides whether the counted time between two edges is more or less than the period of the cross-over frequency and activates the corresponding CMOS-switch (Figure 6.3).

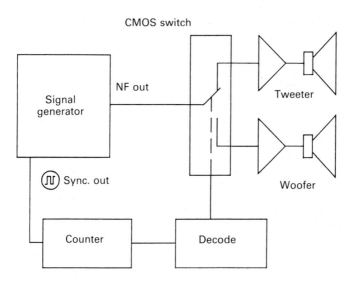

Figure 6.3   *Block diagram of switching cross-over.*

(b) Measurements in a duct

In a Kundt's tube one can generate, in a frequency range determined by the dimensions of the duct, an evenly progressive acoustic wave. On one end of the duct there is a loudspeaker to generate the wave. The other end of the duct must be anechoic. The microphone under test has to be placed on the axis of the duct. The distance from the driving loudspeaker must be at least two diameters of the duct to avoid errors due to near-field effects. The dameter of the microphone should be less then 5% of the diameter of the duct.[3] Figure 6.4 shows the measuring equipment. To measure the frequency response one can use the substitution method as described above.

The anechoic end of the duct can be equipped with absorbing wedges with a length equal to a quarter of the wavelength of the lowest frequency you measure. At 20 Hz this will be about 4.3 m, so the duct will be at least about 15–20 m long. Another possibility is to equip the anechoic end with an active system. In this case, the duct will be terminated by a loudspeaker, which is regulated by an electronic system.[4] Figure 6.5 shows the principle.

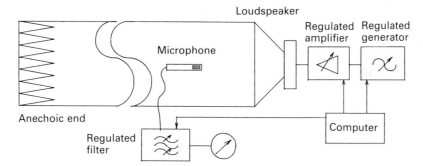

Figure 6.4 *Set-up to measure the frequency response in a duct at low frequencies.*

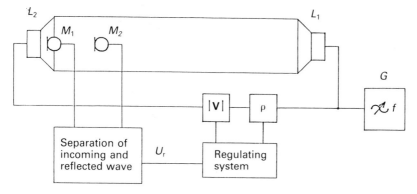

Figure 6.5 *Principle of the structure of an active system to form an anechoic end of a duct.*

In this diagram M_1 and M_2 are microphones and L_1 and L_2 are loudspeakers. The generator output is divided between two paths. One is fed directly to the loudspeaker L_1; the other is connected to a system that regulates the phase and the level. This circuit is controlled by the electrical signal U_r, which is proportional to the reflected wave. Its output signal is connected to loudspeaker L_2. The phase and the level must now be regulated so that the reflected wave becomes close to zero. The circuit required to separate the waves is shown in Figure 6.6.[5]

The system in Figure 6.6 consists of two microphones, three delays and two subtractors. This structure is known in high-frequency applications as a waveguide coupler. The microphone M_1 must be placed at least one diameter of the duct to the loudspeaker L_2, while the microphone M_2 is fixed at the distance Δx from the microphone M_1. The electrical signal of the wave coming from loudspeaker L_1 (P_e) is available at point (1). At point (2) one can measure the electrical signal of the wave Pr coming from the other end. The electrical delay corresponds to the distance of M_1 to M_2.

$$\tau = \Delta x / c_0, \qquad c_0 = 343 \text{ m/s} \tag{6.1}$$

Equation (6.2) describes the sound-pressure distribution in the duct:

$$\mathbf{p}(x,t) = \mathbf{p}_e + \mathbf{p}_r$$

$$= p_e e^{j(\omega t + \omega x/c_0)} + r p_e e^{j(\omega t - \omega x/c_0)} \tag{6.2}$$

The sound pressure at the places of M_1 and M_2 is

$$\mathbf{p}_1(x = 0,t) = p_e e^{j\omega t} + r p_e e^{j\omega t} \tag{6.3}$$

$$\mathbf{p}_2(x = \Delta x,t) = p_e e^{j\omega t} e^{j\omega \Delta x/c_0} + r p_e e^{j\omega t} e^{-j\omega \Delta x/c_0} \tag{6.4}$$

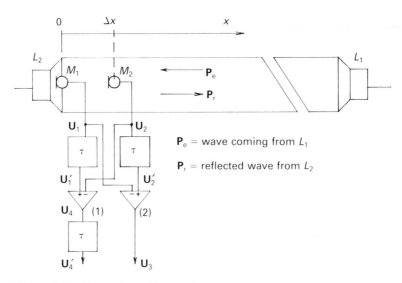

Figure 6.6 *Separation of* $\mathbf{P}_e$ *and* $\mathbf{P}_r$.

The output signal of the microphone is proportional to the sound pressure at this place;

$$\mathbf{U}_1(0,t) \;=\; (1 \,+\, r)u_e e^{j\omega t} \tag{6.5}$$

$$\mathbf{U}_2(\Delta x,t) \;=\; (e^{j\omega \Delta x/c_0} \,+\, r e^{-j\omega \Delta x/c_0}u_e e^{j\omega t} \tag{6.6}$$

$\mathbf{U}_2$ is delayed by τ and $\mathbf{U}_1$ substracted from $\mathbf{U}_2$:

$$
\begin{aligned}
\mathbf{U}_3 &= \mathbf{U}_2(\Delta x,t{-}\tau) - \mathbf{U}_1(0,t) \\
&= r u_e e^{j\omega t}(e^{-2j\omega \Delta x/c_0} - 1) \tag{6.7}
\end{aligned}
$$

A comparision with equation (6.5) shows that the reflected wave at point $x = 0$ is included.

$$\mathbf{U}_3 \;=\; \mathbf{U}_r(0,t)(e^{-2j\omega \Delta x/c_0} - 1) \tag{6.8}$$

The equation (6.5) is formed in the same manner for the point $x = 0$

$$\mathbf{U}_4 \;=\; \mathbf{U}_e(0,t)(e^{-2j\omega \Delta x/c_0} - 1) \tag{6.9}$$

The part in the brackets becomes zero if the exponent of the e-function in the brackets is a multiple of $\lambda/2$. So the distance Δx between the two microphones must be less then $\lambda/2$. This is the resulting limit at high frequencies:

$$f_g \;<\; \frac{c_0}{2\Delta x} \tag{6.10}$$

To get the complex reflection factor one must express U_3 and U_4 as a ratio.

This regulation procedure is very slow. One can calibrate the system once a day and store the parameters (phase and amplitude of the driving signal of L_2 for each frequency) in a computer. In this way, a time-consuming regulation delay is circumvented. In practice such a duct will be about 4 m long for a lower limit of 10 Hz, and for a higher limit of 250 Hz the diameter must be less than 0.5 m.

(c) Diffuse field testing

Testing in diffuse sound fields is carried out to get information about the directional properties of the microphone. On-axis free-field response M_F [dB] and diffuse-field response M_{diff} [dB] are related[3] by the directivity index B:

$$B \;=\; M_F - M_{diff} \quad [dB] \tag{6.11}$$

This parameter indicates the directional behaviour as regards the way the microphone output increases with increasing directivity. It can also be calculated from the polar pattern measured in a plane wave. The advantage of using a diffuse sound field is that the directivity index can be measured even for non-rotational-symmetric microphones (e.g. an artificial head). Moreover, it is a very fast method. Some standard values are given in Table 6.1.

A diffuse field is defined as having the same value probability and the same power as an incident sound wave in all directions. The average energy density and the square of the sound pressure are equal at all points.

Table 6.1 *Directivity index*

Type of directional characteristic	Directivity index B [dB]
sphere	0
cardioid	4.8
supercardioid	5.7
hypercardioid	6.0
figure of eight	4.8

Two different ways of generating such a sound field are used.

(1) *Echoic chamber.*   The uniform statistical distribution of the incident direction is achieved by multiple reflections at the walls, floor and ceiling. Therefore, a hard reflecting and non-vibrating surface has to be used. The lower limiting frequency f_g depends on the enclosed volume $V[\text{m}^3]$:

$$f_g = 125(180/V)^{1/3} \quad [\text{Hz}]^3 \tag{6.12}$$

Care must be taken in choosing the measurement point. It has to be outside the reverberant radius of the omnidirectional sound source and the walls. The minimum distance is

$$r = 0.06(V/T)^{1/2} \quad [\text{m}]^3 \tag{6.13}$$

where T = reverberant time [seconds].

Because of the finite number of reflections that can be achieved, the frequency spectrum at the measurement point consists of many peaks and dips. Therefore random noise (one third octave bandwidth) has to be used as the test signal.

(2) *Artificial diffuse field.*   A set of n loudspeakers is distributed equally on the surface of a hypothetical sphere at some distance around the measurement point. The whole set-up is placed in an anechoic chamber to avoid unwanted reflections that would give rise to an interference field. Each source is now connected to one of n uncorrelated random-noise generators. This ensures random sound incidence from all n directions, which is essential for a constant pressure distribution in a limited region around the measurement point. A total of eight sound sources has proved to give an accuracy of 0.5 dB if the loudspeakers have an almost omnidirectional characteristic and are matched within a tolerance range of ± 0.25 dB for every one third octave band.[6]

It should be stated at this point that because of the statistical nature of diffuse sound fields no phase measurements can be made. Only the amplitude response with the frequency resolution of the narrow-band filter used (one third octave, preferably) can be obtained.

(d) Nearfield measurements with an artificial mouth (proximity effect)

When a cardiod or bidirectional microphone is measured close to an omnidirectional sound source (e.g. a vocalist's mouth), a boost of low frequencies is observed. This effect is used by vocalists to make their voices more deep and warm, or to transmit only the voice of a speaker without disturbing low-frequency environmental noise. In the latter case, the microphone is designed for

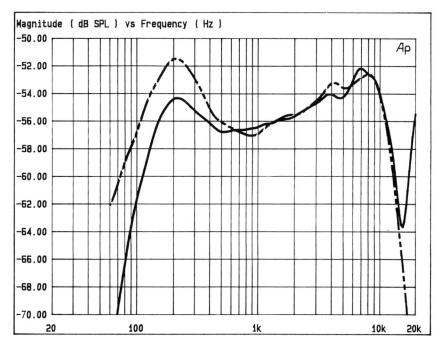

Figure 6.7 *Proximity effect of MD 531.*

minimum bass transmission, so that the speaker must use the proximity effect to boost the lower frequency range of his voice. On the other hand, the sound is sometimes not natural, so the proximity effect is not always welcome. Figure 6.7 shows a typical frequency response of a dynamic microphone at a distance of 1 m (even progressive wave) and 0.1 m (spherical wave).

Pressure microphones do not exhibit this effect. The output of this kind of transducer is

$$\mathbf{U}_P = S\mathbf{p} \qquad \text{when } S = \text{proportional factor} \qquad (6.14)$$
$$\mathbf{p} = \text{pressure.}$$

A cardioid or bidirectional microphone is a velocity transducer. Its output can be written in a similar way as:

$$\mathbf{U}v = T\mathbf{v} \qquad \text{when } T = \text{proportional factor} \qquad (6.15)$$
$$\mathbf{v} = \text{velocity.}$$

If $kr \gg 1$ $\mathbf{p}$ and $\mathbf{v}$ decreases with $1/r$, where

$$k = \frac{2\pi f}{c_0} \qquad c_0 = 343 \text{ m/s} \quad (T = 20°C) \qquad (6.16)$$

$r = $ distance to source, $\qquad f = $ frequency.

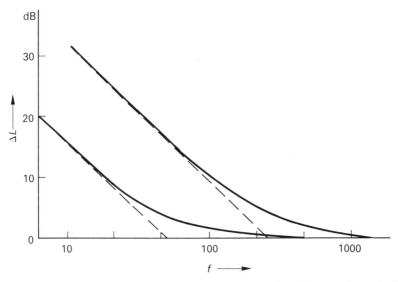

Figure 6.8 *Level difference between a pressure microphone and a velocity microphone exposed to a spherical wave.*

In a spherical wave the phase shift between **p** and **v** is nearly zero. In the near field of the omnidirectional sound source **p** also decreases with $1/r$. Because of the interdependence of the spherical wave impedance to kr, **v** decreases with $1/r^2$ and there is a phase shift greater than zero. The level difference ΔL between a pressure and a velocity transducer is given by

$$\Delta L = 10 \log \left(1 + \frac{1}{k^2 r^2} \right) \quad \text{[dB]} \tag{6.17}$$

In Figure 6.8 you can see the level difference at 1 m and 0.2 m. An ideal velocity and pressure microphone and an ideal spherical wave source are assumed.

The measurement of this effect is similar to the free-field measurement as described above. Only the sound source must be changed. This source is a so-called artificial mouth. It is a spherical wave source that simulates the acoustic conditions of a human mouth. A simple one is shown in Figure 6.9. Other realizations include the complete human head. These are more accurate in evaluating the distance-dependant differences in the near-field frequency response. During the measurement the distance of the microphone to the artificial mouth should be the same as in practice. Usually the distance is 5 to 10 cm for vocalist microphones.

6.2.3 Measurement of directional characteristics

As explained above, the directivity index can be measured in a diffuse sound field. To get some more detailed information about the microphone's directional behaviour it is tested under free-field conditions at different angles. Two kinds of graphical presentation are used.

Figure 6.9   *A simple artifical mouth.*

(a) Polar pattern

The microphone is mounted on a turntable and the loudspeaker is set at constant frequency at 80 dB SPL. Whilst the microphone revolves round its reference axis, its output level is recorded on a piece of circular graph paper with five progressively larger circumferences, each usually indicating five decibels sensitivity difference from the next adjacent circumference (Figure 6.10).

(b) Direction-dependent frequency response

The microphone's frequency response is measured at different angles. Because of the properties of the various types of microphones (omni-, bi-, unidirectional), the following angles are used: 0°, 45°, 90°, 135°, 180° (Figure 6.11).

6.2.4 Phase measurements

Phase measurements are carried out in a plane progressive wave or in a sound-pressure chamber (pressure-type microphones only). As the testing of microphones is always a comparison between the unit under test and the reference unit, the phase responses of both have to be measured. This can be done by using the substitution or simultaneous-comparison method with either a sine wave or an impulse excitation. Care must be taken in adjusting the microphones in space, because a mismatch in the measurement distance results in a time delay which gives an increasing phase shift with increasing frequency. For a 1 mm difference the phase error will be 21° at 20 kHz.

The phase response can be displayed in either wrapped or unwrapped form. The wrapped display shows more detail because the range always covers 360° while time delay effects can be more easily seen in the unwrapped presentation with a linear scaling of the frequency axis.

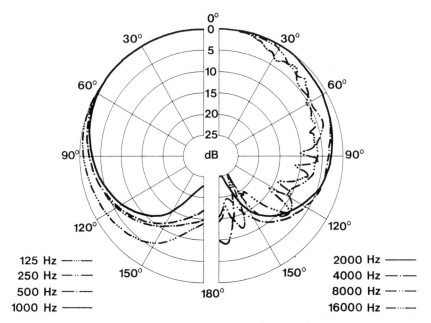

Figure 6.10   *Example of polar plot at different frequencies.*

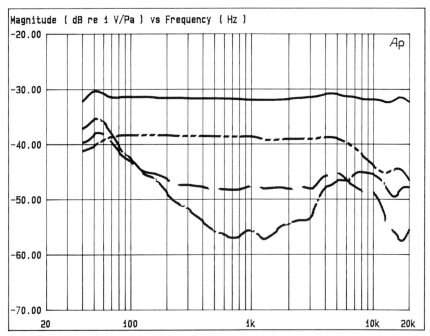

Figure 6.11   *Example of direction dependent frequency response, at 1 kHz, from top to bottom : 0°, 90°, 135°, 180°.*

(a)

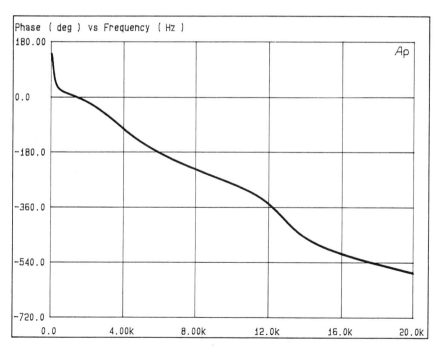

Figure 6.12 *(a) Wrapped and (b) unwrapped phase display.*

Phase measurements should always be made with a linear frequency stepping. This results in better resolution at high frequencies, which is essential to capture every 360° phase shift. Otherwise the data will not be correct. Usually more frequency steps than in magnitude measurements are necessary.

6.2.5 Impulse measurements

(a) Basics

Now that powerful digital signal processors are available which calculate a 1024 point FFT in some milliseconds, impulse measurements become of interest. They are fast because the whole frequency range is transmitted at one time, whereas in conventional sinusoidal sweeps one frequency is measured after another. Especially in production testing, where fast methods are essential, digital equipment is more and more used. It has the further advantage that quality classification and failure statistics can easily be carried out. Because of this, some basics and some special details related to acoustic impulse testing will be given as follows.

The transfer function of a linear, time-invariant system is completely defined by either magnitude and phase in the frequency domain or by the impulse response in the time domain. This means that both kinds of representation describe the system behaviour for both transient and steady-state input signals. So transforming the complete impulse response into the frequency domain by means of the Fourier transform gives the same results for magnitude and phase as would have been measured by means of a slowly swept sine wave signal. First some consideration should be given to the measuring time. As stated above, the minimum sampling time equals the time in which the impulse response dies down beneath a certain value. This threshold is determined by the dynamic range, depending on the A/D converter's resolution, the background noise and the measurement technique, which will be discussed later. The duration of the impulse response is mainly fixed by the low-frequency behaviour of the loudspeaker, the microphone and the reverberant time of the measurement environment, which normally is an anechoic chamber. Therefore most of the room's influence concerns the range below the room's cut-off frequency.

Another parameter that affects the measuring time is the desired resolution in the frequency domain. This is simply the reciprocal value of the sampling time, and gives equal line spacing on a linear frequency scale. On a logarithmic scale the distance between the spectral lines becomes smaller with increasing frequency. Therefore the sampling time is again determined by the low-frequency range. A 1 Hz line spacing for example gives 1 s of measuring time. To economize storage space and computation time during post-processing, only the data needed for a nearly equal line spacing on a logarithmic scale will be saved after the fast Fourier transform (FFT) has been calculated.

There are reasons for dividing the frequency scale into two ranges, one, for example, from 20 Hz to 1 kHz and the other from 1 kHz to 20 kHz:

(1) A 1 Hz line spacing will result in a 40 000 point Fourier transform (40 kHz sampling frequency) to cover the whole audio range up to 20 kHz. These are too many points to calculate in an acceptable time. Using two FFTs with a smaller number of points and different line spacing is much faster. The FFT algorithm requires a number of sampling points that is a power of two. Furthermore, aliasing effects must be avoided. A sampling frequency that equals four times the upper frequency limit is a good choice. In the first range, which reaches up to 1 kHz, a 4096 point FFT with 4096 kHz sampling frequency results in a 1 Hz line spacing. In the second range a sampling frequency of 81.92 kHz is a practical

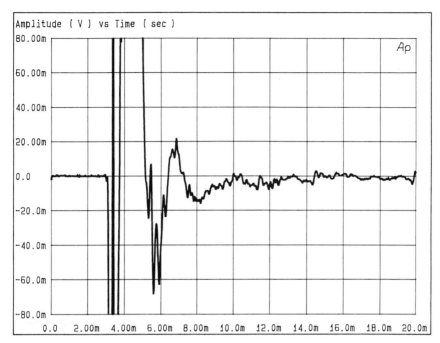

(a)

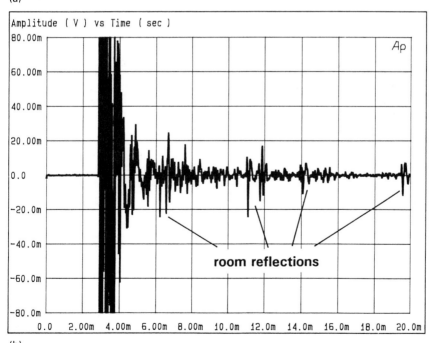

(b)

Figure 6.13 *Microphone impulse response from excitation in (a) low frequency range and (b) high-frequency range.*

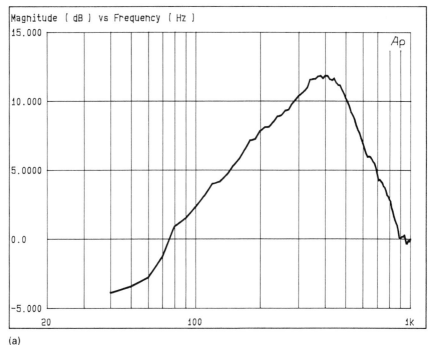

(a)

(b)

Figure 6.14 *Microphone frequency response calculated from (a) whole impulse response at low frequencies; (b) at 6 ms truncated impulse response at low frequencies*

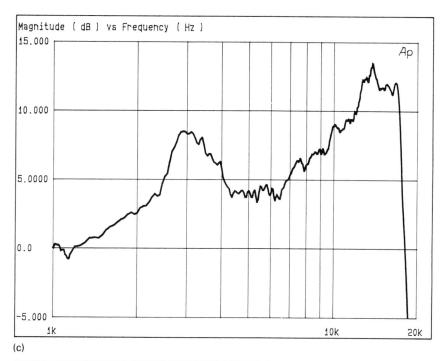

(c)

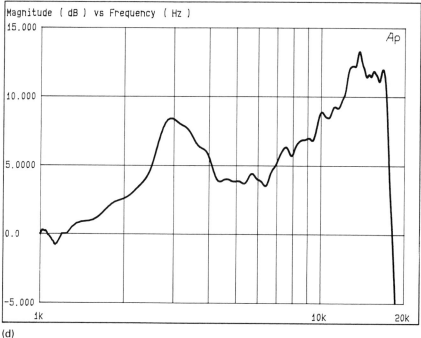

(d)

Figure 6.14   *(c) whole impulse response at high frequencies; (d) at 6 ms truncated impulse response at high frequencies.*

value and gives 20 Hz distance between the spectral lines calculated from 4096 points.

(2) As mentioned above, a two-way loudspeaker system without a crossover should be used. It is a good idea to switch between these two speakers when changing the frequency range. An active high-pass filter must be used in the high-frequency chain to keep low-frequency signal components from the tweeter.

(3) Even in so-called anechoic rooms there are reflections which affect the measurement result. It would now be desirable to eliminate these smaller peaks in the impulse response (see Figure 6.13b). This is possible only if they can be clearly separated from the loudspeaker–microphone answer.

The following example clarifies the way in which microphone free-field testing can be improved:

A microphone having a poor transfer function was chosen because peaks and dips in the magnitude response will lengthen the impulse response. This was done to see if the method will work under practical conditions.

The impulse response was measured separately for the low- and high-frequency range as defined above (Figure 6.13).

From both signals, the magnitude was calculated. The results are presented in Figure 6.14. The frequency of 40 Hz was chosen as the lower limit because measurements were not valid beneath that limit due to environmental noise. The same time period was used for both impulse diagrams in Figure 6.13 and the amplitude scaling is magnified to visualize the significant signal parts. Let us look at the high-frequency range. Most of the energy is concentrated between 3 ms and 4 ms, followed by some ringing up to 6 ms due to the 1 kHz electrical cut-off used in this range. After 6 ms the room reflections are clearly separable. So the first separable reflection arrives after twice the time of the direct signal. Zeroing the amplitude after 6 ms before calculating the FFT will cancel these unwanted reflections. The effect can be seen in comparing Figure 6.14c and 6.14d. The response is smoothed and more correct, especially in the range around 5 kHz. What would happen if we did the same windowing in the low-frequency range? If the signal hadn't died away after 6 ms, calculating the FFT will then give wrong results, which is clearly visible comparing Figure 6.14a and 6.14b. The correct graph will be obtained in combining the responses of Figure 6.14a and 6.14d.

We conclude that at low frequencies the impulse responses of loudspeaker and microphone cannot be separated from the room response. The whole response has to be used by the FFT algorithm to get the same magnitude and phase display as would have been measured with a slowly swept sine excitation. At high frequencies room reflections can be eliminated if the room is big enough. This is an advantage of impulse testing because the magnitude response is more correct than when measured under stationary conditions (Figure 6.15).

(b) Impulse measurement technique

The usual way to carry out impulse testing is to generate a small square pulse as the exciting signal. The amplitude of the impulse is limited by the range of linearity of the system and its duration by the range of frequencies of interest. Therefore it is difficult to deliver enough energy to the system to overcome any noise that is present. Although techniques exist for producing an impulsive acoustical excitation – electronic spark gaps, pistol shots or exploding balloons are often used – it is difficult to ensure that the energy is equally distributed over all frequencies of interest and, furthermore, that the energy distribution is the same each time.

To overcome these problems, pseudo-random noise is often used. The duration of the signal equals at least the length of the impulse response. In this way much

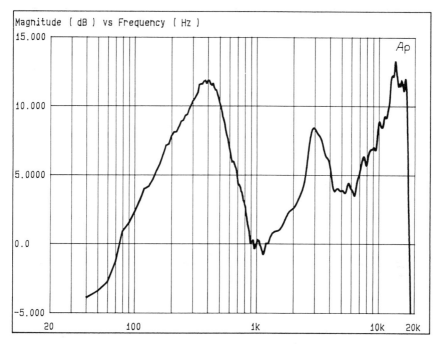

Figure 6.15 *Complete microphone frequency response composed of Figure 14(a) and (d).*

more energy can be sent to the system for a given amplitude of signal, circumventing the dynamic range problem. The total measuring time is the same as if a single impulse had been used. Because the measured response is the convolution of the impulse response with the excitation, this impulse response is identical with the cross-correlation-function between the input and output signal.[7–13]

A special kind of pseudo random noise is the so-called maximal-length sequences (MLS). These are periodic, binary, pseudo-stochastic signals which can be generated in a deterministic process. They have some advantages:

(1) The autocorrelation function of maximal sequences is a perfect impulse. This means that the spectrum is flat everywhere except at direct current.[7]
(2) The MLS provides a signal-to-noise ratio which is only 4 dB lower compared to an unfiltered swept sine excitation.
(3) They are easy to generate in a binary feedback shift register. Also the cross-correlation operation is particularly simple. An efficient algorithm exists, based upon the fast Hadamard transform, which performs only additions and no multiplications, thus saving computation time.[9]
(4) Although these signals have statistical properties like true white noise, they are actually deterministic signals which can be repeated precisely. Differences in the response of the system to successive measurements can be unambiguously attributed to noise which can be furthermore reduced in an averaging procedure.[9]

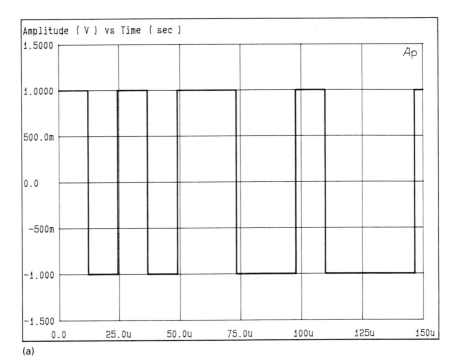

(a)

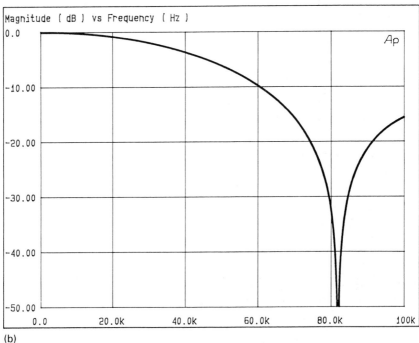

(b)

Figure 6.16   *(a) Time-section and (b) spectrum of maximum-length sequence of n = 4096 points, sampling frequency = 81.92 kHz.*

6.3 Measurement of non-acoustic parameters

6.3.1 Handling noise

The sensitivity to mechanical vibrations is important if the microphone is used on stage. Grasping or fixing on a stand produces mechanical excitations which result in an unwanted electrical output. Therefore a shock mount is used as a mechanical low-pass system to avoid broad-band excitation. This shock mount can be an external device or already built into the microphone's grip.

There are two possibilities in measuring the sensitivity to mechanical vibrations:

(1) The microphone is mounted on a vibration exciter which is set to constant acceleration (Figure 6.17): $1 \, \mathrm{m/s^2}$ has proved to be a practical value. It is large enough to get a good signal-to-noise ratio but small enough to keep most shock mounts in their linear range. The spectral analysis is carried out in the lower frequency range up to 250 Hz.

(2) To obtain the influence of the grip mass a force excitation must be used. This is the more meaningful measurement because the weight of the microphone influences the noise figure, especially in handheld applications. The device under test is hung by two thin threads to allow movement with a very low resonance frequency. Then it is excited in its most sensitive direction by means of an impulse hammer carrying a force transducer. The result is obtained by calculating the

Figure 6.17 *Microphone mounted on a vibration exciter.*

transfer function from the force spectrum as the input and the spectrum of the voice coil voltage as the output signal.

To allow a comparison between different microphones the results should be related to the acoustical sensitivity of the transducers.

6.3.2 The pop effect

When a speaker at a meeting or a vocalist in a band use their microphones close to the mouth, you can sometimes hear explosive noises coming from the loudspeakers. This noise is due to the performer's singing or saying words involving the letters 'p', 'b', or other plosive sounds. When these sounds are spoken, an air impulse reaches the transducer and produces disturbing noise near the capsule. This is called 'pop effect'. Engineers try to reduce this effect when they design their microphones. But first it must be measured. In the following, a method is described, which agrees well with subjective evaluations. No standardized procedure existed until now to measure the pop effect. So the following measurement is an example of a test, which is used in the industry. Figure 6.18 shows the equipment needed to measure the pop effect.

The generator creates a periodical voltage, which you can measure one at a time at the loudspeaker in Figure 6.19 when the switch is pressed. The rising edge of this signal is not so steep that the loudspeaker generates an acoustical output. The electrical input of the loudspeaker is shown in Figure 6.20. This method of stimulation allows automatic measurement.

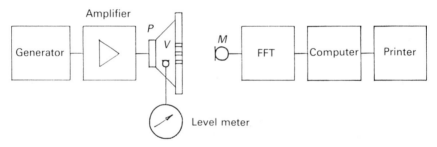

Figure 6.18   *Principle of construction of pop-measuring set-up.*

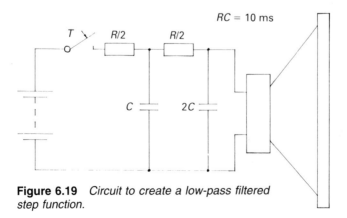

Figure 6.19   *Circuit to create a low-pass filtered step function.*

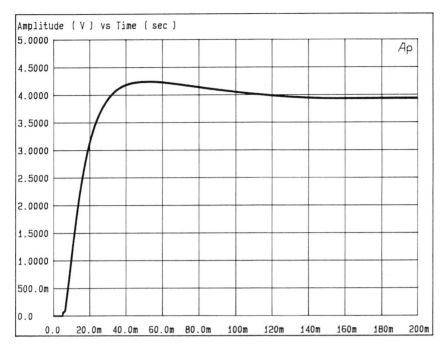

Figure 6.20 *Voltage at the loudspeaker of the pop generator.*

The amplifier boosts the incoming voltage, so that the maximum sound pressure in the volume *V* is 140 dB SPL. This should be controlled with a suitable microphone (see Figure 6.18). The shape of the sound pressure in the volume *V* is given in Figure 6.21.

The compressed air in the volume *V* will be blown through the nine holes in the plate that bounds the volume *V* in front of the loudspeaker.[14] The position of the nine holes can be seen in Figure 6.22. It must be arranged that the boundaries of the holes have no sharp edges.

Many experiments have shown that the holes in Figure 6.22 give the best results simulating a human mouth. The microphone under test should be placed on the axis of the loudspeaker 10 cm before the nine holes.[15,16] This distance is very important, because behind this position the air that is blown out of the volume *V* will be turbulent (Figure 6.23). This turbulence is one reason for the pop effect.

A second measurement should be made at the most usual distance, depending on the type of microphone.

The microphone under test is directly connected to an FFT-analyser. If it is necessary to amplify the microphone signal, the connected amplifier must have a low-frequency limit of about 5 Hz. Otherwise the level at low frequencies will be measured wrongly. The incoming signal should be averaged several times. The resulting spectrum has to be related to the proximity frequency response of this particular microphone. This is the best way to compare different microphones. The different frequency response and sensitivity of the microphones have now no influence on the result. This result must be normalized (0 dB = 94 dB SPL), so that it corresponds to an equivalent pop sound pressure level. The proximity frequency response must be measured at the same distance from the sound

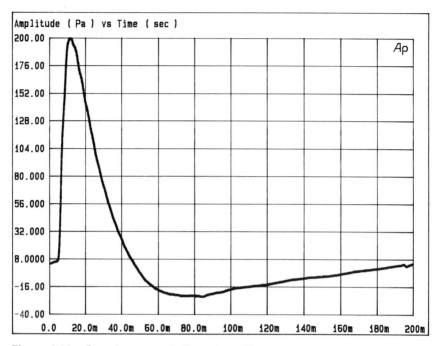

Figure 6.21   *Sound pressure in the volume V.*

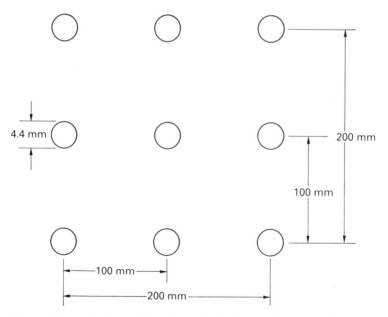

Figure 6.22   *Position of the nine holes in the middle of the plate.*

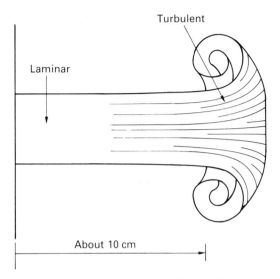

Figure 6.23 *The air-current after leaving the holes.*

source as the pop response. The sound pressure level should be 80 dB SPL. The sound source must be an artificial mouth. The interesting frequency range for the pop effect is about 10 Hz to 1 kHz. The very low frequencies are interesting in relation to the microphone preamplifier because they can overdrive it. The higher frequencies are worse for the sound, because the human ear is more sensitive at these frequencies. Figure 6.24 shows the results of two dynamic vocalist microphones.

For an overview you can use the measurement as follows.[14] The generator is the same as above. The input of the loudspeaker is a sinusoidal signal of 5 Hz. This signal must be amplified until a peak sound pressure level of 140 dB SPL is measured inside the volume of the pop generator. The distance of the microphone to the generator is again 10 cm in the axis of the loudspeaker. The signal coming out of the preamplifier of the microphone must be high-pass filtered at 5 Hz and is weighted by an A-filter. The output of the A-filter is connected to a level meter. The whole setup is shown in Figure 6.25.

Before the pop measurement can be started, the whole set-up must be acoustically calibrated. A suitable set-up is shown in Figure 6.26. It has proved to give sufficient accuracy for this type of testing.

To calibrate this set-up the preamplifier of the microphone under test has to amplify the incoming signal so that the microphone and a calibrated reference microphone have the same level at the same distance and by using the same reference signal (for example a sinusoidal signal at 1 kHz or a double decade noise from 10 Hz until 1 kHz). In Figure 6.26 the pop generator is used as the loudspeaker which emits the calibration signal.

Now the reference microphone and the microphone under test are measured as shown in Figure 6.25. The difference of the level of the microphone under test and the reference microphone is the pop sensitivity of this microphone.

In the two methods described it must be secured that the noise level you can measure at about 5 cm out from the perforated area is 10 dB lower than the measured signal.

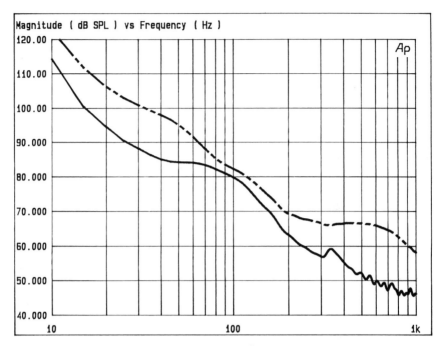

Figure 6.24 *Equivalent pop sound-pressure level of two high-quality dynamic microphones.*

6.3.3 Measurement of wind sensitivity

The noise of the wind on a stormy night that you can hear on TV when a reporter speaks outdoors is very disturbing. So this is an effect which must be reduced as much as possible. The wind noise produced by a microphone is similar to pop noise. Therefore the frequency range of interest for this effect is the same. Because all wind sensitivity measurements show great deviation, a general method of measurement is not known. There are different methods to create the test wind conditions, such as:

● the pendulum method
● the rotating frame method
● the wind tunnel method.

(a) *The pendulum method.*   The microphone under test is connected to the end of a pendulum. When the pendulum is oscillating, the relative velocity of the wind at the point of the microphone varies very much. This is the reason why one cannot measure the wind sensity very well with this method.

(b) *The rotating frame method.*   In this method you create the wind while circling the microphone around. But after the first turn the wind is disturbed. This turbulence is not what you want because it is not natural. To circumvent this the whole set-up must be very large. So this method is not used often.

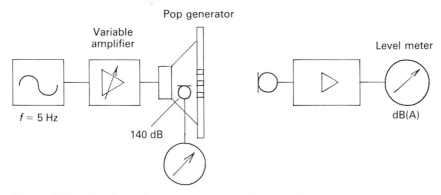

Figure 6.25 *Simple equipment to measure the pop effect.*

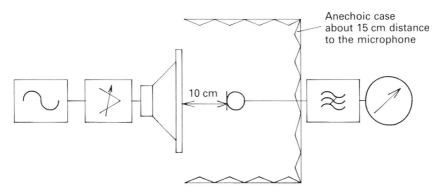

Figure 6.26 *An example to calibrate the pop set-up.*

(c) *The wind-tunnel method.* This is the most practicable and best method to measure the wind sensitivity.[3] In this case, the microphone is measured in a distance of 80 cm from the outlet of the wind tunnel (Figure 6.27).[17] The wind coming out of this wind tunnel is not laminar, but many experiments have shown that in the specified distance it is similar to the wind outdoors.

The microphone is connected to a preamplifier. The wind velocity should be about 4 m/s. For a first overview this is a practicable value. The microphone must be turned to an angle where the wind sensitivity is most intense. Then the output of the microphone is measured with a level meter. The result must be related to the sensitivity of the microphone as described in Section 6.3.2. This result corresponds to an equivalent sound pressure. The angle at which the microphone is least sensitive to the wind must be determined. This measurement should be repeated at several wind velocities. The velocities should correspond to reality. The result is presented in a table or a curve, where the equivalent sound pressure as a function of wind velocity is represented.[3]

More detailed information contains a spectrum of the microphone output. In this case the microphone is directly connected to an FFT analyser. The resulting spectrum must be related to the free-field frequency response of the microphone.

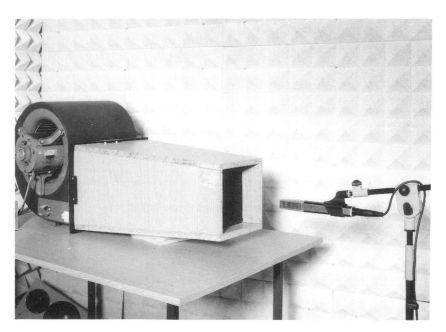

Figure 6.27   *A wind tunnel used in industry.*

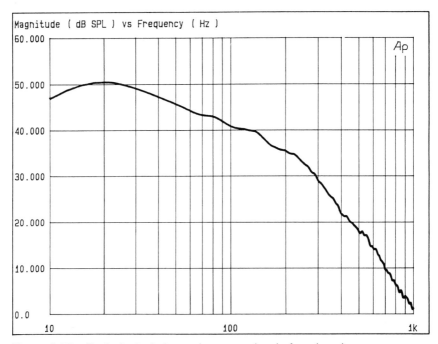

Figure 6.28   *Equivalent wind sound-pressure level of a microphone.*

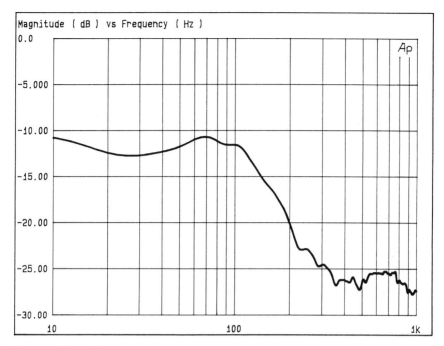

Figure 6.29 *Effectiveness of a wind screen.*

A typical equivalent wind sound-pressure level is shown in Figure 6.28. This is a good base from which to optimize the design of the microphone or windscreen.

To measure the effectivity of a windscreen you should show the deviation between the spectrum with the screen and the spectrum without the screen (Figure 6.29).

References

1. Bruel & Kjaer Type 4143, *Instruction Manual*.
2. *Meß-und Prüfmöglichkeiten der PTB* (1980). Physikalisch-Technische-Bundesanstalt, Braunschweig, p. 59.
3. IEC 268–4 (1972). *Sound system equipment. Part 4, Microphones*.
4. Dicke G. (1991). *Realisierung eines Tieftonmeßsystems für Mikrofone*. Diploma Thesis, University of Paderborn (Meschede).
5. Guicking D. and Karcher K. (1984). Active impedance control for one-dimensional sound. *J. Vibration, Acoustics, Stress, and Reliability in Design*, **106**, 393.
6. Sander H. (1985). *Erzeugung und Anwendung eines diffusen Schallfeldes in einem reflexionsarmen Raum*. Diploma thesis, University of Hanover.
7. Fasbender J. and Günzel D. (1980). An experimental system for computer-supported impulse measurements in acoustics. *Acustica*, **45**, 151.
8. Schroeder, M. R. (1979). Integrated-impulse method measuring sound decay without impulses. *JASA*, **66**, 497.
9. Borish J. and Angell J. B. (1983). An efficient algorithm for measuring the impulse response using pseudorandom noise. *JAES*, **31**, 478.

10. Xiang N. (1988). Messung von Impulsantworten mit Hilfe von Maximalfolgen. *Fortschritte der Akustik DAGA*, 893.
11. Struck C. J. and Biering H. (1991). A new technique for fast response measurements using linear swept sine excitation. *90th AES Convention*. Paris, preprint No. 3038, February.
12. Otshudi L. and Guilhot J. P. (1990). Considerations of the energy properties of pseudo-binary signals and their use in acoustic excitation. *Acustica*, **70**, 76.
13. Niedrist G. (1990). Neues Impulsmeßverfahren für elektroakustische Wandler. *Fortschritte der Akustik DAGA*, 175.
14. Wollherr H. (1991). Meßtechnische Bestimmung der Pop – Empfindlichkeit von Mikrofonen. *Fortschritte der Akustik DAGA*.
15. Wienhöfer W. and Sennheiser J. (1987). Measuring equipment for the estimation of 'pop' sensitivity of microphones. *82nd AES Convention*, London, preprint No. 2423.
16. Werner E. (1989). Dependency of microphone pop data on loudspeaker properties. *86th AES Convention*, Hamburg, preprint No. 2787.
17. Werner E. (1990). Zur Messung der Windempfindlichkeit von Mikrofonen. *Fortschritte der Akustik DAGA*, 305 (1990).

7 Ribbon microphones

Günter Rosen

7.1 History of ribbon microphones

The ribbon microphone was invented around 1930. It is assumed that the ribbon loudspeaker, invented in 1924 by E. Gerlach, was modified by Olson. The first ribbon microphones exhibited a figure-of-eight directivity pattern. They had an advantage therefore over the then commonly used omnidirectional moving-coil microphones, in that they absorbed less reverberation from the environment. Microphones with cardioid or hypercardioid directivity patterns, as in common use today, had not been invented at that time. The first cardioid microphones were formed from a combination of a ribbon microphone with a figure-of-eight pattern and an omnidirectional moving coil microphone. Additional exotic designs were evolved, e.g. an omnidirectional ribbon microphone (Olson and Preston, RCA, approximately 1950).

The early ribbon microphones were very large and heavy. The ribbon was considerably longer than is usual today and was also corrugated, which made it sensitive to being moved by wind or vibration. Even by blowing on it (as part of a functional test) the ribbon could be stretched so much that the microphone no longer functioned according to specification. In 1958, Eugen Beyer introduced a short ribbon microphone. The dimensions for this microphone are no larger than those for a moving-coil microphone. It had a sound characteristic comparable with that of a capacitor microphone, but without its disadvantages, namely sensitivity to moisture and the supply voltage required. The microphones, type M 130, M 160, M 260 are still manufactured in more or less the same way today. The magnetic material used is aluminium–nickel–cobalt (Alnico). The (free) length of the ribbon is still approximately 20 mm. The ribbon is less than 2 mm wide and approximately 3 μm thick. The ribbon microphone has an effective 'anti-popping' windshield and may be used as a hand-held vocalist's microphone with excellent sound quality. The short ribbon has the advantage that it is much more robust. The sensitivity to vibration is considerably less. The M 130 and M 160 are double ribbon microphones. Instead of a single ribbon, two ribbons are used, one directly behind the other. In this way the damping characteristic of the microphone is improved, with simultaneous improvement in the reproduction of treble.

A further improvement in the magnetic material used (Neodym) led, in the 1980s, to the ribbon microphone being further reduced in size. In the case of the HM 560, a microphone which is mounted on a headset boom, the diameter of the system has shrunk to 10 mm. The dimensions of the ribbon have not changed. The microphone was designed in such a way that on speaking close to the microphone (approximately 2 cm from the mouth) a linear frequency response characteristic from 40 Hz to 20 kHz is obtained.

The ribbon microphone is preferred by singing drummers and keyboard players, etc. A further equally important application is the reporting of sports events. Here the ribbon microphone demonstrates its advantages yet again, i.e. good reproduction of speech with few 'popping' and hissing tendencies. A low output voltage is irrelevant here as the headset microphone is mounted close to the mouth, where the useful sound pressure is very high.

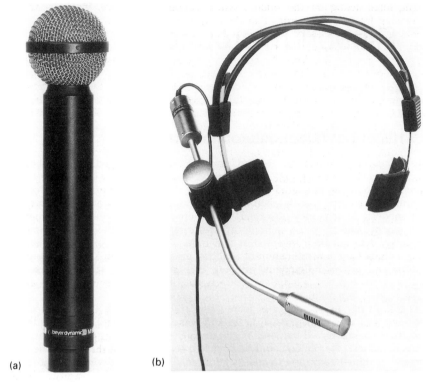

(a)   (b)

Figures 7.1(a)   *A compact high-quality ribbon microphone with construction as in Figure 7.2(b); (b) a miniaturized ribbon microphone constructed as in Figure 7.2(c), which is designed for mounting on a headset boom (Beyer)*

7.2 Function

The ribbon microphone belongs to the category of dynamic microphones. An electrical voltage is produced in the same way as in a dynamic moving-coil microphone, i.e. on the basis of the electrodynamic (or electromagnetic) principle. That is, when an electrical conductor moves in a magnetic field at right angles to the magnetic field lines, an electrical voltage is induced in the conductor. In this case, the electrical conductor is the ribbon, an extremely thin, rolled out metal foil which is moved by the sound waves (movements of air) exciting it.

The induced voltage may be calculated as follows:

where U = induced voltage $U = Blv$
$\quad\quad B$ = magnetic flux density
$\quad\quad l$ = length of the ribbon in the air gap
$\quad\quad v$ = velocity of ribbon at right angles to the field lines of the magnet.

B and l are constants, v is a time-dependent parameter and hence the output voltage U is also dependent on time.

The force acting on the ribbon, which causes it to move, increases with increasing frequency. In order that the output voltage remains independent of frequency, it is necessary to ensure that the resonant frequency due to the flexibility and mass of the ribbon lies at the lower end of the transmitted frequency range. Therefore, the system operates in the mass-controlled range. This phenomenon is also present in dynamic loudspeakers and moving-coil microphones. Friction in the ribbon is very low. It is important that the ribbon is not held taut like a stretched string. If this were the case, the resonant frequency would be too high and a poor transient characteristic would be obtained. Ideally, the transient response of the ribbon is aperiodic.

The advantage of a ribbon microphone is that the mass moved is very small. In a dynamic moving-coil microphone, the moving part comprises a diaphragm with an oscillating coil. In a ribbon microphone it is reduced to the thin aluminium ribbon, which moves in the field of a strong magnet. In this case, the ribbon represents both membrane and oscillating coil. The small mass produces sound characteristics very similar to those of a capacitor microphone. For speech and singing, the ribbon microphone gives very clear reproduction and little tendency for producing 'popping' and hissing sounds.

The schematic construction of a ribbon microphone, as often illustrated in books, is shown in Figure 7.2. It is possible to recognize both the magnet with its magnetic field and the ribbon which moves in that field. The ribbon is shown here corrugated as in an accordion. This corrugation was used in the very early constructions, but modern constructions no longer use it.

The directivity pattern for such a microphone forms a nearly perfect figure-of-eight. Therefore the maximum sensitivity lies at 0° and 180°. Sensitivity is at a

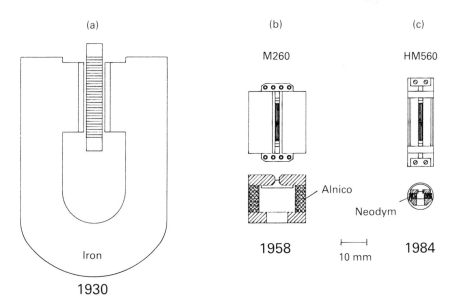

Figure 7.2 *(a) An older type of ribbon microphone using a large iron magnet; (b) a modern type of compact ribbon microphone with a smaller short ribbon and a smaller Alnico magnet with a short magnetic return path (Beyer); (c) a miniature ribbon microphone using a modern high-efficiency type of magnet and a vestigial return path (Beyer).*

minimum at 90°. Deviations from the ideal figure-of-eight characteristic are produced only at very high frequencies. On the one hand, these are due to reflections and diffraction by the magnet structure. On the other hand, the finite dimensions of the ribbon are relevant. If the directivity pattern is measured in the horizontal axis, the ideal pattern is produced when the ribbon is vertical. The direction of the ribbon is usually shown by two red dots on the case of the microphone which, to a certain extent, represent the end points of the ribbon. The microphone may therefore be oriented so that the ideal directivity pattern is produced.

In the studies of stereo recording this topic is considered in more detail regarding both its importance and why ribbon microphones are especially suitable for M–S stereo recording.

By modifying the capsule appropriately, other directivity patterns than the figure-of-eight pattern may also be obtained. As with other microphone types this is achieved because the sound at the back reaches the ribbon via phase shifting or transit time generating elements. The diagrams of Figure 7.3 clearly illustrate the effect.

The way in which the desired directivity pattern is created may be understood by imagining that the direction of the maximum suppression is deflected from 90° to other angles. Without this modification, the microphone has the figure-of-eight characteristic mentioned above. A sound incident at less than 90° reaching the ribbon has the same transmission time to both sides of the ribbon (as shown in (a))

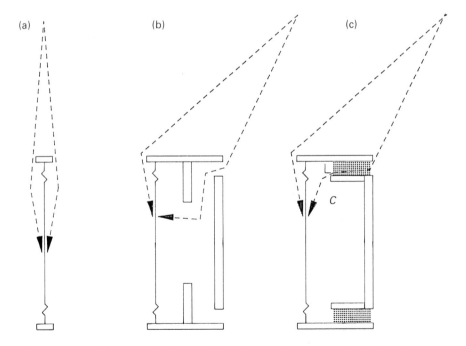

Figure 7.3   *(a) A sound incident at 90° with equal access to the front and rear of a microphone ribbon; (b) as in (a), but with rear access via an open acoustic labyrinth; (c) as in (b) but with a rear labyrinth incorporating a finite acoustic impedance designed to introduce a phase shift, thus giving a cardioid type of polar curve.*

and hence reaches both sides of the ribbon at the same time. The ribbon is not deflected by sound reaching it from the direction shown.

In (b) a diverted sound entry path is provided at the rear of the microphone. The sound must then travel over a larger distance in order to reach the rear of the membrane. The direction of sound for which both paths of sound are the same is therefore moved back a little. Hence, a figure-of-eight pattern is no longer obtained, but instead a hypercardioid.

In (c) the sound bypass is replaced by an acoustic element (resistance-capacitance element). This acts similarly to the sound bypass and delays the sound reaching the rear and hence also causes the direction of the maximum suppression to be moved back. The directivity pattern in this case is also a hypercardioid.

7.3 Directivity pattern at high frequencies

A further aspect to be considered is the directivity pattern at high frequencies. At 10 kHz the wavelength is 3 cm, at 20 kHz it is 1.5 cm. At these small wavelengths, the dimensions of the microphone membrane (in this case the ribbon) are already of great significance. Pressure build-up effects are produced both in front of the membrane and in the shadow behind it. These effects are directly influenced by the mechanical dimensions of the membrane. Ideally the largest membrane dimension should be less than half the wavelength, i.e. less than 7.5 mm.

In capacitor microphones, the diaphragm diameter can be kept this small, so that directivity effects at high frequencies are negligible. However, with these small membrane diameters, electrical noise may increase considerably.

The ribbon microphone does not have a circular diaphragm, but a rectangular ribbon instead. The width of the ribbon at 2 mm is clearly below the required 7.5 mm. However, the length of the ribbon at 21 mm is greater. What does this mean in practice? If the microphone is upright, so that the ribbon is vertical, then the horizontal directivity pattern produced is ideal. In the vertical axis, a small narrowing of the pattern is obtained due to the length of the ribbon.

It is seen that the ribbon microphone maintains an ideal directivity pattern even at very high frequencies. Hence, even at these high frequencies it collects sound over a very wide angle. The recording of sound requires, in particular, that the initial strong sound reflections that are reflected from side walls are also recorded, even at high frequencies. The recording thus contains more treble. An electrical accentuation of the treble in this case would also directly accentuate the sound entering the microphone and would lead to 'sharp, clear' tonal reproduction.

The sound characteristics of the ribbon microphone are in this context considered as 'airy' and also has 'soft highs'.

7.4 Pulse characteristics

As already mentioned, an advantage of the ribbon microphone is that the mass moved is very small. Therefore, a very good impulse characteristic can be expected. A small mass on its own, however, does not automatically result in a good impulse characteristic, as the microphone operates in the mass-controlled range.

The diaphragm of a moving-coil microphone definitely cannot be considered to be a rigid plate at high frequencies, but instead it has the property of vibrating in

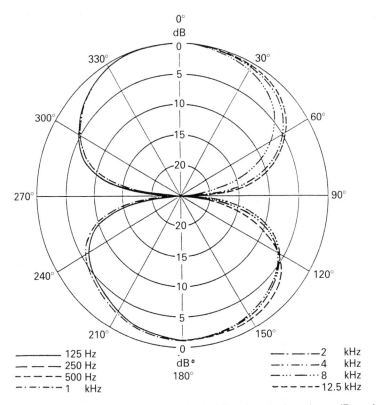

Figure 7.4   *Polar curves of a typical bidirectional microphone (Beyer).*

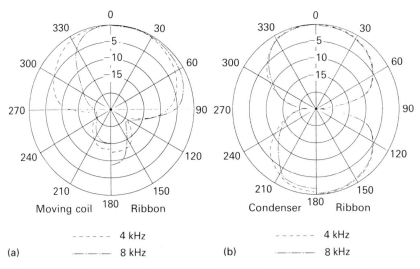

Figure 7.5   *Comparisons between polar curves of different types of microphone: ribbon, condenser and moving coil.*

several modes. Oscillations occur within the membrane itself. In an extreme case, this may mean that on excitation one part of the diaphragm moves to the inside and one part to the outside, without any axial movement of the coil. In addition, due to a certain degree of elasticity between the diaphragm and the coil, the coil does not always follow the oscillations of the diaphragm exactly, especially at very high frequencies. In practice, this is noticeable by a drop in the frequency response at high frequencies. In order to compensate for this drop, a high frequency acoustic resonator is located in front of the diaphragm. The transmitted frequency range of the microphone is thus increased, but the gradient of the drop at the upper end of the transmitted frequency range also increases. When excited with an impulse, this resonator is incited to oscillate. These oscillations are present in the impulse response. These acoustic resonators, under certain conditions, may be heard by trained ears, as the sound is slightly degraded.

The ribbon microphone does not have these resonators, since the membrane and oscillating coil are not separate. A movement in the ribbon results in a corresponding output voltage. The frequency response of a ribbon microphone is characterized by a wide linear range, which drops only slowly at high frequencies. Therefore there is no need for a treble boosting resonator. This is also clear from the impulse response.

In practical applications, ribbon microphones are preferred when an impulse sound or a sound containing strong harmonics occurs, as, for example, in

speech ('s' sound)
brass instruments
pianos, grand pianos
or very complex tonal reproduction such as concert recordings.

These are typical applications, where on the one hand, sufficiently high sound pressure exists so that the ribbon microphone produces sufficient voltage, and on the other hand, very strict requirements are made with respect to the quality of the recording. In contrast to the capacitor microphone, the ribbon microphone has

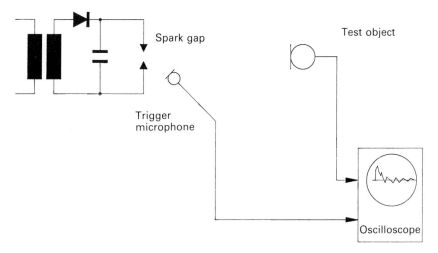

Figure 7.6 *The schematic of equipment for measuring the impulse response of a microphone (Beyer).*

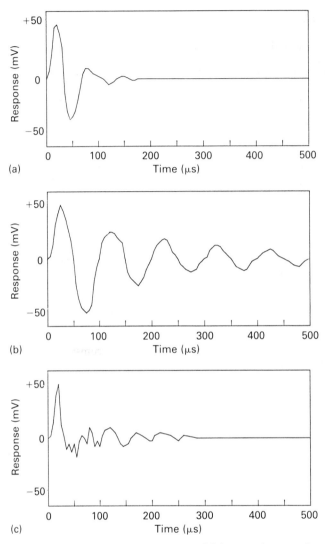

Figure 7.7   *The impulse response of (a) a condenser microphone, (b) a moving coil microphone and (c) a ribbon microphone (Beyer).*

the additional advantage that it is more or less incapable of being overloaded, as indeed are all dynamic microphones.

If the pulse characteristic or transient response of various microphones is measured (e.g. the impulse response to a 'spark detonation'), the ribbon microphone exhibits an impulse characteristic which is very similar to that of a good capacitor microphone. In contrast, the impulse characteristic of the dynamic moving coil microphone clearly exhibits transient and decay processes. These transient and decay processes are superimposed on the recording of an instrument, which of course changes the sound of that instrument.

7.5 Transformers

An important component of the ribbon microphone has not yet been mentioned, i.e. the output transformer. The ribbon microphone, as described hitherto, has an extremely low impedance. The impedance of the ribbon is about 0.2 Ω. The output voltage is also very low. In moving-coil microphones there are a considerable number of turns on the coil (e.g. 50) in a narrow air gap with a high field strength. In ribbon microphones there is only one 'turn'. This disadvantage is only partially compensated by the low mass of the ribbon. The output voltage must therefore be increased by means of an output transformer.

The output transformer has a transformation ratio of approximately 1:30, i.e. the output voltage is about 30 times higher than the voltage induced in the ribbon. Since there is no power gain in the transformer, Figure 7.8 applies, therefore the impedance is transformed by 900 times, which results in an output impedance of 180 Ω for an input impedance of 0.2 Ω. This is approximately the value of the impedance for standard 200 Ω microphones.

In addition, the impedance of a ribbon microphone is dependent upon frequency. In particular, in the vicinity of the natural resonance of the ribbon the impedance may double. Therefore, care must be taken that the terminating impedance of the microphone (i.e. the input impedance of the microphone amplifier) is at least 5 times as large as its nominal impedance, which in this case is 1000 Ω.

In order to keep losses in the primary circuit low, it is important that it is designed to have an especially low resistance. Consequently, only a few turns made from thick wire are used.

In order to avoid damaging the microphone, under no circumstances must external electrical voltages be applied to it. If such voltages were applied, the output transformer would operate in such a way that it increased the currents by the transformation ratio. An ohmmeter which sends a current of 1 mA through the secondary winding of the microphone causes a current of 30 mA to be

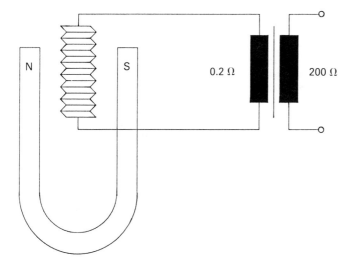

Figure 7.8 *A ribbon microphone and an incorporated step-up transformer to raise the impedance from 0.2 Ω to the line value of 200 Ω (Beyer).*

produced in the primary winding. This may not be sufficient to melt the ribbon, but together with the high field strength of the magnet, considerable forces are created which act on the ribbon. As a consequence the ribbon may be pushed out of the air gap and snap. Therefore, never connect an ohmmeter to the output terminals for carrying out measurements in a functional test. (In a moving-coil microphone such a test does not cause any damage.)

An additional bad habit is to blow into the microphone as part of a functional test. With some microphone designs this can lead to the ribbon being stretched, with the result that it fouls the magnet. However, in most modern designs with short ribbons, there is little danger of this.

7.6 Stereo recording using ribbon microphones

Ribbon microphones are usually used for intensity stereophony. Two common processes exist for this: $X-Y$ and $M-S$ stereophony.

In the $X-Y$ process, two directional microphones are oriented at a certain angle (approximately 30° to 80°) with respect to the recording axis. The output signals from the microphones produce directly the signals L (left) and R (right). The stereo effect is obtained by differences in intensity. Two M 160 or M 260 microphones could, for example, be used for this. A special version of the $X-Y$ process is the Blumlein technique where two microphones with a figure-of-eight characteristic are oriented at ± 45° with respect to the main recording direction. The bass characteristic of a microphone with a figure-of-eight characteristic is very good and almost comparable with that of a pure pressure microphone. The uniformity of the figure-of-eight directivity pattern, which is especially marked in modern ribbon microphones, produces a very accurate directivity image.

$M-S$ stereophony is also an intensity stereophony. $M-S$ means middle–side stereophony. A directional microphone is used as the M system, which is aimed in the main recording direction. (In special cases, an omnidirectional microphone is used as the M microphone). The S microphone with figure-of-eight characteristic is oriented at 90° to the main recording direction. The L and R signals are derived by means of a matrix from the $M-S$ signals. This matrix produces the sum or difference of the M and S signals. In $M-S$ stereophony the uniformity of the figure-of-eight pattern for the S microphone is of particular importance. If the figure-of-eight is asymmetrical, this leads immediately to the image being moved in a sideways direction and hence a less exact stereo image is produced. Here the characteristics of the ribbon microphone, with its ideal directivity pattern over a wide frequency range, are advantageous.

8 Microphone amplifiers and transformers

Peter Baxandall

8.1 Introduction

In this chapter the primary aim is to present the *fundamental principles* that underlie the design of microphone amplifiers and transformers as lucidly as possible. However, attention will also be given to engineering aspects which practical experience has shown to be important.

Section 8.2 has been included largely because it is felt that some appreciation of how the present state of the art has come about helps to make work in this field more interesting and satisfying. The Section also provides a convenient way to introduce some topics that are dealt with in greater detail in later parts of the chapter.

8.2 History

A basic microphone amplifier is nowadays a diminutive device of great reliability, tiny power consumption and low cost – but how very different things once were!

The recording industry had no requirement at all for microphone amplifiers until about 1925 when electrical disc recording came in, and the significant history of such amplifiers really starts at about the same time as that of regular broadcasting, say 1920 or just after – although Wente's capacitor microphone and amplifier for sound-intensity measurement,[1] based on work at Bell Telephone Laboratories, appeared in the literature as early as 1917. This latter date is in fact only about five years after the vital notion first began to dawn that the triode valve, invented in 1907 by Lee de Forest, and regarded at first merely as an improved radio detector, had potent amplifying properties and could be cascaded to make high-gain amplifiers.

The BBC (British Broadcasting Company, as it then was) took over responsibility, in November 1922, for running three broadcasting stations that had previously been owned and operated for a short period by independent electrical firms.

One of these firms was Western Electric, whose 0.5 kW London station was soon moved (by steam lorry!) to Birmingham. It had equipment of American design, including a Western Electric Type 373 double-button stretched-diaphragm carbon microphone, which gave much better quality than ordinary telephone microphones, and a Western Electric Type 8A microphone amplifier. This amplifier, which was powered by storage batteries, is shown in Figure 8.1, and cost £160, equivalent to several thousand pounds today. It had three choke–capacitance coupled stages with sizable input and output transformers.

Surprising as it may now seem, a microphone amplifier as such was not used at all in other BBC installations in the very earliest days, the output of the studio

Figure 8.1   *Western Electric Type 8A three-valve microphone amplifier as installed, with added feed-current milliammeters, in Birmingham control room, 1923. There was a separate filament rheostat for each valve and choke – capacitance interstage coupling was employed. Storage batteries were used for the power supply.*

carbon microphones – of the telephone variety – being fed directly to the input of the modulation amplifier in the transmitter. This was the initial state of affairs at the original London station 2LO as installed by the Marconi Company at Marconi House in the Strand, though it did not last long.

The much superior quality of the Birmingham transmissions was noticed by listeners – perhaps the first faint glimmering of a hi-fi outlook! – and before long Western Electric carbon microphones and amplifiers were introduced at other stations. Nevertheless it became evident that even these had fairly serious performance shortcomings, and this, together with their very high cost and the fact that microphones and amplifiers were obviously going to be needed in much larger numbers as the broadcasting service expanded, resulted in a good deal of effort being put into finding other solutions.

Captain H. J. Round at Marconi's, who was a key figure in the early days of broadcasting in Britain (and with whom I once spent a most enjoyable day when he came to see some work we were doing at the Royal Signals and Radar Establishment about 1948), soon evolved the Marconi–Sykes magnetophone, the essential notion coming from a 1920 patent of Adrian Sykes. This was a form of moving-coil microphone whose diaphragm was a flat annular coil of aluminium wire suspended on cotton wool in the field of a large pot-type electromagnet consuming 4 A at 8 V from a battery.

An early prototype of the magnetophone, and the amplifier which Round had designed to go with it, were brought into regular service in May 1923, just after the BBC London studios at Savoy Hill were occupied. The designs were finalized within a few months and became the standard equipment in most BBC studios for several years. The microphone was widely known as the meat-safe!

Figure 8.2 *Marconi Type GA1 single-channel five-valve microphone amplifier, having resistance – capacitance interstage coupling. The metal screening boxes over the valves have been removed, but can be seen in Figure 8.3. This was the standard BBC microphone amplifier for several years. It operated from a 6 V filament supply and a 300 V HT supply.*

The production version of the microphone amplifier, Marconi Type GA1, is shown in Figure 8.2, with the metal screening covers over the valves removed. The amplifier weighed over 45 kg, but was considerably smaller than the original experimental prototype. There were five R–C coupled stages of triode amplification, with input and output transformers mounted behind the valve panel. The price was only about half that of the Western Electric Type 8A.

Each valve, connected by gold-plated contacts, was mounted in a sub-assembly suspended by rubber bands, with flexible leads to the main circuit. This was done to reduce microphony, or 'ponging' as it was often called, which occurred if the valves were subjected to the slightest vibration – this was indeed a major problem with directly heated valves of the types then employed.

The GA1 amplifier had provision for adjusting the high-frequency response 'to suit the acoustic properties of the hall in which the microphone is used', to quote from the leaflet supplied.[2]

The magnetophone and GA1 amplifier were used not only in studios but also for outside broadcasts, as also were the Western Electric items previously mentioned. Figure 8.3, taken in 1925, shows two of the Marconi amplifiers rigged for one of the famous early OBs of the song of the nightingale from a wood in Oxted, Surrey.[2] Notice the boxes of batteries and the checking radio receiver.

Though the above Marconi equipment was in widespread BBC use by 1924, Birmingham continued to use Western Electric products, soon changing from the double-button carbon microphone to a capacitor microphone, which was retained there until 1926. It was the only capacitor microphone in BBC service. The microphone and its amplifier 'were both very susceptible to the least trace of dampness and so were apt to be "temperamental" '. Though capacitor microphones were employed to a very limited extent after 1926, it was many years before their early BBC reputation for making frying noises and exhibiting general unreliability was largely overcome.

Figure 8.3 *Two Marconi GA1 microphone amplifiers installed for a 1925 outside broadcast of the song of the nightingale from a Surrey wood. A 6 V battery and six wooden boxes containing 50 V batteries may be seen.*

A very significant event in the history of microphones and microphone amplifiers in the BBC was the introduction of the Marconi–Reisz transverse-current carbon microphone – the familiar marble-block device of octagonal shape that is seen in so many early broadcasting photographs. It was invented in Germany by Georg Neumann, founder of the present-day firm of that name, while employed by the Reisz company.[3] It was brought into service towards the end of 1925, and was in almost universal use throughout the BBC by 1927. Many of these microphones, together with a BTH version of the same type but round in shape, were to be seen in action until 1935, and to a limited extent even later.[2,4]

The Reisz microphone was robust, extremely reliable, acceptably small by the standards of those days, it had a convenient (resistive) impedance of about 300 Ω, and, above all, it was highly sensitive, giving well over 10 mV r.m.s. on the grid of the input valve even on speech. The bass response extended without loss down to 30 Hz and below. The axial high-frequency response exhibited a very broad peak of over 10 dB centred around 4 kHz, much of this being due to diffraction. However, with suitable equalization, which was incorporated in the later amplifier designs, and with appropriately positioned artists, quite good sound quality was obtainable.

The microphone took about 20 mA at 6 V DC to polarize it, and generated considerable noise, of $1/f$ spectrum. The noise, expressed as an equivalent sound pressure level (SPL), though a good deal higher than for a normal modern microphone, was not intolerably great, but bearing in mind the high sensitivity, it corresponded to quite a high electrical output noise level, which over-rode the thermal noise in the amplifier input circuit by a substantial factor. This greatly eased the problems of amplifier design from a noise and microphony viewpoint and it also reduced the audibility of interference picked up by microphone wiring. The microphone unfortunately gave considerable non-linearity distortion at very high SPLs, but its limitations were well understood by the engineers of the day and by-and-large it gave very satisfactory results on medium-wave radio.

Figure 8.4 shows, in redrawn and slightly simplified form, the circuit of a Reisz microphone amplifier as given in Reference 5 of 1928. The first three valves are 2 V filament valves of fairly low gain as then used in ordinary radio sets, the output valve being a small power valve with 6 V filament. Note the arrangement of three flashlamp bulbs whose relative brightness, in association with the meter readings, gave an immediate indication, in the event of a filament failure, of which valve was faulty – such failures were considerably less rare than in later times.

Note too the way the 6 V supply for the filaments serves also for energizing the microphone, and that the microphone circuit is not a balanced one. This latter feature would be frowned upon today, but apparently it was then found good enough, largely because of the high output level of the Reisz microphones.

Obviously gain adjustment effected right at the input as in Figure 8.4 leaves the amplifier output due to valve noise and microphony at its full level when the gain is reduced, but because of the high microphone sensitivity, and consequently only moderate maximum-gain requirement, this scheme was found to be quite satisfactory, and it had the great virtue of keeping the amplifier distortion to a minimum even at the low gain settings required with loud programme sources – negative feedback was not known about when this amplifier was designed.

The purpose of the input gain control, as normally used, was to set the gain to suit the type of programme involved, so that the level at the microphone amplifier output would be suitable for easy handling on the faders at the control desk.

Before the advent of Reisz carbon microphones, when magnetophones and GA1 amplifiers were employed, these amplifiers and their batteries were usually located close to the associated studios, but when Reisz microphones came in, the amplifiers were installed in the control room, allowing all batteries to be located nearby and shared between amplifiers. This new scheme, unlike the old one, involved long cable runs at microphone level, and these, as already mentioned, were operated in an unbalanced mode.

Another feature of the Figure 8.4 circuit is the use of grid stoppers in the first two stages. These markedly reduced the tendency of the circuit to respond, by rectification, to radio-frequency interference, and their use for this purpose has been credited to H. J. Round. This dodge, of course, is still used in the semiconductor era, though sometimes primarily for preventing VHF oscillation.

After 1928, the design of BBC microphone amplifiers, otherwise called 'A' amplifiers, evolved for several years along lines fairly closely related to the Figure 8.4 type of circuit, but by 1933 indirectly heated triodes, which became available about 1930 with the advent of mains-operated radio receivers, had displaced the filament type and were less microphonic.[6]

Though Reisz microphones were still in very widespread use in 1933, they were beginning to be replaced by more modern and sensitive forms of moving-coil microphone, and, to a much lesser extent, by capacitor microphones which had individual preamplifiers.[7,8]

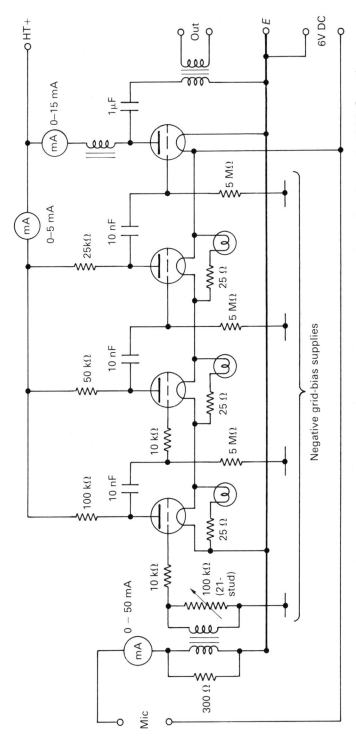

Figure 8.4 *BBC microphone amplifier as used with Marconi-Reisz transverse-current carbon microphones in 1928. Sometimes two microphones in parallel were connected to the input. The low internal resistance of the storage-battery HT supply made early-stage decoupling unnecessary.*

With the new valve types and related transformer designs, the new 'A' amplifiers required only three stages. To compensate for the over-emphasized high-frequency response of the Reisz microphones, a low-Q series tuned circuit, resonating at just under 4 kHz and having adjustable series resistance, could be switched across the output of the first stage.

A shelving bass loss below about 300 Hz was also switched in, because of a belief that the Reisz microphones had a lifting bass response of at least +6 dB at 50 Hz. This belief must, I think, have been mistaken; a recent measurement on a microphone of this type, using a B & K Type 4133 measuring microphone, showing it to have a low frequency response flat down to 30 Hz within ±1 dB, as would be expected from its principle of operation. (It is noticeable that around 1932/1933 published BBC microphone response curves,[7,9] not only for carbon microphones but also for stretched-diaphragm and Voigt slack-diaphragm capacitor microphones, all exhibited a fairly similar bass rise, suggesting that the measuring set-up then in use probably had an unsuspected bass loss. This is perhaps less surprising when one bears in mind that acoustic measurement techniques were in their infancy at the time.)

The gain adjustment, for accommodating different types of programme source, came, as before, at the input, though the change had been made to a normally connected potentiometer rather than a shunt 'rheostat', and the primary of the input transformer was floating so that it could be used in either a balanced or an unbalanced manner.

Despite the use of indirectly heated valves in the new amplifiers, the heater, high-tension and grid-bias supplies were all still obtained from central batteries – there seems to have been a profound distrust of mains supplies for amplifiers by broadcasting engineers in those days, presumably on grounds of less reliability and propensity to mains hum, the latter being a characteristic of much badly designed public-address equipment at the time!

The next major change was the bringing into service in 1935 of the BBC/Marconi Type A ribbon microphone, after a short trial period. By 1936 the majority of BBC microphones were of this type. The ratio of the internal transformer was chosen to give an output impedance of 300 Ω as for the Reisz microphones, which was convenient. Though the new microphones were of much better fidelity and had polar characteristics of a usually beneficial type, the lower output level nevertheless caused acute problems at first.

The cable runs from studios to control room not then being of the balanced variety, studio preamplifiers were initially found necessary with ribbon microphones in order to avoid troubles from switch-click interference etc. However, by changing to balanced twin-core lead-covered cable, plus balanced input transformers with screens, long microphone cable runs, even of 300 m, were found to be quite satisfactory.

As already mentioned, outside broadcasts in the earliest days were mostly done using cumbersome amplifiers designed for studio or control-room purposes, accompanied by massive boxes of batteries, but in later years amplifiers designed specifically for OB use were introduced, of which the OBA/8, shown in Figure 8.5, which appeared in 1938, is the one of by far the greatest technical significance.[4,10,11]

The OBA/8 amplifier was an outstandingly successful design which, for several reasons, had a great influence on later developments in both OB and studio equipment. It remained in service for several decades and was manufactured in rack-mounting form (Type APM1) as well as in the box form used for normal OBs. Large numbers of these amplifiers were used as the basis of temporary wartime studio installations. The main new features, apart from reduced size and weight, were:

Figure 8.5   *OBA/8 outside-broadcast equipment. Two amplifiers are on the table, surmounted by a pair of power units between which is a four-channel microphone mixer. The monitoring loudspeaker, in a case unfolding to form a baffle board, contained a power amplifier consisting simply of a pair of ACSP/3 valves in push-pull – 'extensive tests indicated that the maximum output of about 1 ½ watts was adequate'.*

(1) Only two valves for signal amplification, these being AC/SP3 high-slope television-type pentodes. The 4 V 1 A heaters were run on 50 Hz AC and the mains-derived HT was at 250 V.
(2) Total HT current only 25 mA, allowing ordinary radio-receiver type dry HT batteries to be used for emergency operation in the event of mains failure, plus a storage battery to provide the total heater current of 3.5 A at 4 V.
(3) Negative feedback employed, with combined feedback and passive gain control.[10,11]
(4) Equalization incorporated to correct for falling response of ribbon microphones at high frequencies.
(5) Built-in peak programme meter provided.

The amplifier was capable of supplying a power output of 25 mW, with not more than about 1% harmonic distortion, to any resistive land-line impedance between 75 Ω and 600 Ω , and had sufficient available gain to be able to do this, using ribbon microphones, for any likely kind of programme material. At reduced settings of the single-knob gain control, the amplifier could handle inputs corresponding to ribbon microphone SPLs of about 133 dB* with similarly low distortion and without the sacrifice of signal-to-noise ratio which the simple input gain-control scheme of Figure 8.4 would have produced. The distortion under most operating conditions was substantially less than 1% and subjectively negligible.

*The later Type AXBT version of the Marconi ribbon microphone had a Ticonal magnet and was several dB more sensitive,[12] giving a lower maximum SPL for 1% distortion.

On outside broadcasts and in studios it is, of course, frequently required to mix the outputs of several microphones. Nowadays, since microphone amplifiers are small and relatively cheap, the almost universal practice is to have a separate amplifier for each microphone and do the mixing at a moderately high signal level. But in earlier days the mixing was often done at low level, with an amplifier shared between several microphones. Thus a standard adjunct to the OBA/8 amplifier was a four-channel mixer or series fader. Each control knob on this unit varied the amount of series resistance inserted between the associated 300 Ω microphone and the OBA/8 input, the input impedance also being 300 Ω.

If one microphone was fully faded up, with the others faded out, there was no sacrifice of signal-to-noise ratio, nor was there if more than one was fully faded up, but with all knobs at less than the maximum setting, the signal-to-noise ratio became significantly inferior to what can be achieved by using a separate amplifier for each microphone and doing the mixing at a higher level. However, by judicious use of the main gain control on the OBA/8 amplifier in conjunction with the mixer controls, excellent results could be obtained, though this procedure was less than ideally convenient for the operator.

A further feature of this low-level mixing scheme is that the level of the contribution from one microphone is reduced when further microphones are faded up.

Before the advent of the OBA/8 equipment, rather surprisingly in retrospect, the control of the dynamic range of outside broadcasts had normally been done in the main control room, the gain of the OB amplifier being kept constant. This method of working was far from optimum with respect to OB amplifier distortion and land-line signal-to-noise ratio, and the much sounder policy of effecting the dynamic-range control at the OB point was introduced when the OBA/8 amplifier came into use. The built-in peak programme meter enabled this to be done in an optimum manner.

The development of mains-operated equipment of the OBA/8 type, plus the very reassuring experience gained by using it in numerous temporary war-time studios, resulted in the abandonment after the war of the centralized control-room location of microphone or 'A' amplifiers and the post-fader 'B' amplifiers. The studio then became a self-contained unit so far as amplifier equipment was concerned. In the earlier post-war years of austerity, OBA/8 and APM1 amplifiers continued to be widely used, pending the introduction of new and more versatile equipment specifically intended for studio installations. Thought had in fact been given to the design of such apparatus during the later part of the war, leading to what became known as Type A studio equipment.[13,14] Though this began its service trials in December 1944, it was not until the mid 1950s that it had replaced virtually all earlier equipment in studios.

The Type A equipment, which was still based on the pre-war ACSP/3 pentode valve, had relatively small amplifier modules, mounted on sliding shelves in lockable cabinets in such a way as to be easily removable for servicing.

A new feature was that a separate 'A' amplifier was used for each microphone or gramophone source, low-level mixing being abandoned. A jack field, accessible without unlocking the main cabinet doors, enabled the output of each 'A' amplifier to be fed to any chosen fader, to suit operational requirements.

With the amplifiers at a distance of some metres from the control desk, the feeds to and from the faders had to be at low impedance, balanced stud-type constant-impedance bridged-T faders being employed, with 600 Ω input and output transformers in all amplifiers – a clean technique, although rather expensive.

The use of constant-impedance faders fed in this way from 600 Ω source impedances allowed passive mixing to be achieved simply by parallel-connecting the fader outputs, and the scheme avoided the interaction between controls that was a characteristic of the simple series-fader arrangement as carried out with OBA/8 amplifiers.

To secure a good output signal-to-noise ratio, with a considerable number of inputs and allowing some control range to spare – so that none of the faders would normally need to be set at maximum – the individual microphone or 'A' amplifiers had to have sufficient gain to supply a fairly high input level to the faders.

The power-supply switching for the amplifiers was controlled by push-buttons on the control desk and standby 'A' and 'B' amplifiers were provided which, by operating the appropriate desk keys, could be instantly substituted for any amplifier unit that had gone faulty.

Many changes have occurred in the design of microphone amplifiers and associated equipment since the first post-war designs mentioned above came into service. These changes have resulted not only from far-reaching developments in the general field of electronic technology but also from changing attitudes concerning the style of presentation of many kinds of sound programme. The introduction of stereo broadcasting, which began on a regular programme basis in July 1966, has also of course had a major influence on equipment design.

The general trend, for better or for worse, has been in the direction of far greater system complexity, coupled with a requirement that the equipment should be of more flexible design than in the past, allowing it to be more readily adapted to changing operational needs.

The first of the new and more economical equipment introduced into BBC studios from about 1954 to replace Type A was known as Type B,[15] whose main new features were:

(1) Miniature components used throughout, including CV455 (inter-services equivalent of ECC81 or 12 AT 7) 9-pin glass-based double-triode valves in all stages. Each pair of triodes required less than half the heater power of a single AC/SP3 pentode, permitting economical push–pull output stages to be employed. This eliminated DC polarization of the cores of the transformers used to provide balanced 600 Ω outputs, leading to a substantial reduction in the size and cost of these transformers.

(2) The amplifiers were built as small plug-in rack-mounting units of standardized size, so easily removable for servicing that the jack sockets or meters for anode-current checking incorporated in earlier amplifiers were deemed unnecessary and were omitted.

(3) Plugging of specific microphone lines to specific faders carried out on a jack field preceding the microphone amplifiers instead of coming after them as in Type A equipment.

(4) Control-desk modules built on standard-sized sub-panels, permitting a variety of functionally different control desks to be quickly assembled to suit particular operational requirements, and altered if these requirements changed.

This Type B equipment still followed the philosophy of accommodating the electronics on racks or cabinets separate from the control desk.

The pace of development in electronic technology towards the end of the 1950s was such that both the above studio equipment and also a fairly new valve design of outside broadcast equipment, the OBA/9,[16] began to seem obsolescent fairly soon after going into service. This was mainly because, by about 1955, germanium PNP junction transistors had reached the stage at which their characteristics were good enough to enable microphone and other amplifiers to be designed with a performance that would be acceptable for high-quality broadcasting purposes.

The BBC reacted quite quickly to this development and by May 1957 prototype circuits for transistor outside broadcast amplifiers and peak programme meters were undergoing service trials. The same circuit design work also led to what became known as Type C studio equipment,[17,18] which underwent trials in two drama studios in Broadcasting House Extension, opened in 1961. The units operated off a single mains-derived stabilized supply voltage of 24 V, positive earthed. The microphone amplifier noise figure was initially about 5 dB, which is at least 3 dB worse than for a good valve amplifier, but this situation soon improved as lower-noise transistors became available.

A further most significant feature of this new Type C studio equipment was that it adopted the now almost universal practice of accommodating the amplifiers within the control desk. One of the arguments against doing this in the valve era was that control desks are subjected to a certain amount of mechanical vibration when in use, which may include accidental kicking, and this would be liable to produce audible sounds due to valve microphony – transistors, however, are not appreciably microphonic.

Incorporating the amplifiers in the desk, in addition to saving floor space, had the advantage of eliminating the long cable runs to faders that were previously necessary, and this made it possible to use good-quality carbon potentiometers in place of expensive stud-type constant-impedance balanced faders and associated transformers.

The Type C equipment included 'means of manipulating the frequency response of a microphone channel to obtain the effects required in the production of some kinds of light music'.[18] A bass and treble negative-feedback tone-control circuit, which I published in 1952,[19] was employed, plus a tuned circuit to give an optional 'presence boost' of 3 or 6 dB centred on 2.8 kHz.

The provision of the above controls was indicative of a new trend in attitudes, largely instigated by the pop recording industry, and it has continued to the present day. Such frequency-response manipulation is often called equalization or 'EQ'. Logically, however, the word 'equalization' implies correction of a non-flat frequency response, for example of a land-line, or a tape replay head, or the bass rise given by a directional microphone on close speech, so that its use in relation to tone controls, presence circuits etc., whose function is artistic (sometimes inartistic!) seems rather unfortunate.

Another trend in control desk design, for both mono (largely TV) and stereo, has been to have much larger numbers of microphone channels than in the past – often well in excess of thirty – and this has been facilitated by the development of small channel modules, using either discrete transistors or integrated circuits and each including the basic microphone amplifier, plus the frequency-response manipulating controls, the routing switches etc. and the slider-type fader.

The BBC Type C equipment, which was monophonic and had rotary faders, came before the above-mentioned modern type of channel-module format had been evolved, but it represented a most praiseworthy pioneering stage in the development of simple yet elegant circuit and constructional techniques for transistor sound equipment.

Figure 8.6 shows a Type C microphone amplifier unit and its circuit diagram as given in Reference 18. The diagram is drawn on a 'collectors upwards' basis, even though this necessitates having the negative supply line at the top and conventional current flowing upwards. An issue is here involved that was highly controversial at the time – I strongly favoured the 'positive upwards' convention.[20] However, this issue largely resolved itself later when the NPN silicon planar transistor became the dominant active device – diagrams could then be both 'positive upwards' and 'collectors upwards' and yet have the earth line at the bottom.

The 10 kΩ potentiometer shown in Figure 8.6(b) was mounted externally to the unit and connected by pins on the plug-in connector at the back. In some applications it might be the channel fader, whereas in others it might be replaced by an external pair of resistors to give a fixed gain, or by a switched network of gain-setting resistors.

The work on the Type C equipment was followed by further BBC work which led to new transistor studio equipment with even better performance, known as Type D,

The advent of stereo during this period had a profound effect on BBC studio equipment activities, as might be expected. At first the fairly small number of stereo programme contributions were handled by making a few special stereo desk units for use in association with Type B valve amplifiers. These had quadrant-type faders mounted as 'outrigger units' on the desk surface in front of the other controls.

Though the above-mentioned Type D equipment was initially conceived on a monophonic basis, it was nevertheless adapted to full stereophonic operation and was installed, for example, in the sound radio studios at Pebble Mill, Birmingham, which opened in 1971.

The arrival of stereo, in combination with other influences, both technical and political, has tended to shift most of the detailed design work on new sound equipment in recent years from the BBC to industry.

The proliferation of independent recording studios employing multi-track recording techniques, allied with the great increase in the number and complexity of the sound facilities connected with TV and sound broadcasting, both national and local, has brought into being a large and flourishing industry for the

(a)

Figure 8.6 (a)   *Type C microphone amplifier unit.*

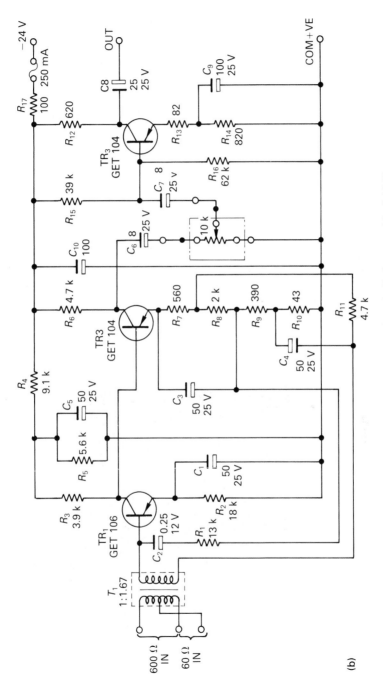

Figure 8.6 (b) Circuit of Type C microphone amplifier unit as given in Reference 18 of 1963.

(b)

manufacture of control desks, or mixer desks as they are now more frequently called.

The faders in these commercial desks normally have conductive-plastic elements, giving smooth and continuous variation of gain, and are of the linear-motion rather than quadrant type. The mixing is done on a virtual-earth basis,[21] which effectively eliminates any interaction between controls.

The BBC, of course, still has complex and exacting requirements for new sound equipment, both for studios and for outside broadcasts, but its role in fulfilling these needs has tended to become more one of drawing up specifications and carrying out approval testing on the resultant commercial products than in doing the major part of the detailed design work.

Though the above account has been closely related to BBC practice – largely because this is so well documented – the trends outlined are broadly representative of developments throughout world broadcasting and recording organizations.

8.3 Some facts about random noise

Though man-made interference is sometimes included within the meaning of the term noise, only truly random noise of natural origin – usually perceived as a gentle hiss or mush in the background – is considered in this Section, which presents some basic noise concepts underlying the design and use of microphone amplifiers.

8.3.1 Noise waveforms

Random noise waveforms normally have Gaussian statistical characteristics, and one consequence of this is that for 99.7% of the time the peak instantaneous value does not exceed 3 times the r.m.s. value.

Another fact about Gaussian noise is that if it is measured on an ordinary mean-rectifier a.c. voltmeter, the true r.m.s. voltage is $2/\sqrt{\pi}$ or 1.128 times the scale reading. (If this voltmeter is of the electronic variety, care must be taken to ensure that its circuitry is not overloaded by instantaneous inputs up to at least 3 times the r.m.s. input.)

An entirely separate attribute of random noise is its frequency spectrum, which shows how the mean squared noise voltage or current per unit bandwidth varies with frequency. If this spectrum is flat over the frequency band of concern, then the noise is called white noise, by analogy with white light.

The digitally generated white noise tracks on test CDs are liable to be non-Gaussian, the probability density function being somewhere between Gaussian and rectangular.

When two random noise voltages V_{N_1} and V_{N_2} are produced by totally independent systems or components, they are said to be uncorrelated, and when they are added electrically the total r.m.s. value V_{N_T} is given by

$$V_{N_T} = \sqrt{(V_{N_1}^2 + V_{N_2}^2)} \tag{8.1}$$

If, on the other hand, V_{N_1} and V_{N_2} are derived from the same original source, via equal time delays if these are appreciable, then they are said to be 100% correlated and their sum is obtained by simple algebraic addition, as for any other pair of voltages with identical waveforms.

8.3.2 Johnson noise

One cause of random noise in microphone amplifier equipment is thermal agitation or Johnson noise, produced in every resistor or other resistive element. It may be represented by a noise voltage V_N acting in series with the resistance and given by

$$V_N = \sqrt{(4kTBR)} \tag{8.2}$$

where V_N = r.m.s. noise voltage (V)
$\quad\quad\quad k$ = Boltzmann's constant, 1.38×10^{-23} (J/°C)
$\quad\quad\quad T$ = absolute temperature (K)
$\quad\quad\quad B$ = effective bandwidth over which the noise is measured (Hz)
$\quad\quad\quad R$ = resistance (Ω)

It is helpful to remember one specific result, such as that $V_N = 1.0\,\mu\text{V}$ r.m.s. for $R = 3\,\text{k}\Omega$, $B = 20\,\text{kHz}$ and $T = 300\text{K}$ (27°C).

The Johnson noise current that flows when a pure resistance R is short-circuited is given by

$$I_N = \sqrt{\left(\frac{4kTB}{R}\right)} \tag{8.3}$$

where I_N = r.m.s. noise current (A), and the other quantities are as before.

8.3.3 Shot noise

Another cause of random noise in valves and semiconductor devices is shot noise, which arises because the currents flowing within them involve the rapid* and mutually independent passage of very large numbers of identical electronic charges, each producing a tiny output pulse of very short duration. Like Johnson noise, shot noise is normally both Gaussian and white, and its magnitude is given by

$$I_N = \sqrt{(2qI_{dc}B)} \tag{8.4}$$

where I_N = r.m.s. shot-noise current (A)
$\quad\quad\quad q$ = electronic charge, 1.60×10^{-19} (C)
$\quad\quad\quad I_{dc}$ = d.c. current (A)
$\quad\quad\quad B$ = noise bandwidth (Hz)

8.3.4 Flicker noise

In addition to Johnson noise and shot noise, practical amplifying devices also produce flicker noise, whose spectral density is inversely proportional to frequency – in other words, the mean squared noise voltage or current per unit bandwidth, instead of being constant as for white noise, varies as $1/f$, rising, therefore, at 10 dB/decade or 3 dB/octave with falling frequency. Flicker noise is otherwise known as $1/f$ noise, pink noise, or excess noise, and for such noise it is the mean-squared noise voltage per octave that is independent of frequency.

*In great contrast, the charges in copper conductors normally move at less than 1 mm/s.

Flicker noise is greatly influenced by slight imperfections in manufacturing techniques and varies much more from sample to sample than does shot noise. Advances in technology, however, have brought flicker noise under much better control than was once the case.

The noise output component from a microphone amplifier due to flicker noise, though not white, is normally Gaussian.

8.3.5 Popcorn noise

Another type of noise sometimes exhibited by semiconductor devices is so-called popcorn or burst noise, whose waveform is far from being Gaussian – as the name suggests, it is of a sporadic or pulse-like nature. With earlier transistors and op. amps. it was fairly frequently encountered, but improved manufacturing techniques have made it comparatively rare nowadays.

8.3.6. Representation of total amplifier noise

Ignoring popcorn noise, the random noise generated inside a microphone amplifier is due to a combination of Johnson, shot and flicker noise, but may always be correctly represented by equivalent voltage and current noise generators placed right at the input of an imaginary noiseless amplifier with other characteristics identical to those of the actual amplifier, as shown in Figure 8.7.

With the input terminals short-circuited, only V_N produces any noise output from the amplifier; with the input open-circuited, V_N has no effect but I_N now gives rise to a noise output by feeding its current into the amplifier input impedance Z_{in}, thus producing an input voltage.

With a signal source of finite impedance connected, both V_N and I_N produce noise output, and their fixed values, plus the degree of correlation between them, can always be chosen to give perfect simulation of the manner in which the noise output due to the amplifier's internal noise mechanisms is influenced by the impedance of the signal source. Additional Johnson noise is, of course, produced by the signal source itself – unless this is purely reactive, or at zero absolute temperature. The correlation between the V_N and I_N noise generators is normally very slight in audio-frequency amplifiers and can be ignored.

V_N and I_N at the input of a well-designed microphone amplifier are mainly due to the input transistors themselves, augmented as little as possible by Johnson noise from associated resistors and later amplifying stages.

Just as the noise produced by a complete amplifier may be accurately represented by voltage and current noise generators right at the input, so also may be the noise produced by an isolated transistor or op. amp.

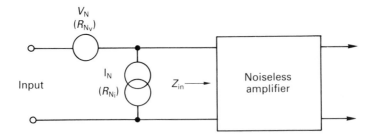

Figure 8.7   *Equivalent voltage and current noise generators representing the total noise produced within an amplifier.*

8.3.7 The R_{N_v} and R_{N_i} concept[22,23]

In data sheets for semiconductor devices, information on voltage and current noise is often given in nV/√Hz and pA/√Hz respectively. These data-sheet quantities are the values of $V_N/\sqrt{B}$ and $I_N/\sqrt{B}$ at the device input, and are sometimes denoted by the symbols e_n and i_n.

As will become apparent, there are very real advantages to be gained by expressing device noise, and also total amplifier noise, in the alternative form of equivalent room-temperature Johnson-noise-producing resistance values R_{N_v} and R_{N_i}, and from equations (8.2) and (8.3) it follows that

$$R_{N_v} = \frac{[V_N/\sqrt{B}]^2}{4kT} \tag{8.5}$$

$$R_{N_i} = \frac{4kT}{[I_N/\sqrt{B}]^2} \tag{8.6}$$

Figure 8.8 presents the relationships (8.5) and (8.6) in a convenient form, enabling the R_{N_v} and R_{N_i} values, corresponding to given data sheet nV/√Hz and pA/√Hz figures for $V_N/\sqrt{B}$ and $I_N/\sqrt{B}$, to be quickly read off.

R_{N_v} and R_{N_i} do not represent actual resistances appearing in the circuit – they merely constitute a convenient method for expressing the magnitudes of the noise voltage and noise current generators, as shown in Figure 8.7. V_N is the room-temperature open-circuit Johnson-noise voltage produced by R_{N_v} and I_N is the short-circuit Johnson-noise current produced by R_{N_i}. Representing voltage noise by an equivalent resistance R_{N_v} was a well-established notion in the valve era, but the use of R_{N_i} to represent semiconductor current noise seems to have been first proposed in 1966 by Dr E. A. Faulkner of Reading University,[23] and its more widespread adoption is much to be recommended.

If only Johnson noise and shot noise were present within the devices, then R_{N_v} and R_{N_i} would be independent of frequency throughout the whole audio band,

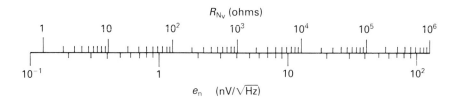

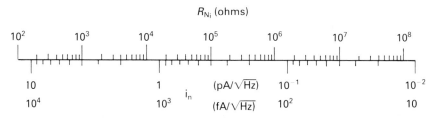

Figure 8.8 *Conversion scales (20°C).*

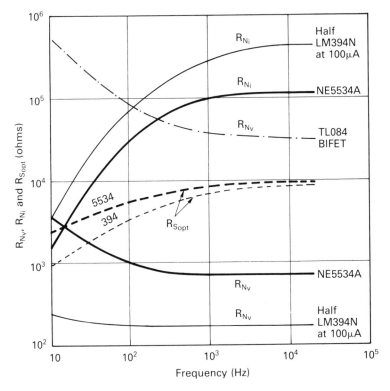

Figure 8.9 R_{N_v}, R_{N_i} and $R_{S_{opt}}$ *curves for representative devices.*

but, as shown in Figure 8.9, flicker noise in ordinary bipolar transistors causes a reduction in R_{N_i}, corresponding to an increase in the pA/√Hz figure, below a certain frequency, and at a normally much lower frequency still, a rise in R_{N_v}.

With field-effect transistors (FETs) on the other hand, flicker noise shows up mainly on the R_{N_v} curve, with negligible reduction, even at 20 Hz, in the very high R_{N_i} value. See TL084 curve in Figure 8.9.

8.3.8 Noise figure

An important concept when amplifiers are fed from passive resistive signal sources, to which moving-coil and ribbon microphones approximate, is that of noise figure, which is usually expressed in decibels and defined as in equation (8.7):

$$NF = 10 \log_{10} \left[\frac{\text{Total noise output power from amplifier}}{\text{Noise output power due to source Johnson noise alone}} \right]$$

(8.7)

Since the output power is proportional to the mean-squared output voltage, an alternative definition is

$$NF = 10\log_{10}\left[\frac{\text{Total mean-squared noise output voltage}}{\text{Mean-squared noise output voltage due to source Johnson noise alone}}\right]$$

(8.8)

Neglecting any slight correlation between V_N and I_N, it may be shown that*

$$NF = 10\log_{10}\left[1 + \frac{R_{N_v}}{R_S} + \frac{R_S}{R_{N_i}}\right]$$

(8.9)

In this equation R_S is the source resistance, and R_{N_v} and R_{N_i} are the values applying to the complete amplifier – they may or may not be approximately equal to the values for the input device in isolation, depending on the circuit design details.

It is evident from the form of equation (8.9) that there must be an optimum value of R_S which will give a minimum noise figure, and this is given by

$$R_{S_{opt}} = \sqrt{(R_{N_v} R_{N_i})}$$

(8.10)

(Alternatively the value of $R_{S_{opt}}$ in kilohms may be obtained by dividing the data-sheet nV/√Hz figure by the pA/√Hz figure.)

It is also apparent from equation (8.9) that for this optimized noise figure to be low, as normally desired, R_{N_i} must be much larger than R_{N_v}, the actual relationship being

$$NF_{opt} = 10\log_{10}\left[1 + 2\sqrt{\frac{R_{N_v}}{R_{N_i}}}\right]$$

(8.11)

These facts may be related to features of Figure 8.9 as follows:

(1) The broken-line curves representing $R_{S_{opt}}$ come exactly midway between the corresponding R_{N_v} and R_{N_i} curves.
(2) Wide spacing of the R_{N_v} and R_{N_i} curves makes the noise figure very low when $R_S = R_{S_{opt}}$, and tolerably low over a very wide range of R_S values.

For the NE5534A, NF_{opt} at 1 kHz, from equation (8.11), is 0.7 dB, corresponding to $R_{S_{opt}} = 8.7\,k\Omega$, and from equation (8.9) the noise figure is 2 dB or better for any value of R_S from 1.3 kΩ to 57 kΩ. For comparison, the values for one half of an LM394N super-matched pair, at $I_{dc} = 100\,\mu A$, are $NF_{opt} = 0.2$ dB for $R_{S_{opt}} = 6.8\,k\Omega$, and NF = 2 dB or better for R_S values from 310 Ω to 150 kΩ.

For early germanium transistors, the R_{N_i} and R_{N_v} curves were much less widely spaced, making it essential, for a tolerably good noise figure, to operate with R_S values not departing very far from $R_{S_{opt}}$. One reason for the relatively poor performance of these transistors is that the collector-to-base leakage current, I_{c_o},

---

*The presence of Z_{in} (noiseless) in Figure 8.7 affects signal and noise equally, so has no influence on the noise figure. It is convenient to assume either $Z_{in} = 0$ or $Z_{in} = \infty$ in deriving equation (8.9).

often gave a substantial increase in current noise. With silicon transistors, however, such leakage currents are normally quite negligible.

With ordinary op. amps. the operating current of each input transistor is, of course, predetermined (it is about 160 μA for the NE5534A), but when discrete transistors are used, a suitable current must be decided upon by the circuit designer, and this has a large influence on the values of R_{N_v} and R_{N_i} obtained.

8.3.9 Theoretical prediction of transistor noise

Noise information, sometimes only very scanty, and unfortunately often rather erroneous, is given in various forms in manufacturers' data sheets, but it is helpful to appreciate that at frequencies high enough for flicker noise to be neglected, the approximate values of R_{N_v} and R_{N_i}, and the manner in which they vary with operating current, can often be satisfactorily estimated on a theoretical basis.

Although the expressions involved are a little cumbersome at radio frequencies, where operating currents are liable to be quite high, nevertheless at the relatively low currents and frequencies involved in microphone amplifiers, the following very simple equations are applicable with adequate accuracy:

$$R_{N_v} = r_{bb'} + \frac{1}{2g_m} \tag{8.12}$$

$$R_{N_i} = \frac{2\beta_{dc}}{g_m} \tag{8.13}$$

where R_{N_v} = equivalent resistance representing voltage noise (kΩ)
R_{N_i} = equivalent resistance representing current noise (kΩ)
$r_{bb'}$ = base spreading resistance (kΩ)
g_m = mutual conductance of the intrinsic transistor (mA/V)
β_{dc} = d.c. current gain, I_c/I_b.

Figure 8.10 illustrates the significance of equations (8.12) and (8.13). The base spreading resistance $r_{bb'}$ comes between the base lead b and the effective base terminal b′ of the 'inner ideal transistor' or 'intrinsic transistor' as it is usually called. The mutual conductance of the intrinsic transistor is given by

$$g_m = \frac{qI_c}{kT} \tag{8.14}$$

where I_c = collector current (A), the other quantities being as given under equations (8.2) and (8.4). g_m is 40 mA/V at $I_c = 1$ mA.

The terms $1/2g_m$ and $2\beta_{dc}/g_m$ in equations (8.12) and (8.13) account for the collector-current and base-current shot noise respectively. (Though shot noise is given basically by equation (8.4), which involves q, q has been eliminated from equations (8.12) and (8.13) by exploiting equation (8.14), thus allowing (8.12) and (8.13) to be given in a form often more convenient for practical purposes.)

The Johnson noise produced by $r_{bb'}$ has been allowed for in Figure 8.10 by increasing the value of R_{N_v} correspondingly, $r_{bb'}$ then being regarded as a noiseless resistance.

R_{N_v} and $r_{bb'}$ can obviously be shown in the opposite sequence if preferred, with R_{N_v} immediately next to the b terminal, but the R_{N_i} noise generator, unlike that shown in Figure 8.7, inevitably does not come right at the input. However, the

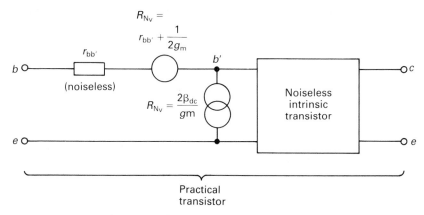

Figure 8.10 *Voltage and current noise generators as given by equations (8.12) and (8.13).*

presence of the noiseless resistance $r_{bb'}$, whose value seldom exceeds about 300 Ω, in series with the input, has a quite negligible effect on the noise performance under all normal microphone amplifier conditions, since the noise voltage-drop across it due to current from R_{N_i} is insignificant compared with the R_{N_v} noise voltage. This is true even at very low frequencies, where flicker noise reduces the value of R_{N_i} and therefore increases the current it produces.

Thus both R_{N_v} and R_{N_i}, as given by the simple equations (8.12) and (8.13), can in practice be taken as coming right at the input as shown in Figure 8.7, without significant error, the noiseless resistance $r_{bb'}$ being simply ignored.

The curves in Figure 8.11 are based on equations (8.12), (8.13) and (8.14), taking $r_{bb'} = 300\,Ω$ and $β_{dc}$ (or h_{FE}) = 300, these values being fairly typical for an input transistor such as the BC 109.

An alternative equation to (8.13) for R_{N_i}, which may sometimes be more useful in practice, may be obtained by putting I_{dc} in equation (8.4) equal to the d.c. base current I_b, or bias current as it is usually called with op. amps. This gives I_N, which may then be expressed as an equivalent R_{N_i} value using equation (8.3), the basic result being equation (8.15):

$$R_{N_i} = \frac{2kT}{qI_b} \qquad (8.15)$$

A convenient version of this equation, for $T = 293\,K$ or 20°C, is

$$R_{N_i} = \frac{50.5}{I_b} \qquad (8.16)$$

where R_{N_i} is in kilohms and I_b is in microamperes.

The NE5534A provides an illustrative example of the application of equation (8.16). The data sheets give a bias current value of 0.5 μA, and substituting this in equation (8.16) gives $R_{N_i} = 101\,kΩ$, which agrees well with the high-frequency end of the R_{N_i} curve in Figure 8.9, where the effect of flicker noise is negligible. This curve was derived from the data sheet curve for input noise current density.

A warning is called for in relation to equations (8.15) and (8.16), because some op. amps. employing bipolar input transistors achieve low bias current values, not by operating these transistors at very low base current, but by means of additional bias-current-neutralizing circuitry. This scheme, however, does not reduce the base-current noise – indeed it adds further uncorrelated noise current – so that the value of R_{N_i} for such op. amps. is very much less than is given by the above equations. Op. amps. of this type are not ideally suited to microphone amplifier applications.

8.3.10 'Low-noise transistors'

The term 'low-noise transistor' is used in the literature with two distinct meanings.

According to the first meaning, a low-noise transistor is one whose voltage noise is low, i.e. which has a low value of R_{N_v}, implying that $r_{bb'}$ is also low. Such a transistor might be used without an input transformer to amplify the output of a 30 Ω microphone, for example, and yet give an acceptable noise figure. Under such conditions the source resistance is usually much less than $R_{S_{opt}}$, so that only R_{N_v} rather than R_{N_i} is of real significance.

The requirement for low $r_{bb'}$ can be met by choosing a fairly large-area type of transistor, as used in the driver stages of audio power amplifiers. For example, types BFX85 (NPN) and BC461 (PNP) have an effective noise $r_{bb'}$ value in the

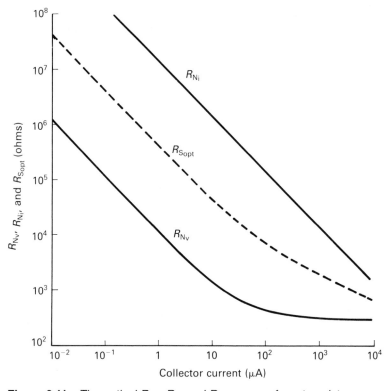

Figure 8.11 *Theoretical R_{N_v}, R_{N_i} and $R_{S_{opt}}$ curves for a transistor assumed to have $r_{bb'} = 300\ \Omega$ and a constant β_{dc} (or h_{FE}) of 300.*

region of $15\,\Omega$, compared with maybe $200\,\Omega$ for small transistors such as the BC109.

An alternative is to connect n small transistors in parallel, which divides the effective values of $r_{bb'}$, R_{N_v} and R_{N_i} by n if the transistors are identical and each is operated at the same current as the original single transistor. Some transistors, such as the National Semiconductors LM394N supermatched pair, exploit integrated-circuit technology and have a large number of paralleled transistors built in – 50 per section in the example just quoted.

For any single transistor, flicker noise causes the value of R_{N_i} to fall off below a certain corner frequency – see Figure 8.9 – and increasing the working current of such a transistor in order to reduce R_{N_v} always raises this corner frequency, thus degrading the overall noise performance for $R_s = R_{s_{opt}}$. However, if the total working current is increased by adding more transistors in parallel, each passing the same current as the original single transistor, then R_{N_v} and R_{N_i} are reduced without adversely affecting the flicker noise corner frequency.

Flicker noise is not normally much of a problem in the design of microphone amplifiers, largely because the ear has reduced sensitivity at the low frequencies involved, and because the flicker noise is usually well below the studio background noise, whose spectrum also rises at low frequencies. However, liberties should not be taken and it is wise to avoid unnecessarily high current densities.

According to the second meaning of the term a 'low-noise transistor' is one which, when operated at an appropriate current and fed from a source of optimum resistance, gives an exceptionally low noise figure. It is evident from equation (8.11) that the criterion in this case is the ratio R_{N_i}/R_{N_v}, which should be as high as possible. This is equivalent to the requirement that the product (nV/√Hz) × (pA/√Hz) should be as small as possible. Table 8.1 gives some typical values for a selection of transistors and op. amps. R_{N_i} often varies much more

Table 8.1 *Parameters for a range of transistors and op. amps.*

Type	Pol	Case	I_{dc} (mA)	R_{N_v} (Ω)	R_{N_i} (kΩ)	$2\beta_{dc}/g_m$ (kΩ)	R_{N_i}/R_{N_v}	$R_{S_{opt}}$ (Ω)	NF_{opt} (dB)
BC109	NPN	T018	0.1	300	100	150	330	5480	0.45
LM394	NPN	8-pin DIL	0.1	180	260	275	1440	6840	0.22
BFX37	PNP	T018	1.0	90	8.0	10	89	850	0.84
GET106*	PNP	–	0.3	190	3.8	8.0	20	850	1.6
2N4401	NPN	T092	1.0	31	5.5	8.0	177	413	0.61
2N4403	PNP	T092	1.0	28	8.0	11	286	473	0.49
BFX85	NPN	T039	2.5	25	1.5	2.2	60	194	1.0
BC143	PNP	T039	2.5	13	1.6	2.4	123	144	0.72
BC143	PNP	T039	1.0	21	3.2	6.0	152	259	0.65
NE5534	NPN	8-pin DIL		760	100		132	8720	0.70
LT1028	NPN	8-pin DIL		45	16		356	850	0.44

*Input transistor used in 1962 in BBC microphone amplifiers.

from sample to sample than R_{N_v} and is sometimes much lower than the ideal β_{dc}/g_m value because of the influence of $1/f$ noise. Noise below 1 kHz and above 10 kHz was attenuated in making the noise measurements.

From equations (8.12) and (8.13) it is evident that to give the desired high R_{N_i}/R_{N_v} ratio, β_{dc} should be as high as possible and well maintained down to low working currents. The base spreading resistance $r_{bb'}$ should be reasonably low, but at a typical current of 100 μA there is not much advantage in having a value of $r_{bb'}$ below, say, 50 Ω.

An important point to appreciate is that when the source impedance can be correctly optimized, by means of a transformer or otherwise, no improvement whatever in noise figure can be obtained by adding further transistors in parallel – a point of which the designer of at least one commercial microphone amplifier appears to have been unaware!

An optimized noise figure of 1 dB or better is quite readily obtainable without resort to costly input devices.

The description 'low noise' in device data sheets, in addition to the above meanings, usually implies that the devices have been subjected to production tests to eliminate specimens with high flicker or popcorn noise.

8.3.11 Noise criteria for moving-coil, ribbon and capacitor microphones

Modern moving-coil microphones usually have a nominal impedance within the range 150 to 300 Ω, with a typical open-circuit voltage sensitivity of about 0.13 mV/μbar or 1.3 mV/Pa. (1 Pa = 1 N m^{-2}, which corresponds to a sound-pressure level or SPL or 94 dB.)

The passive resistance of the coil constitutes the main component of the impedance over most of the audio spectrum, though this is augmented in mild degree by motional impedance and, at the very highest audio frequencies, by passive inductive reactance.

The motional impedance above about 100 Hz, for good omnidirectional moving-coil microphones, is fairly frequency-independent and resistive, and typically represents about 20% of the total impedance – see curve (a) in Figure 8.12.*

For cardioid moving-coil microphones, the motional component of the impedance varies much more with frequency, and the total impedance usually rises to a peak of at least twice the d.c. resistance at some frequency in the region of 100 Hz – see curve (b) in Figure 8.12.

Despite the above detailed effects, no error that is of much consequence in normal circumstances is introduced if it is assumed that the noise output of a moving-coil microphone, in truly silent surroundings, has a white spectrum with a magnitude corresponding to Johnson noise in the nominal resistive impedance, and the same applies to ribbon microphones – see curve (c) in Figure 8.12.

Assuming a resistive microphone impedance of 200 Ω and a noise bandwidth of 12.5 kHz (this relates to an A-weighting network with unity gain at 1 kHz and for

---

*A level acoustic frequency response requires that the diaphragm should be resistively controlled over most of the audio-frequency band, so that frequency-independent acoustic pressure gives frequency-independent velocity and hence output voltage. If, now, a frequency-independent current is fed to the coil, this will produce frequency-independent force, velocity, and hence motional voltage – in other words a frequency-independent resistive motional impedance. At the highest frequencies the effects of diffraction have to be allowed for in the design, and the use of a Helmholtz resonator technique to augment the bass output by applying pressure to the rear of the diaphragm also complicates matters.

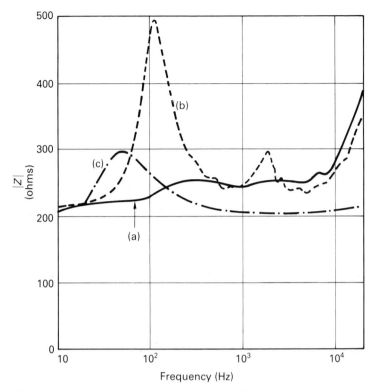

Figure 8.12 *Impedance characteristics for three types of microphone: (a) omnidirectional moving-coil; (b) cardioid moving-coil; (c) ribbon.*

frequencies up to 20 kHz*), then the A-weighted noise voltage at 293 K (20°C) is 0.20 μV r.m.s. With the above-mentioned typical microphone sensitivity of 0.13 mV/μbar, this noise level corresponds to an equivalent noise SPL of 17.7 dBA. Hence if used with an amplifier which achieves a 1 dB noise figure, the actual noise SPL obtained would be 18.7 dBA.

For capacitor microphones the much higher voltage sensitivity of 1 mV/μbar is about average, with a typical output impedance from the internal active circuitry of around 150 Ω. However, the output noise level in silent surroundings is much higher than that of Johnson noise in 150 Ω, and the noise spectrum, unlike that for moving-coil and ribbon microphones, is far from white. The noise output arises mainly from the following causes:

(1) The action on the rear surface of the diaphragm of a frequency-independent thermal agitation noise pressure originating within the viscous air-damping resistance between the diaphragm and the fixed electrode.
(2) Shot and flicker noise associated with the FET drain current, as represented by R_{N_v} in series with the gate lead.
(3) Gate current shot noise as represented by the current generator R_{N_i} between gate and source.

*The 12.5 kHz equivalent noise bandwidth figure is an accurately computed value supplied by John Vanderkooy.

(4) Noise current fed to the gate due to the Johnson noise voltages acting in series with the high-value leak resistors, if present.

A further small amount of electrical noise may in practice be produced by additional transistors or op. amps. and their associated resistors when these are incorporated.

Before considering the quantitative noise performance of practical capacitor microphones, it is perhaps worth pausing to contemplate the fact that the ultimate and irreducible noise level of an imaginary ideal capacitor microphone is determined by the thermal agitation noise pressure associated with the radiation resistance seen by the diaphragm. If this were the only source of noise, the noise performance would be very much better than that of any existing practical microphone and, moreover, it would be independent of the microphone diameter provided this was small compared with the wavelength at all frequencies.* The achievable signal-to-noise ratio with such an ideal microphone would be a characteristic of the acoustical field itself rather than the microphone.[24] The calculated A-weighted noise SPL for frequencies up to 20 kHz is then about –4 dBA.

The above result emphasizes the point that in microphone capsules, just as in low-level electrical circuits, the presence of resistance always degrades the noise performance, and that the amount of degradation present in even the best commercial microphone designs is quite large, since these seldom have a noise SPL as good even as +14 dBA.

In practical omnidirectional capacitor microphones, cause (1) in the above list results in a noise output component that has the same spectrum as the acoustical pressure response, which rolls off at high frequencies if the axial free-field response is flat. This source of noise is normally the dominant one at the higher audio frequencies in a good design, but is liable to be overridden at lower frequencies by noise from the other causes listed. However, with a good enough FET and a leak resistance value of some thousands of megohms – or no leak at all as in many electret-polarized microphones – the noise due to cause (1) may predominate down to frequencies not very much higher than 100 Hz.

Turning now to cardioid and figure-of-eight capacitor microphones, the actuating acoustic pressure difference between the two sides of the diaphragm now becomes progressively smaller as the frequency falls, so that to maintain a level frequency response the diaphragm stiffness has to be made very much less than in an omnidirectional microphone – by a factor well in excess of a hundred in practice. The damping resistance, which has about the same value as in omnidirectional versions, then exerts a dominant control over the diaphragm motion down to quite low audio frequencies, though there is usually an appreciable reduction in response at the extreme bass end.

Though the total signal force is much less at low frequencies than in an omnidirectional capacitor microphone, the noise force due to the viscous damping is about the same and therefore produces a much larger diaphragm amplitude, and corresponding noise output voltage, than in the omnidirectional version. The spectrum of the output noise due to cause (1) in the list now rises with falling frequency at 20 dB/decade, maintaining its dominance over the other causes listed at all frequencies in a good design.

---

*For practical microphones of small size, the equivalent noise SPL, if due predominantly to the internal viscous damping resistance, is reduced by 3 dB on doubling the diameter, assuming the Q-value is kept constant. The noise performance can also be improved by under-damping the diaphragm and correcting the resultant high-frequency peak by electrical equalization.

Hence it is evident that in directional capacitor microphones with the usual type of capsule design, a substantial rise in the noise spectrum with falling frequency is quite unavoidable no matter how good the associated circuit design may be.

The inherently lower low-frequency noise given by omnidirectional capacitor microphones may be reduced even further by employing RF circuitry of enlightened design in place of the ordinary FET scheme.

It is interesting to notice that whereas a rising low-frequency noise spectrum is inherent in practical cardioid and figure-of-eight microphones of the capacitor type, because their diaphragms are resistively controlled, this is not an inherent feature with ribbon microphones, which are basically mass-controlled.

To put the above matters in their proper perspective, it should be added that although obtaining extremely low noise levels at low frequencies is undoubtedly a matter of some academic interest, it is nevertheless of no real practical concern in an ordinary music context, for even the relatively high low-frequency noise level of a good commercial cardioid or figure-of-eight capacitor microphone is normally quite swamped by low-frequency studio background noise, and may be below the threshold of hearing at normal listening levels.

The noise performance requirements for a microphone amplifier are much less exacting if it is to be used only with capacitor microphones than if it is also required to give first-class results, with an input transformer if necessary, when moving-coil or ribbon microphones are employed. That this is so becomes evident from the following example.

As already mentioned, a very good present-day capacitor microphone may possibly have a noise level, expressed as an equivalent SPL, as low as 14 dBA. At a sensitivity of 1 mV/µbar this corresponds to a noise output voltage of 1 µV r.m.s., which is the Johnson noise voltage, in the A-weighting bandwidth of 12.5 kHz, for a resistance value of 5.2 kΩ.

In the above situation, amplifier noise will worsen the overall noise performance by 3 dB if the amplifier R_{N_v} value is 5.2 kΩ, and by 1 dB if R_{N_v} is 1.35 kΩ. The R_{N_i} noise, for any reasonable value of R_{N_i}, will certainly be fairly negligible if the microphone output impedance has the typical value of 150 Ω mentioned previously.

For a good noise performance with a 200 Ω moving-coil microphone used without an input transformer, R_{N_v} would have to be much less than 200 Ω and R_{N_i} would have to be much greater than 200 Ω. For a 1 dB noise figure, possible values would be $R_{N_v} = 26$ Ω and $R_{N_i} = 1.5$ kΩ. If, as might well turn out to be the case, R_{N_v} was not as low as this, then a suitable input transformer would be required for achieving the desired low noise figure – but with or without the transformer, the above 58:1 ratio of R_{N_i} to R_{N_v} is a minimum requirement for achieving a 1 dB noise figure. A very much smaller ratio would suffice with the capacitor microphone.

Thus with capacitor microphones only the voltage noise of the amplifier is likely to be of any real significance, whereas with other types of microphone both voltage and current noise must be taken into account if the best possible performance is to be obtained.

Though microphone amplifier voltage noise can be quoted in nV/√Hz, or as an equivalent R_{N_v} value, a widely adopted practice is to express it in dBu, i.e. in decibels relative to 0.775 V r.m.s.* for some stated or assumed bandwidth such as 20 kHz.

---

*This voltage across 600 Ω gives 1 mW. The 600 Ω figure is a hangover from the early days of telephony, being the characteristic impedance of a typical open-wire telephone line as carried on old-style telephone poles. In America dBv has the same meaning as the European dBu. Another unit is the dBV, or dB(V), which means decibels relative to 1 V r.m.s.

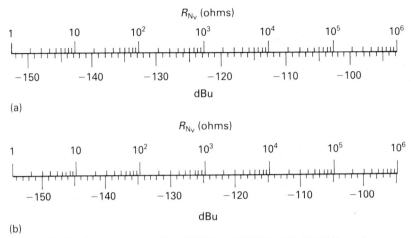

Figure 8.13 *Conversion scales (20°C): (a) 20 kHz, (b) 12.5 kHz noise bandwidth.*

Figure 8.13 provides a convenient means for converting R_{N_v} values to dBu or vice versa, and makes it evident, in relation to the above capacitor microphone example, that if the amplifier voltage noise is −125 dBu or less, it is unlikely to degrade the noise performance of any capacitor microphone by more than 1 dB. Most commercial mixer desks achieve a noise performance at least as good as this.

Though dBus are here being used to express just the voltage noise of the amplifier itself, they are also frequently used to express the total noise in a microphone input circuit with a passive microphone resistive impedance present – this usage is fully explained in Section 8.3.13.

8.3.12 Input impedance, negative feedback and noise

It is important to appreciate the following two points.

Firstly, the application of overall negative feedback to an amplifier does not in itself affect the noise figure at any specific frequency, though the resistors introduced for applying the feedback will generate some extra Johnson noise, which is kept to a minimum in good designs.

Secondly, though the input impedance of an amplifier is often greatly influenced by the application of negative feedback – feedback applied in shunt with the input lowering it and feedback applied in series raising it – nevertheless the value of source resistance, $R_{S_{opt}}$, for optimum noise performance is not affected by the feedback as such.

8.3.13 The 'equivalent input noise' (EIN) concept

As already expounded, the noise performance of a microphone amplifier may be specified very satisfactorily by quoting its R_{N_v} and R_{N_i} values, or the corresponding nV/√Hz and pA/√Hz figures.

Given such information, equation (8.9) then enables the noise figure for any passive resistive source impedance to be easily calculated, and equation (8.10) gives the source impedance required for optimum noise performance. This optimum performance is then as indicated by equation (8.11).

For complete information the frequency dependence of R_{N_v} and R_{N_i} should ideally be included, though the variation in R_{N_v} is usually negligible. In practice, however, single broad-band or spot-frequency values are normally quite adequate.

An alternative and widely used method for presenting the noise performance of a microphone amplifier, or amplifying device, is based on the notion that with some specific value of passive source resistance R_S connected to the input, the total noise output voltage magnitude may be accounted for in terms of a single equivalent input noise (EIN) voltage of white spectrum acting in series with the source.

A change in the value of R_S alters the magnitude of the EIN voltage, both because of the different Johnson noise voltage generated by R_S and also because the value of R_S affects the magnitude of the contribution from the amplifier's current-noise generator – see Figure 8.7. However, the total output noise may still be regarded as being due solely to a single EIN voltage, now of modified value, acting in series with R_S.

The EIN voltage itself may be expressed in several different ways.

It may be given in $nV/\sqrt{Hz}$, or as a noise voltage in a stated bandwidth. Figure 8.14(a) and (b), based on the data sheets for the NE5534 op. amp., embody these two methods, the descriptions above them being those used by the manufacturer.

Figure 8.14(c) is taken from the data sheets for the Trans-Amp balanced-input microphone amplifier module made by Valley People Inc., the EIN voltage here being expressed in dBu, though labelled 'dBv (dB re 0.775 V)' on the American data sheet.

It is evident from the definition of EIN voltage that if the amplifier itself was noiseless, then the EIN voltage would be just the Johnson noise voltage in the source resistance R_S, and it is helpful to include graphs, shown in broken-line, depicting this ideal performance. The vertical spacing between the device curve and the corresponding broken-line graph, at any given value of R_S, then gives the noise figure for that R_S value.* The excellent noise performance of the Trans-Amp when fed directly from a 200 Ω moving-coil microphone will be noticed, this unit having a considerable number of transistors in parallel in each half of its input stage.

8.4 Electronically balanced input circuits and noise

Though microphone amplifier circuits with transformerless balanced inputs differ considerably in design details, the circuit shown in Figure 8.15(a) is representative of good contemporary practice.[25] The value of R_G is varied to provide gain adjustment, sometimes in 10 dB steps. Additional components, here omitted for clarity, are often included to reduce the sensitivity to r.f. interference and to control the rate of attenuation of loop gain so as to give good stability margins.

The resistors labelled R' serve to hold the input circuit at the right d.c. level, but are normally made of sufficiently high value to avoid significant shunting of the signal source.

The negative feedback from each op. amp. output, via R_{fb_1} or R_{fb_2} to the emitter of the associated transistor, ensures that the two collector voltages are at all times

---

*For (a) and (b) in Figure 8.14, NF = 20 log (noise voltage ratio), whereas in (c) the spacing may be related directly to the decibel scaling.

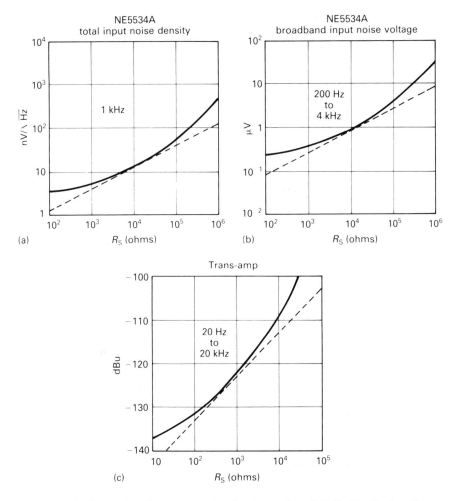

Figure 8.14 *Equivalent Input Noise (EIN) voltage data (20° C). The broken-line graphs represent the source Johnson noise alone.*

held very close to the d.c. bias voltage level $+V_B$, thus keeping the instantaneous collector currents very nearly constant.

When a common-mode input voltage V_{com} is applied to input terminals 1 and 4, voltage changes virtually equal to V_{com} are caused to appear also on points 3 and 6, thus keeping constant the voltages across R_{fb_1} and R_{fb_2} and hence the transistor currents.

The equal voltages on 3 and 6 are applied as a common-mode input to the balun stage involving op. amp. 3, giving zero output from it if the resistor values are accurately matched. Thus the complete Figure 8.15(a) circuit has ideally zero response to common-mode inputs. No voltage appears across R_G for such inputs.

When a balanced input voltage V_{bal} is applied between terminals 1 and 4, the feedback again holds the transistor currents virtually constant, and to achieve this it is now necessary for the feedback circuit to produce a voltage change between the emitters equal to V_{bal}. Hence

$$V_{3\text{-to-}6} \times \frac{R_G}{2R_{fb} + R_G} = V_{bal} \tag{8.17}$$

where $R_{fb_1} = R_{fb_2} = R_{fb}$.

The voltage between points 3 and 6 is subjected to an inverting gain of unity in the balun stage, the final output voltage therefore being given by

$$V_{out} = -V_{bal} \times \frac{2R_{fb} + R_G}{R_G} \tag{8.18}$$

The circuitry within each broken-line enclosure in Figure 8.15(a) may be regarded as constituting an op. amp. with modified characteristics, terminal 1 of the left-hand circuit being the non-inverting input, terminal 2 the inverting input and terminal 3 the output. Representing these modified op. amps. by simple broken-line triangles enables the Figure 8.15(a) circuit to be drawn as at (b).

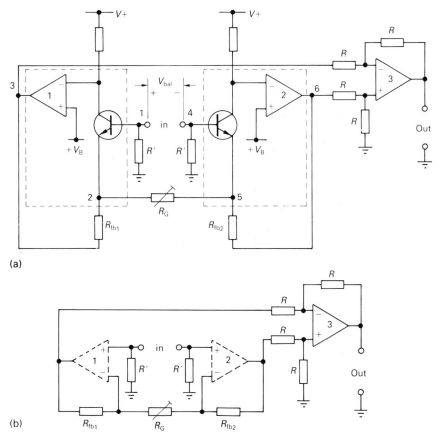

(a)

(b)

Figure 8.15 *(a) Basic circuit of a modern transformerless microphone amplifier with balanced input; (b) simplified version of (a).*

A satisfactory practical microphone amplifier, suitable for feeding straight from $200\,\Omega$ moving-coil and ribbon microphones, could indeed be directly based on Figure 8.15(b), op. amps. such as the LT1028 (see Table 1), which have low voltage-noise, being used in positions 1 and 2.

In recent times, however, a single op. amp., type SSM2015P, intended specifically as a balanced-input microphone amplifier, has become available and has internal circuitry in broad conformity with Figure 8.15(a). This product provides a neat, economical and generally highly satisfactory solution to the microphone amplifier problem, tending, indeed, to render most other solutions somewhat obsolescent.

The noise aspect of electronically balanced microphone amplifiers is interesting and demands careful consideration.

Referring again to Figure 8.15(a), it will be assumed that the input transistors provide sufficient gain to make it reasonable to ignore the noise produced by the op. amps. Thus the significant noise sources are the transistor voltage and current noise generators, Johnson noise in the feedback network, and source Johnson noise.

The high-value resistors R' exert a minor shunting effect across the source, which only very slightly degrades the noise performance, and they also produce a common-mode noise input, to which, however, the system is non-responsive. Hence the influence of these resistors on the noise output can be neglected in practice.

Figure 8.16 represents the essential elements of Figure 8.15 from a noise point of view.

Considering voltage noise first, it is evident that there will be introduced in series with the input circuit, in a balanced manner, a total Johnson noise voltage corresponding to a resistance value of $R_{N_{v1}} + R_{N_{v2}} + R_S + R'_G$, where R'_G is the parallel value of R_G and $(R_{fb_1} + R_{fb_2})$.

R_{fb_1} and R_{fb_2} are made as low in value as is practicable, consistently with the ability of the op. amps. to drive them to a sufficient output level when R_G is set to a low value. Thus the Johnson noise contribution from the feedback network is kept minimal.

The total Johnson noise voltage from the above-mentioned four resistances in series is subjected to a gain of $(R_{fb_1} + R_{fb_2} + R_G)/R_G$.

The noise current generators I_{N_1} and I_{N_2} together produce both a common-mode component of current noise, to which the complete circuit is non-responsive, and also a balanced component flowing round the input circuit. This latter component produces voltages across R_S and R'_G which give rise to a noise

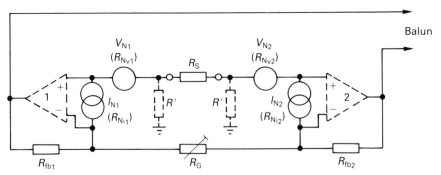

Figure 8.16   *Diagram for explanation of noise performance of Figure 8.15*

output. It is therefore necessary to consider the derivation of these separate noise current components.

Let the instantaneous values of I_{N_1} and I_{N_2} be i_{n_1} and i_{n_2} respectively, both being taken as positive when flowing upwards, say. Whatever the values of i_{n_1} and i_{n_2} may be, another pair of current components i_a and i_b can always be chosen so that, at the time instant concerned, the following relationships are satisfied:

$$i_{n_1} = i_a + i_b \tag{8.19}$$

$$i_{n_2} = i_a - i_b \tag{8.20}$$

These are two equations with two unknowns, so that given the i_{n_1} and i_{n_2} values at any instant, i_a and i_b can be determined.

It is evident that i_a constitutes the common-mode current component, flowing in the same direction in both current generators, whereas i_b constitutes the balanced component which flows round the circuit through R_S and R'_G.

Adding and subtracting equations (8.19) and (8.20) gives

$$i_a = \frac{i_{n_1} + i_{n_2}}{2} \tag{8.21}$$

$$i_b = \frac{i_{n_1} - i_{n_2}}{2} \tag{8.22}$$

Now the noise waveforms of which i_{n_1} and i_{n_2} in equation (8.22) are instantaneous values each have an r.m.s. value of I_N, it being assumed that $I_{N_1} = I_{N_2} = I_N$ in Figure 8.16. However, since I_{N_1} and I_{N_2} are uncorrelated, the minus sign in equation (8.22) is now of no consequence and the r.m.s. value of the right-hand side of the equation is $\sqrt{2}I_N/2$, i.e. the r.m.s. value of the balanced noise current flowing round the circuit due to the noise current generators is $I_N/\sqrt{2}$. The effective value of equivalent current-noise resistance for the complete circuit is therefore $2R_{N_i}$, where R_{N_i} is the value for a single transistor.

From the above, assuming equal R_{N_v} and R_{N_i} values for the two transistors and ignoring the small noise effect of the R' resistors, the noise performance of the Figure 8.16 balanced circuit is the same as that of the unbalanced circuit shown in Figure 8.17(a). Comparison with Figure 8.7 and reference to equations (8.9), (8.10) and (8.11) shows that, for the Figure 8.17(a) circuit

$$NF = 10 \log_{10} \left[1 + \frac{2R_{N_v} + R'_G}{R_S} + \frac{R_S}{2R_{N_i}} \right] \tag{8.23}$$

$$R_{S_{opt}} = \sqrt{[(2R_{N_v} + R'_G)(2R_{N_i})]} \tag{8.24}$$

$$NF_{opt} = 10 \log_{10} \left[1 + 2 \sqrt{\frac{2R_{N_v} + R'_G}{2R_{N_i}}} \right] \tag{8.25}$$

When, as is often the case, the feedback network resistance R'_G is negligibly small, we then have

$$NF \approx 10 \log_{10} \left[1 + \frac{2R_{N_v}}{R_S} + \frac{R_S}{2R_{N_i}} \right] \tag{8.26}$$

$$R_{S_{opt}} \approx 2\sqrt{(R_{N_v} R_{N_i})} \tag{8.27}$$

$$NF_{opt} \approx 10 \log_{10} \left[1 + 2 \sqrt{\frac{R_{N_v}}{R_{N_i}}} \right] \tag{8.28}$$

It is seen from equations (8.27) and (8.28) that though $R_{S_{opt}}$ is twice as high as it would be for a single transistor of the same type, nevertheless use of this higher R_S value gives the same noise figure as for the single transistor.

It is of interest to determine what happens to the noise performance of the Figure 8.15 type of circuit if, instead of operating with a balanced signal source, one end of the source is earthed to give unbalanced operation.

For simplicity it will again be assumed that R'_G is negligibly small. In other words it is being assumed that the currents in the bottom leads of the two current generators in Figure 8.16 flowing into the feedback network produce negligible noise output, the significant effect in normal balanced operation being produced by the balanced component of current in their top leads flowing through R_S. However, when the left-hand end, say, of R_S is earthed, I_{N_1} is virtually prevented from contributing anything at all to the noise output, the observed current-generator-originated noise output now being due almost entirely to the voltage across R_S produced by the full value of I_{N_2} flowing through it.

Voltage noise from $R_{N_{v1}}$, R_S and $R_{N_{v2}}$ is of course still applied between the non-inverting inputs as before. The noise performance of the system is therefore approximately as for the circuit of Figure 8.17(b), and we now have

$$NF \approx 10 \log_{10} \left[1 + \frac{2R_{N_v}}{R_S} + \frac{R_S}{R_{N_i}} \right] \tag{8.29}$$

$$R_{S_{opt}} \approx \sqrt{(2R_{N_v} R_{N_i})} \tag{8.30}$$

$$NF_{opt} \approx 10 \log_{10} \left[1 + 2 \sqrt{\frac{2R_{N_v}}{R_{N_i}}} \right] \tag{8.31}$$

Comparisons of equations (8.29), (8.30) and (8.31) with (8.26), (8.27) and (8.28) shows that earthing one end of R_S to give unbalanced operation has the following effects:

(1) If the value of R_S remains unchanged, the noise figure is worsened.
(2) The value of R_S required for optimum noise performance is reduced by a factor of $\sqrt{2}$.
(3) Even with R_S reduced as in (2), NF_{opt} is still not as good as with balanced operation.

These effects are borne out experimentally, but if the value of R_S used is well below $R_{S_{opt}}$, as is frequently the case in practice, the differences become quite negligible since voltage noise then tends to become dominant and is the same for balanced and unbalanced operation.

An interesting sidelight on the above relates to the ordinary use of op. amps., for example in the virtual-earth manner shown in Figure 8.18(a). The long-tailed-pair input stage of the op. amp. is here being operated in the unbalanced manner referred to above, and the noise performance obtainable is commensurate with the device's R_{N_v} and R_{N_i} (or nV/$\sqrt{\text{Hz}}$ and pA/$\sqrt{\text{Hz}}$) values as normally quoted. However, by changing to the (b) scheme, op. amp. 1 input stage is now operated under proper balanced conditions and a noise performance somewhat better than given by the data sheet figures is obtained.

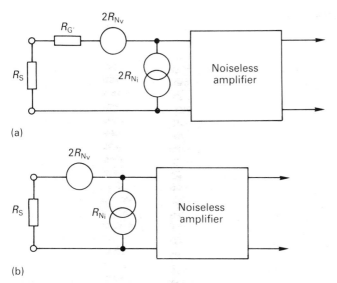

(a)

(b)

Figure 8.17 *(a) Circuit having the same noise performance as the circuit of Fig 8.16; (b) as for (a) but with one end of R_S in Figure 8.16 earthed. $R_{G'}$ in (a) is the small parallel value of R_G and $(R_{fb1} + R_{fb2})$, and was neglected in deriving (b).*

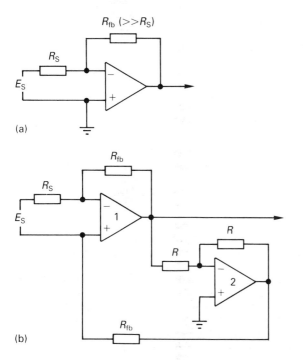

(a)

(b)

Figure 8.18 *Diagrams illustrating a point about op. amp. noise characteristics.*

8.5 Transformers

8.5.1 Introduction

Very few present day professional audio engineers have any detailed under-standing of the theoretical principles and practical design problems relating to audio transformers, and this often creates difficulties in the liaison between users and manufacturers.

While the properties of a textbook 'ideal transformer' are generally well appreciated, difficulties arise in trying to understand the numerous ways in which practical transformers depart from this simple ideal.

When looked into in detail, the behaviour of an audio transformer is actually quite complex – and fascinatingly interesting – but the skilled designer knows, largely from experience, which complications are likely to be significant in a specific context and which can be ignored. Though the complications really constitute shortcomings, which no manufacturer can completely avoid, never-theless in a well-designed transformer, used in the intended manner, none of them produce any audible quality degradation, even under the most searching conditions of subjective assessment. A good transformer is a virtually immaculate device.

Microphone transformers for use with transistor amplifiers do not normally have to step the source impedance up to nearly such a high value, for good noise performance, as was necessary in the valve era, and this has greatly eased the design problem with regard to the effects of leakage inductance, secondary shunt capacitance and inter-winding capacitance.

Indeed, the conditions under which a modern microphone transformer is required to operate are sometimes such that a first-rate performance can be achieved using only the very simplest design procedure, and a design falling into this category will first be considered.

8.5.2 A simple microphone transformer design

Suppose that it is desired to use a pair of $30\,\Omega$ 4038 A ribbon microphones – the commercial equivalent of the BBC 'PGS' design[12,32] – to feed via balanced lines to the unbalanced inputs of a Sony PCM-F1 digital recorder system.

A simple noise test may first be made by connecting resistors of various values to one of the microphone inputs of the PCM-F1 and plotting, on the vertical axis, the square of the amplified noise output voltage against the resistor values horizontally. (The lower audio frequencies, say below $500\,\text{Hz}$, should be attenuated by a simple output $C–R$ filter to subdue possible hum and/or flicker-noise contributions. Alternatively an A-weighting or CCIR filter may be employed.)

The graph obtained is found to be a good straight line below about $2\,\text{k}\Omega$, with an intercept on the horizontal axis at about $-400\,\Omega$, so that the value of R_{N_v} for this microphone amplifier is approximately $400\,\Omega$.

Above $2\,\text{k}\Omega$, however, the graph starts to bend over towards the horizontal, because internal $10\,\text{k}\Omega$ resistors have unfortunately been incorporated across the inputs, considerably worsening the noise performance potentially obtainable from these otherwise excellent microphone amplifiers, which have a junction-FET long-tailed-pair input stage in combination with a separate op. amp. The $10\,\text{k}\Omega$ resistors, R_{103} and R_{203} – incidentally of a type looking more like a ceramic capacitor – should preferably be changed to $330\,\text{k}\Omega$, to give the amplifiers a respectably high R_{N_i} value. I have made such a modification to several PCM-F1 units. The modification is quite pointless, of course, if only capacitor microphones are to be used – see Section 8.3.11.

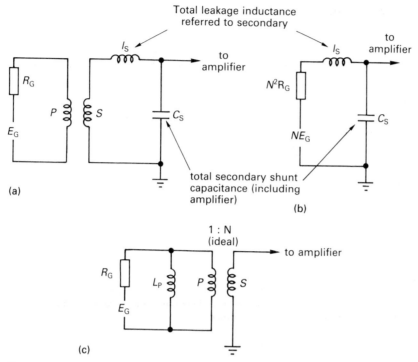

Figure 8.19 *Approximate equivalent circuits for microphone transformer, omitting copper and core losses; (a) and (b) apply at high frequencies, (c) at low frequencies. The ratio N of the ideal transformer is preferably made precisely equal to the actual turns ratio, though some authors depart slightly from this usage.*

If the input transformer has a ratio 1:N, then so far as medium- and high-frequency behaviour is concerned, the effective circuit is approximately as shown in Figure 8.19(a).* When the primary circuit impedance is low, as in the present instance, the effect of primary shunt capacitance may be completely ignored.

By transferring all quantities to the secondary side, the equivalent circuit of Figure 8.19(b) is arrived at. Now the number of turns required on the primary side is determined by low-frequency considerations, as discussed later, so that increasing the step-up ratio N can be achieved only by increasing the number of turns on the secondary winding. This increases the value of the leakage inductance l_S shown in the diagram, which, for a given winding geometry, is proportional to the square of the number of secondary turns.† The value of C_S, on the other hand, is almost unaffected by the number of turns.

*The source resistance and e.m.f. are here given the suffix G, for 'generator', since in a transformer context the suffix S, used for 'source' in the earlier part of this chapter, is now more conveniently employed to denote 'secondary'.

†The concept of leakage inductance is a convenient way to represent the fact that the magnetic coupling between the windings of a practical transformer is always slightly less than 100%, i.e. not quite all the flux that links with the turns of one winding also links with those of the other.

Thus, as the ratio N is increased, the resonance frequency of l_S and C_S falls, and would ultimately come right down into the audio band if N was made excessively high. But even if the resonance occurs at a frequency somewhat higher than the top of the audio band, it may still have an adverse effect on the levelness of response at high audio frequencies, to an extent dependent on the Q-value of the series-tuned circuit formed by the elements shown in Figure 8.19(b).

By careful design with regard to the values of l_S and C_S, and maybe with the addition of further damping in shunt with the secondary, N^2R_G may be given quite a high value without sacrificing uniformity of audio response up to about 20 kHz. Values of N^2R_G up to at least 100 kΩ can sometimes be achieved if C_S can be kept low enough. Such designs are considered in more detail later.

However, if N^2R_G is made not more than several kilohms, even very limited practical experience will make it obvious to the designer that the resonance will then occur at a frequency so far above the top of the audio band that its influence on the response at audio frequencies will probably be negligible. The above complications may then be ignored and the values of l_S and C_S allowed to look after themselves while other aspects of the design are attended to.

First, in the present design, the type and size of core to be employed must be decided upon, and Mumetal, Permalloy C or other commercially equivalent nickel–iron laminations, having about 78% nickel content, are almost universally used in such transformers, because of their very high value of initial permeability.[26] A lamination thickness of 0.38 mm or 0.015 in is normally adopted; thinner laminations would give the transformer a higher value of shunt eddy-current loss resistance (see Section 8.5.6), but the advantage of this in a normal audio transformer context is very small, so that the extra cost is seldom justified.

With regard to the influence of core size on performance, this is considered more fully in Section 8.5.12. Obviously the larger the core size selected, the higher will be the cost of the transformer and its associated screening can. The Type 187 size of lamination shown in Figure 8.20 is economical and enables a good performance to be obtained, especially if a stack thickness of twice the width of the middle limb is adopted, as for the present design.

The core dimensions l_m and A have the meanings shown in Figure 8.21, but in determining the effective cross-sectional area A_{eff}, it is usual to multiply the geometrical area shown by a stacking factor, often taken as 0.9, to allow for surface oxidation of the laminations – paint insulation is now much less used than was once the case. For the present '2 × square' design, the values taken are A_{eff} = 0.72 cm^2 and l_m = 5.1 cm.

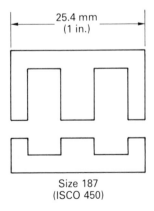

25.4 mm
(1 in.)

Size 187
(ISCO 450)

Figure 8.20 *A widely-adopted lamination size for microphone transformers. Other sizes in use are illustrated in Reference 1.*

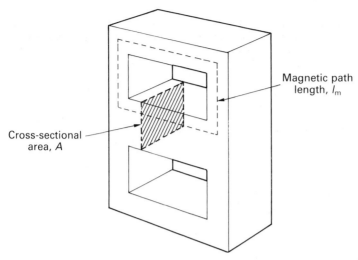

Figure 8.21 *Significance of the quantities A and l_m in a normal transformer core. In some sizes, e.g. Size 21 (ISCO 421), which is sometimes used for larger microphone transformers, the width of the centre limb is not quite double that of the other parts, requiring 'corrected' values of A and l_m.*

The equivalent circuit, at its simplest, for determining the low-frequency performance, is shown in Figure 8.19(c).

If R_G is 30 Ω and the response is to be not more than 1 dB down at 20 Hz, this corresponds fairly closely to –3 dB at 10 Hz, so that the reactance of L_P should be at least 30 Ω at 10 Hz, requiring $L_P = 0.48$ H or more.

Adopting the rationalized MKS, or SI (Système International) system of units, as is now the normal practice, the inductance is given by

$$L = \frac{4\pi n^2 A_{\text{eff}} \, \mu}{10^7 \, l_m} \tag{8.32}$$

where n = number of turns
A_{eff} = effective core cross-sectional area (m²)
l_m = magnetic path length (m)
μ = effective relative or specific permeability.*

To use equation (8.32) to determine the number of turns required, a value for μ must be assumed – that applicable at low frequencies and low-signal levels to a small transformer core of the normal type using interleaved laminations of 0.38 mm Mumetal. This is called the effective initial permeability, μ_i.

Several decades ago it was customary to assume a value of 7000 for the μ_i of an ordinary Mumetal core, but much progress has since been made in improving these nickel–iron alloys by the addition of small amounts of other constituents

---

*It is here convenient to use μ without a suffix, though in more fundamental work μ_s is often used for specific permeability. The quantity $4\pi/10^7$ in equation (8.32) is the permeability of free space, often denoted by μ_0. The product $\mu_0\mu_s$ is then called the absolute permeability and denoted by μ without a suffix. But since only specific or relative permeability is normally referred to in practical transformer design work, it is widespread engineering practice to use μ for this. Other suffixes of a more practical nature may then be added, e.g. μ_i for initial (specific) permeability.

and by improvements in the production technology. Effective values of μ_i well in excess of 20 000 are now regularly found.

In the present design μ_i = 15 000 was assumed. Then substituting the above mentioned values L_P = 0.48 H, A_{eff} = 0.72 cm^2 and l_m = 5.1 cm in equation (8.32) gives just over n = 134 turns. A winding of 136 turns of 32 s.w.g. (0.27 mm) self-fluxing enamelled wire was in fact adopted, and for experimental purposes this was put on as a centre-tapped bifilar winding requiring 68 bobbin revolutions – see Section 8.5.15.

The secondary winding, outside the primary, and similarly occupying the full bobbin width, was made 1150 turns of 42 s.w.g. (0.10 mm) enamelled wire, giving a ratio of 1:8.46 and thus ideally stepping up the nominal 30 Ω primary source resistance to 2.14 kΩ.

A copper-foil screen was placed between the two windings, being separated from each by a layer of approximately 0.1 mm insulation tape. Care was taken to avoid forming a shorted turn – a vital point. For comments on the effects of such a screen, reference should again be made to Section 8.5.15.

Another practical point is that when assembling a laminated Mumetal core, the laminations should be treated gently, rather as if they were made of glass. Stressing them beyond the elastic limit, so as to leave a permanent bend, inflicts serious magnetic damage and can cause a large reduction in permeability.

8.5.3 Simple tests on the above transformer

Resistance of value 31.6 Ω was shunted across the 600 Ω output of a Levell R–C oscillator, to provide a 30 Ω source for feeding the transformer primary. The secondary was fed to an oscilloscope on 10 mV/cm sensitivity, with the oscillator and oscilloscope earth terminals joined. The *outside* of the secondary was taken to the HI input of the oscilloscope, whose input capacitance is approximately 100 pF.

The frequency response was then determined in terms of oscillator attenuator readings for a constant output level of 10 mV peak-to-peak, and was found to be within the limits ±0.25 dB from 40 Hz to 20 kHz, being –1.0 dB at 20 Hz and –2.9 dB at 10 Hz. Thus all seemed well with regard to audio-frequency response.

The d.c. resistances of the windings were measured and found to be r_p = 1.9 Ω and r_s = 151 Ω* Hence, assuming a 30 Ω resistive source and referring all quantities to the secondary side, gives the equivalent circuit of Figure 8.22 when the transformer is connected to the PCM-F1 amplifier. The noise figure is given by

$$ NF = 10 \log_{10} \left[\frac{2150 + 687}{2150} \right] = 1.2 \, \text{dB†} $$

*If such a resistance measurement is made on a transformer after inserting the core, it can leave the core in a sufficiently polarized state to give some reduction in low-level inductance and a noticeable increase in microphony. Ideally it is then desirable to depolarize the transformer by applying a signal current, at say 30 Hz, of sufficient level to give core saturation, indicated by a highly distorted voltage waveform, and then to turn the level smoothly down to zero over a period of many seconds.

†This simplified calculation ignores the noise contribution (≈0.1 dB at significant frequencies) from the eddy-current loss and the PCM-F1 shunt input resistance.

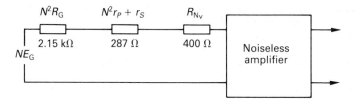

Figure 8.22   *Resistance values for simplified noise calculation of Section 8.5.3. The very small noise contributions from the shunt eddy-current loss, and the amplifier R_{N_i}, are here ignored.*

8.5.4 Tests of a more probing nature

Extending the above frequency-response measurement to higher frequencies gives curve A of Figure 8.23.

A much flatter response for frequencies well above the audio band may be obtained if a $10\,k\Omega$ resistor is shunted across the secondary, as for curve B, which was obtained with the same input level as for curve A. There is also a small improvement in the low-frequency response, the loss at $10\,Hz$ being reduced from $2.9\,dB$ to $2.2\,dB$.

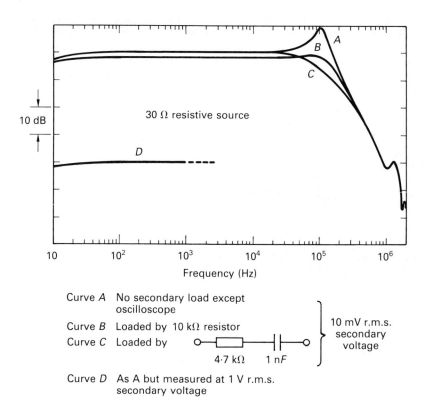

Curve A No secondary load except oscilloscope

Curve B Loaded by $10\,k\Omega$ resistor

Curve C Loaded by $4\cdot7\,k\Omega$ $1\,nF$

Curve D As A but measured at $1\,V$ r.m.s. secondary voltage

$\left.\rule{0pt}{4em}\right\}$ $10\,mV$ r.m.s. secondary voltage

Figure 8.23   *Microphone transformer frequency responses with 30Ω resistive source.*

A disadvantage of such a shunt resistor, which appears between the 287 Ω/400 Ω junction and earth in Figure 8.22, is that it reduces the signal by more than it reduces the noise, increasing the calculated noise figure from 1.2 dB to 2.3 dB. However, this worsening of the noise performance can be avoided if the *equivalent* of a 10 kΩ resistor is obtained by shunt negative feedback via a resistor of much higher value – see Section 8.5.16.

An alternative technique for levelling the high-frequency response, which is widely used, is to shunt a series combination of C and R across the secondary, and curve C was obtained with 1 nF and 4.7 kΩ. This produces negligible worsening of the audio-frequency noise performance with the values here involved, since it is the parallel equivalent resistance value of this series combination that must be compared with the above 10 kΩ value from a noise viewpoint – the parallel resistance is 18 kΩ at 20 kHz, 59 kΩ at 10 kHz, and increases rapidly as the frequency falls further.

The response curves in Figure 8.23 conform closely to what would be expected from the simple equivalent circuits of Figure 8.19 – with damping additions in the case of curves B and C – for frequencies up to about 600 kHz, but above this frequency the rate of attenuation begins to depart from the simple theoretical 40 dB/decade, and above 1 MHz complex behaviour with multiple resonances sets in, as occurs with all transformers if the frequency is made high enough. These resonances involve a combination of leakage inductances between one part of a winding and another, and distributed capacitance. In higher-impedance transformers, complex behaviour is liable to start at a lower frequency.

Figure 8.24 shows 10 kHz square-wave responses corresponding to the three response curves of Figure 8.23. Though waveform A may not be thought to 'look nice', I do not believe that ringing at such a high frequency is ever of any direct subjective significance. Reports, which are not infrequent, that such effects can be perceived, are usually explicable in terms of the amazing power of the human imagination!

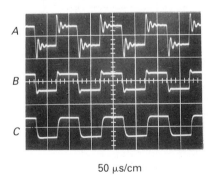

50 μs/cm

A No secondary load except oscilloscope

B Loaded by 10 kΩ resistor

C Loaded by   4·7 kΩ   1nF

Figure 8.24 *Microphone transformer 10 kHz square-wave responses with 30 Ω resistive source.*

However, the addition of secondary damping is sometimes beneficial in ensuring an adequate negative-feedback stability margin in the associated amplifier, and should be regarded as good general practice. See Section 8.5.16.

The response curves *A*, *B* and *C* of Figure 8.23, as already mentioned, were measured at a secondary voltage of only 10 mV peak-to-peak. If the measurement is repeated at a higher level, it is found that the shape of the response curve at low frequencies is altered, curve *D* having been obtained at a constant level of 1 V r.m.s. This happens because the mean slope of the *B–H* loop for the core material is amplitude dependent, as considered in more detail in Section 8.5.7.

The equation from which the peak flux density under sine-wave-voltage conditions may be obtained is

$$\hat{B} = \frac{E_{rms}}{4.44 A_{eff} \; nf} \tag{8.33}$$

where $\hat{B}$ = peak flux density (T)
E_{rms} = induced e.m.f. (V)
A_{eff} = effective core cross-sectional area (m^2)
n = number of turns
f = frequency (Hz)

For 1 V r.m.s. at 10 Hz, with A_{eff} = 0.72 cm^2 or 0.72 × 10^{-4} m^2 and n = 1150, this equation gives $\hat{B}$ = 0.272 T or 2720 G. Waveform distortion is visible below about 13 Hz with a 30 Ω source; 1 V r.m.s. secondary voltage corresponds to a sound pressure level, using a 4038A microphone, of approximately 140 dB.

So far, in accordance with widespread practice, the response curves presented have been obtained using a resistive signal source to simulate the microphone. Microphone impedances, however, as shown in Figure 8.12, are liable to depart fairly markedly, in both magnitude and phase angle, from the nominal resistive value, so that it is really more meaningful to make measurements by injecting the test voltage at very low impedance, preferably in a balanced manner, in series with actual microphones. Section 8.5.18 gives details of a 1000:1 transformer for conveniently carrying out such tests. It is suggested that regular use of such a device would now and then reveal frequency responses by no means as good as had been supposed!

Figure 8.25 shows response curves obtained as above for ribbon and moving-coil microphones, of nominally 30 Ω impedance, feeding the 1:8.46 microphone transformer.

The moving-coil microphone impedance at frequencies below 100 Hz is an almost pure resistance of about 22 Ω, giving an impressively level measured response. The ribbon microphone, on the other hand, rather similarly to that in Figure 8.12, has a broad impedance peak at about 50 Hz, whereas at the lowest frequencies its impedance becomes that of a fairly pure inductance because of the internal transformer; by 10 Hz a simple inductive potential divider effect has become established between this and the primary inductance of the 1:8.46 transformer.

Comparing Figures 8.23 and 8.25, it is seen that the high-frequency resonance comes at a lower frequency when measured with microphones than when measured with a simple resistance source, and this is because of the passive series inductance of these electromagnetic microphones.

Also shown in Figure 8.25 are low-level response curves obtained with a 100 Ω resistive source, under the same three secondary loading conditions as for the microphones; the source e.m.f. was the same for all three curves.

The 100 Ω curves A and C show an interesting feature in that the response from about 200 Hz to 1 kHz shelves upwards by nearly 1 dB. This is because, with a

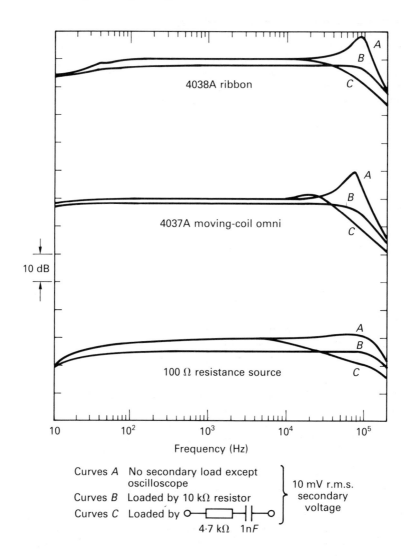

4038A ribbon

4037A moving-coil omni

10 dB

100 Ω resistance source

Frequency (Hz)

Curves *A* No secondary load except oscilloscope
Curves *B* Loaded by 10 kΩ resistor
Curves *C* Loaded by

4·7 kΩ   1nF

10 mV r.m.s. secondary voltage

Figure 8.25 *Microphone transformer frequency responses with microphone and 100Ω resistive sources.*

high source resistance, significant attenuation is produced, at the lower frequencies only, by the shunt eddy-current loss resistance. As discussed in Section 8.5.6, this resistance starts to rise, at a rate proportional to the square root of frequency, above a corner frequency which is in the 200 Hz region for the type of core here used. Such a step is always evident with source resistances high enough to give several decibels of loss at 20 Hz, unless laminations thinner than the normal 0.38 mm are used.

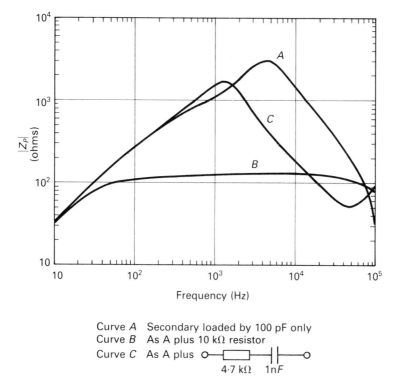

Figure 8.26 shows the results of measuring the primary impedance of the 1:8.46 microphone transformer under the same three secondary loading conditions as before – a 100 pF capacitor was added across the secondary in place of the oscilloscope. The primary was fed with a current of 30 µA r.m.s. via a 100 kΩ resistor, the voltage across the winding being measured via a laboratory amplifier with a flat frequency response.

Curve A actually dips, just off the diagram, to 11 Ω at 110 kHz, due to resonance involving the leakage inductance and the total secondary shunt capacitance. This resonance is brought down to about 50 kHz when the 1 nF plus 4.7 kΩ loading is used.

If a frequency-independent input impedance is desired – and it was vital in the days when low-level series faders were used – then the advantage of simple resistance loading of the secondary is very evident, and Figures 8.23 and 8.25 show that this also gives a somewhat better frequency response under practical working conditions. It is certainly the best scheme to use when the transformer and amplifier can be designed as one entity, enabling the resistance loading to be provided by shunt negative feedback via a resistor of much higher value, thus avoiding the degradation of the noise figure that is experienced when passive resistive loading is employed.*

*An additional passive shunt, consisting of C and R in series, with a very high corner frequency, may also be desirable, in the interests of achieving a good negative-feedback stability margin.

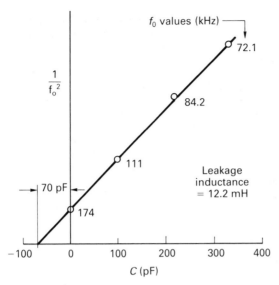

Figure 8.27   *Microphone transformer winding-capacitance and leakage determination.*

A practical method for determining the leakage inductance and effective secondary shunt winding capacitance of transformers such as that here being considered, is to connect several known values of capacitor, C, across the secondary and determine, for each value, the frequency f_0 at which the primary impedance dips to a low value. The core, the screen if fitted, and the primary, should be earthed to the end of the secondary that will be earthed in normal use.

A graph of $1/f_0^2$ (in any convenient units) against capacitor value is then plotted, that for the present transformer being given in Figure 8.27. The winding capacitance is given by the intercept shown. The leakage inductance may then be calculated from the total capacitance corresponding to a point on the graph – well up the graph for best accuracy – and the value of f_0 at that point.

In principle only two points are necessary, and a formula for leakage inductance which avoids the need to draw a graph can then, of course, be easily derived. However, plotting several points and checking that a good straight line is obtained provides reassurance that all is going straightforwardly – complications can sometimes arise with high-impedance transformers.

8.5.5 Modification to provide microphone equalization

Though microphone transformers, and their associated amplifier input circuits, are usually designed with the aim of achieving a flat frequency response at audio frequencies, this approach may sometimes be modified to obtain equalization for the non-ideal acoustic frequency response of a microphone. This turns out to be particularly appropriate with the 4038A ribbon microphone, whose frequency response, though good, falls off somewhat at very low and very high frequencies. The original BBC PGS microphones[12] have a published average axial response, for a large source distance, which is approximately –3 dB at 40 Hz, –2 dB at 10 kHz and −4 dB at 15 kHz. The 4038A commercial version[32] appears to have a high-frequency response not quite as good as this, both from published data and from my own measurements; figures of −2 dB at 10 kHz and −7 dB at 15 kHz are believed to be more nearly relevant.

Referring to the equivalent circuits of Figure 8.19 and the response curves of Figures 8.23 and 8.25, it is evident that what is required for high-frequency equalization is to modify the values so as to bring the resonance peak down to a frequency a little below 20 kHz. Merely increasing the capacitance across the secondary will not do, for the Q-value then turns out to be too low – less than unity. The need is to increase the leakage inductance so that the wanted resonance frequency can be obtained with a smaller value of shunt capacitor. In theory this could be done by increasing the spacing between the primary and secondary windings, and maybe also increasing the number of turns, but both of these changes would increase the winding resistances and hence worsen the noise performance. An alternative is to add extra series inductance externally to the transformer, in series with either the secondary or the primary. Since the latter requires a much smaller inductor value, it is more convenient, and Figure 8.28 shows a design based on this notion. The inductors were made by winding 20 turns of 40 s.w.g. enamelled wire on an R–S components anti-parasitic (ferrite) bead. The flux density in these inductors, even at very high sound levels, is less than 0.1 mT, so that the distortion introduced is negligible. The d.c. resistance is 0.5 Ω each.

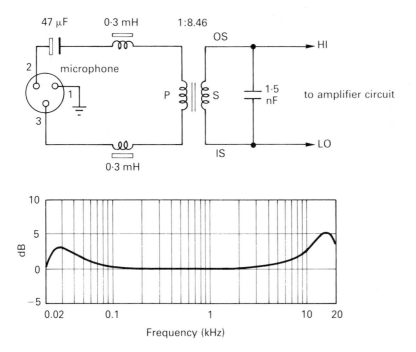

Figure 8.28 *Equalizing scheme for 4038A ribbon microphone. The response curve was measured using the 1000:1 test transformer of Section 8.5.18, at a microphone transformer secondary voltage of 2 mV r.m.s.*

Low-frequency equalization may be achieved by inserting a tantalum electro-lytic capacitor of suitable value to give series resonance with the primary shunt inductance and microphone inductance at about 30 Hz. Though the inductance value is somewhat amplitude dependent, the scheme has been found very satisfactory in practice.

The specimens of 4038A microphone actually employed have a resistive impedance at medium and high frequencies which is somewhat higher than the nominal 30 Ω figure, and using the more realistic figure of 42 Ω in the noise calculation at the end of Section 8.5.3 yields a noise figure of 0.9 dB. With a 50 m microphone cable having a go-and-return resistance of 3.5 Ω, the noise figure becomes 1.2 dB. Adding 0.1 dB in accordance with the footnote in Section 8.5.3 gives a practical noise figure of 1.3 dB. With a verified microphone sensitivity of 0.07 mV/μbar, the A-weighted noise SPL is 17.5 dBA. A slightly larger trans-former of higher step-up ratio would improve this to 17 dBA.

8.5.6 Eddy-current loss[27]

When a low-frequency a.c. voltage of constant magnitude is applied to a winding on a laminated transformer core, the flux in the core induces a frequency-independent voltage (back e.m.f.) not only into the winding itself, but also into the cross-sectional area of each lamination. This causes eddy currents to circulate round little loop paths within the thickness of each lamination, with accompany-ing I^2R power losses. The total eddy-current loss for all the laminations may be represented by an equivalent resistance R_e in shunt with the winding.

At low frequencies, where the main flux for a given applied voltage is relatively large, the flux generated by the eddy currents themselves is very small in comparison. But as the frequency is increased, the magnitude of the main flux density falls, until a frequency is reached where it is of the same order as that caused by the eddy currents. The direction of the eddy-current flux is opposed to that of the main flux. The result is that at high frequencies, near the inner region of the cross-sectional area of each lamination, which is inside most of the eddy-current paths in the lamination, the main and eddy-current fluxes nearly cancel each other, whereas no such cancellation effect occurs at the outside of each cross-sectional area. Consequently, at these higher frequencies, there is very little resultant flux in the inner regions of the laminations, the flux being concentrated near the outside – 'skin effect'. The total eddy-current loss is reduced, so that the value of the shunt resistance R_e representing it is increased.

The flux-neutralizing effect of the eddy currents also reduces the shunt inductance of the winding. Because of the distributed nature of the elements involved, the rise in R_e and fall in shunt L with increasing frequency both follow square-root or 10 dB/decade laws. Both effects come in at approximately the same critical frequency, or corner frequency, which is proportional to $\rho/\mu\delta^2$, where ρ is the specific resistance of the magnetic material, μ is the relative permeability and δ is the lamination thickness.

For 0.38 mm Mumetal, with $\mu = 20\,000$, the critical frequency is a little over 200 Hz. The effect of this on frequency response with high source impedances was mentioned in Section 8.5.4. The low-frequency value of R_e for the 1:8.46 microphone transformer, referred to the secondary, is 56 kΩ.

When the frequency is raised well above the critical frequency, the Q-value, if eddy currents were the only source of loss, would tend to unity.

With ferrite cores the specific resistance is so enormously higher than for Mumetal that eddy-current loss is normally quite negligible in transformer applications.

8.5.7 B–H loops and hysteresis loss[27]

In the rationalized MKS system, the magnetizing field strength H is expressed in ampère-turns per metre. For a core of constant permeability, the relationship between B and H is

$$B = \frac{4\pi}{10^7} \times \mu H \qquad (8.34)$$

where B = flux density (T)
  μ = relative permeability
  H = nI/l_m, in which
  n = number of turns
  I = current (A)
  l_m = magnetic path length (m).

At high flux densities the relationship between B and H follows a hysteresis loop of the familiar shape shown by the larger trace in Figure 8.29(a), in which the approach to saturation is evident.* In driving the material round this loop – upwards on the right and downwards on the left – an amount of energy proportional to the loop area is dissipated per cycle.

At lower peak flux densities, such as for the smaller trace in Figure 8.29(a), the shape of the loop is much simpler and follows quite closely a relationship established by Lord Rayleigh well before the advent of radio broadcasting. The Rayleigh equation is

$$B = \frac{4\pi}{10^7} \left[\underbrace{(\mu_i + \alpha\hat{H})H}_{A} \pm \underbrace{\frac{\alpha}{2}(\hat{H}^2 - H^2)}_{B} \right] \qquad (8.35)$$

where B = flux density (T)
  μ_i = initial relative permeability
  H = magnetizing field strength   (AT/m) ($\hat{H}$ = peak value)
  α = a constant for the magnetic material

For a given peak H excursion, denoted by $\hat{H}$, term A by itself gives a linear relationship between B and H, with a slope that increases as $\hat{H}$ is made larger. It is term B that opens the line out to form a pointed loop; the – sign is taken while H is increasing positively, and the + sign while it is changing in the opposite direction.

Consideration of equation (8.35) makes it evident that as $\hat{H}$ is reduced, the width of the loop diminishes more rapidly than its length, so that the loop tends more and more closely towards being a simple straight line at very low signal levels. The slanting loop in Figure 8.29(b) has about a quarter of the B value that applies to the small loop in (a) and its slimmer proportions are obvious.

To enable the true shape of this slim loop to be more satisfactorily discerned, a signal proportional to H was mixed into the vertical deflection circuit so as to cancel out the mean slope, and the vertical gain was then increased by a factor of 10 to reveal the true shape more effectively. It is seen that the shape of the loop is in very good accord with the Rayleigh representation as 'two pieces of parabola'.

---

*The vertical voltage for these displays was obtained by using a Blumlein integrator to integrate the voltage, proportional to dB/dt, induced in a secondary winding.

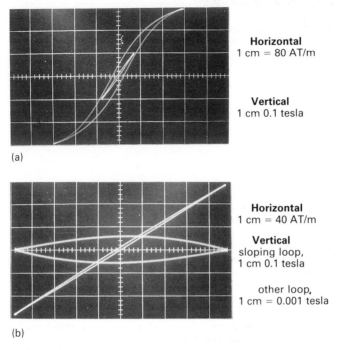

(a)

Horizontal
1 cm = 80 AT/m

Vertical
1 cm 0.1 tesla

Horizontal
1 cm = 40 AT/m

Vertical
sloping loop,
1 cm 0.1 tesla

other loop,
1 cm = 0.001 tesla

(b)

Figure 8.29 *These B–H loops were obtained using a Grade A13 (3H1) ferrite transformer core, to avoid misleading effects caused by eddy currents.*

A consequence of this parabolic shape is that the area of the loop is proportional to $\hat{H}^3$, so that the energy lost per cycle is also proportional to $\hat{H}^3$. But for a constant voltage across the winding, both B and H, at low levels and low frequencies, are inversely proportional to frequency, so that the energy lost per cycle is then proportional to $1/f^3$ and the hysteresis power loss is therefore proportional to $1/f^2$. This, as stated above, is for a constant voltage V across the winding. If V is varied at a fixed frequency, then the energy or power loss is proportional to V^3. Hence when both f and V are varied, we have

$$\text{hysteresis power loss} \propto \frac{V^3}{f^2} \tag{8.36}$$

If the hysteresis loss is represented by a shunt resistance R_h, then it follows that, at low levels and low frequencies,

$$R_h \propto \frac{f^2}{V} \tag{8.37}$$

Hysteresis loss thus tends to become negligible at high frequencies and/or small voltages. It may be added that if the loops had had constant proportions, then R_h would be independent of voltage and directly proportional to frequency.

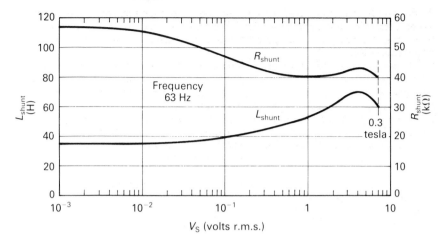

Figure 8.30 *Shunt inductance and resistance measured at the secondary of the 1:8.46 microphone transformer.*

At higher frequencies the eddy-current effects referred to in the previous Section interact with the hysteresis mechanism and matters become more complex, though this is seldom of any practical concern.

Figure 8.30 shows the result of a fixed-frequency measurement on the secondary of the 1:8.46 microphone transformer, made with a simple admittance bridge employing op. amps. The shunt resistance curve virtually represents the total core loss, though strictly speaking a small contribution to it, about 4%, comes from the copper loss.

At low levels the core loss is almost purely eddy-current loss, but as the level is increased there is an increasing hysteresis-loss contribution.

The rise in shunt inductance with measuring voltage reflects the increasing steepness of the B–H loops, which continues until offset by the beginning of the saturation process.

8.5.8 Barkhausen noise

Ferromagnetism is a subtle, complex and very fascinating subject[28,29,30]. Normal polycrystalline magnetic materials, such as Mumetal, are believed to contain a large number of magnetic domains, of many different sizes, which are effectively little magnets, each involving very large numbers of atoms. The magnetic effects are basically produced by moving electrons associated with these atoms.

In unmagnetized material the domains are orientated in a random manner, producing zero resultant magnetic flux.

The application of a magnetizing field of smoothly increasing magnitude produces two kinds of effect:

(a) A smooth and continuous movement of the domain boundaries, resulting in a smooth increase in B.
(b) Sudden 180° reversals in the orientation of some domains. These are known as Barkhausen jumps, and every time one occurs a little impulse voltage is induced in the exciting winding.

If the winding is connected to an amplifier and loudspeaker, these jumps become audible as separate clicks if the d.c. is changed very slowly, but with more usual rates of change they merge together to produce a rushing sound, called Barkhausen noise.

The effects may be convincingly demonstrated by the following experiment, which is well worth actually carrying out. Put a winding of about 1000 turns on a core consisting of just two or three small Mumetal E laminations, overlapping and held together to give a closed magnetic circuit. Feed a variable d.c. voltage of ±15 V, very well smoothed, via about 10 kΩ to this winding, the winding being connected by means of a suitable value of blocking capacitor to the input of a low-noise transistor amplifier and loudspeaker. There should preferably be some bass cut in the system, to avoid overloading on the more rapid rates of current change.

The sound heard is very much like that produced by tilting a cardboard box containing dry gravel one way and the other, with the same feature that on reversing the direction of tilt there is silence at first.

From such an experiment the following facts emerge:

1. No Barkhausen noise at all is audible for $\hat{B}$ excursions of less than about ±2.5 mT (or ±25 gauss). If any jumps do occur below this level, the domains are evidently so small that their effect is drowned by amplifier noise.
2. The loudest Barkhausen noise, involving reversals of the largest domains, occurs while the operating point is traversing the steepest parts of a B–H loop such as the larger one in Figure 8.29(a).

From an audio engineering point of view, however, the significant practical conclusion is that Barkhausen noise in microphone transformers is never audible at all, with a good margin to spare. With a 4038A microphone and the transformer design already described, it turns out that an SPL of 100 dB at 50 Hz gives $\hat{B}$ < 1 mT.

8.5.9 Residual loss

In materials such as Mumetal, eddy-current and hysteresis losses, already discussed, account for very nearly all the core loss – but not quite all, and the remaining small loss is called residual loss. Residual loss, though it gives rise to a very small phase displacement between B and H under sinusoidal conditions, is different from hysteresis loss in that there is no non-linearity, so that it produces no distortion. It involves a time-dependent mechanism within the magnetic material, which may be of a thermal nature.

If residual loss is represented by a resistance R_r in shunt with the winding, the value of R_r is independent of voltage level and directly proportional to frequency. (See conclusion of paragraph under equation (8.37).)

In ferrite materials the residual loss is also small, but because the eddy-current loss is so very much less than for nickel–iron alloys, the residual loss nevertheless becomes the dominant core loss at very low signal levels, for which the hysteresis loss has become vanishingly small. Thus, whereas residual loss can be forgotten about when using Mumetal etc., it is normally given prominence in trade literature on ferrite materials.

8.5.10 Distortion

When, as is normally the case, a transformer winding is driven by a sine-wave source of much lower internal impedance than its own reactive impedance, an approximately sinusoidal voltage is forced to exist across the winding, so that the waveform of the flux density B is also nearly sinusoidal. But from the pointed shape of the B–H loops discussed in Section 8.5.7, it is evident that the waveform of H, and the magnetizing current, will be much more highly distorted than this.

It is the flow of this distorted magnetizing current in the source impedance that gives rise to the voltage distortion that appears across the windings.

Distortion data for design purposes can be given in different ways, sometimes in the form of a family of curves showing the variation in percentage distortion with flux density for a number of different ratios of source resistance to winding reactance.[31]

The unusual method here presented involves only a single curve, which is essentially a characteristic of the core material itself and is independent of the associated circuit conditions.

Referring to Figure 8.31, the procedure for determining the percentage distortion when a transformer primary is fed from a signal source of internal resistance R_G is

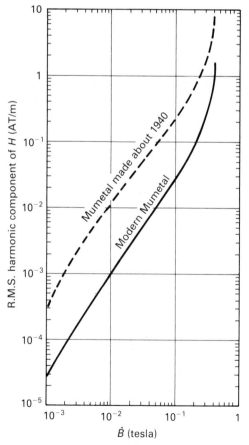

Figure 8.31 *Universal distortion curve for present-day Mumetal laminated cores, plus broken-line curve for early Mumetal.*

(a) From equation (8.33) and the transformer parameters, determine the value of $\hat{B}$ at the frequency and level of interest.
(b) From the curve determine the value of the r.m.s. harmonic component of H, and multiply this by l_m/n to obtain the corresponding r.m.s. harmonic current in amps. (l_m = magnetic path length in metres, n = number of turns in the primary winding.)
(c) Multiply the current value obtained in (b) by R_G to yield the harmonic voltage across the primary winding.
(d) Divide the figure obtained in (c) by the primary fundamental voltage, and multiply by 100 to obtain the percentage voltage distortion.

If the copper resistance of the driven winding is significant, it should be counted as part of R_G. The percentage distortion figure then obtained will apply more accurately to the secondary than to the primary voltage.

The curve of Figure 8.31 is applicable at any frequency, provided this is below the critical frequency mentioned in Section 8.5.6. However, because $\hat{B}$ falls with rising frequency in a transformer handling constant voltage, so also does the distortion; the usual requirement is therefore to know its low-frequency value. This reduction in distortion with rising frequency is an attractive feature of transformers not possessed by amplifying devices such as transitors.

If a small gap is inserted in a transformer core, for example by inserting all Ts one way and all Us the other, this will markedly increase the fundamental component of the magnetizing current for a given applied voltage, but it will hardly affect the harmonic component of this current and will therefore hardly affect the percentage distortion. The Figure 8.31 curve is therefore still fully applicable, and this also applies to the effect of the small unwanted gaps that are inherent in ordinary laminated core assemblies but which are absent in toroidal cores. The curve may thus be regarded as a true characteristic of the material itself, except that slight complications arise at very high flux densities, as discussed in Section 8.5.11, and as a result of unequal magnetic properties along the rolling direction and at right angles to it as mentioned in Section 8.5.13.

The square-law slope of the Figure 8.31 experimental curve at very low levels is consistent with predictions based on the Rayleigh equation (8.35).

However, the curve is best regarded more as a guide to orders of magnitude than as highly accurate design information, since quite large variations occur in magnetic material production and between the products of different manufacturers.

Application of the above method to the 1:8.46 microphone transformer design of Section 8.5.2 gives a distortion figure of 0.14% for 1 V r.m.s. at 40 Hz on the secondary. The distortion at low and medium levels is mainly third harmonic.

8.5.11 Saturation flux-density values

The saturation flux density for Mumetal is usually given in the literature as about 0.8 T, but if a test is made on an ordinary laminated Mumetal transformer core, it will be found that saturation appears to set in at around 0.4 T.

The reason is that when the flux in one lamination approaches a slight gap, it is diverted into the two neighbouring laminations, which therefore experience a doubling of their flux density in this region. Thus a mean flux density in the core of about 0.4 T results in a flux density near the joints of around 0.8 T, and saturation therefore occurs.

This effect is shown very clearly in Figure 8.32, from which it is evident that if a large enough magnetizing field is applied, saturation can be made to occur throughout the whole core at about 0.8 T.

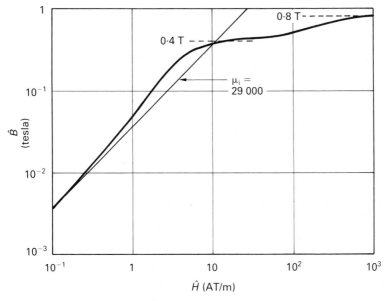

Figure 8.32 *50 Hz measurement on high-μ specimen of Mumetal core (0.38 mm laminations), showing the double-saturation effect described in Section 8.5.11.*

8.5.12 Effects of altering the input transformer size

To give a satisfactory performance with a specific value of source resistance, the number of primary turns, n_p, required, as already discussed, is normally determined by low-frequency considerations. It is of interest to find the effect on this, and on other parameters, of altering the size of the transformer. This is a straightforward problem related to the transformer equations (8.32) and (8.33) and other basic formulae for copper resistance and winding capacitances.

If all linear dimensions are multiplied by the same factor m, then it turns out that, for a constant primary inductance:

(a) $n_p \propto m^{-1/2}$
(b) $\hat{B}$ for a given voltage $\propto m^{-3/2}$
(c) Low-level harmonic distortion at constant voltage $\propto m^{-3}$
(d) Copper resistances $\propto m^{-2}$
(e) Winding capacitances $\propto m$
(f) The leakage inductance and the shunt eddy-current resistance are unaffected.

Thus doubling the linear dimensions reduces the winding resistances to a quarter of their original values and makes the distortion eight times smaller. By also using thin laminations (0.1 mm) for reduced eddy-current loss, and adopting an optimum step-up ratio, a system noise figure of less than 0.5 dB can be achieved in practice.

8.5.13 Toroidal transformers

Toroidal transformers employing spiral or 'clock-spring' cores have certain advantages.

Firstly, the direction of the flux coincides everywhere with the direction of rolling of the magnetic material in manufacture, the highest permeability always occurring in this direction.

Secondly, the absence of slight air gaps such as occur in ordinary transformer cores where the laminations butt together also contributes to enhanced effective permeability values, and avoids the saturation effect discussed in Section 8.5.11.

When these features are combined with the use of superior magnetic materials such as thin Supermumetal or Supermalloy, effective initial a.c. permeability values of 200 000 or more can be realized.

A third advantage of toroidal transformers is that the magnetic hum pick-up is much reduced. However, even if a simple winding occupies the full circumference quite uniformly, it is actually still equivalent to a single circumferential turn so far as hum pick-up is concerned. Consequently, when only a few turns are involved, the advantage of toroidal construction in this respect is not as great as might at first be expected.

A toroidal transformer enclosed in a Mumetal case is employed in the 4038A ribbon microphone.[32]

A disadvantage of toroidal construction, of course, is that winding is less easy, increasing the cost.

8.5.14 Hum pick-up and Mumetal cans

The stray magnetic field produced by mains transformers, both in theory and in practice, varies inversely as the cube of distance, except when very close.

With the normal type of single-bobbin microphone input transformer, a Mumetal screening can is always necessary, and is normally of the deep-drawn cylindrical type with an overlapping lid.

The amount of hum reduction obtained is very significantly influenced by the orientation of the transformer within the case, the least effective condition being when the axis of the transformer bobbin coincides with the axis of the can.

The appreciable reluctance of the junction between the lid and the body of the can results in the dominant magnetic flux inside the can tending to be in the axial direction, but even in the absence of the lid, the axial flux density falls off with increasing distance from the open end at a rate tending towards the theoretical piston-attenuator slope of 41.8 dB per diameter, as illustrated in Figure 8.33.

Thus the transformer should be mounted as far from the lid end as possible, with the bobbin axis at 90° to the can axis. Non-magnetic spacing material, of about 1 mm thickness, should be used to prevent the transformer core from coming into direct contact with the can.

Holes or slots of normal sizes in the can or lid exert negligible influence on the screening effectiveness when the transformer is positioned as above, but should be drilled fairly gently to avoid permanent deformation of the Mumetal.

When the above points are attended to, a properly annealed deep-drawn Mumetal screening can will be found to give at least 50 dB of hum reduction at 50 Hz.

Many capacitor microphones contain diminutive output transformers which can give rise to hum, though the problem is much less acute than with amplifier input transformers fed from electromagnetic microphones – for three reasons. Firstly, because the distance from mains transformers is normally much greater;

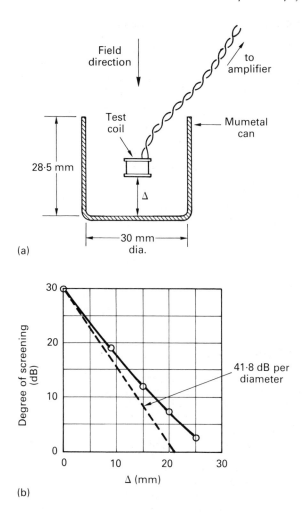

Figure 8.33 *Measurement on Mumetal screening can.*

secondly, because the signal level is usually higher than with other types of microphone; and thirdly because these internal transformers are fed from the low output impedance of an emitter-follower or other feedback circuit – an ideal transformer fed from an active circuit with zero output impedance would give no hum pick-up at all.

Mumetal screening for such internal microphone transformers is therefore frequently omitted.

8.5.15 Bifilar windings and electrostatic screens

When a microphone feeds an amplifier input transformer via a balanced cable, the set-up is as shown diagrammatically in Figure 8.34, and troubles are sometimes experienced due to both audio-frequency interference, for example

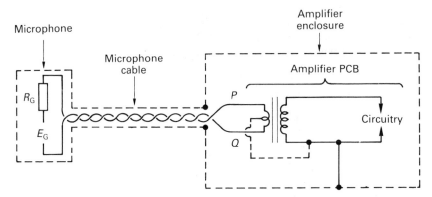

Figure 8.34 *Interconnection of microphone and amplifier.*

from nearby power cables, and RF interference. The latter is made audible by demodulation at the semiconductor junctions.

Audio-frequency balanced-mode interference can be induced magnetically into the small loop areas formed by the inner conductors, but this effect is reduced to a very low level indeed by the use of good twisted-pair or preferably quad cables.

The effects now to be considered involve longitudinal or common-mode interference appearing in the same phase at P and Q, and it is the RF components of such interference that usually constitute the main problem. For such interference the cable behaves as if it was a coaxial one.

One mechanism by which common-mode interference can get onto the inner conductors involves the fact that braided or lapped cables have tiny gaps in their screening – lapped cables, particularly those of the double-lapped or Reusen type, are generally much better than braided ones in this respect.

Another very significant mechanism involves the resistance of the cable.[33] Consider the simpler case of a length of coaxial cable, short-circuited at one end. This will behave as an aerial, in which, in the absence of any resistance, equal voltages would be developed along the lengths of both the outer and inner conductors, giving zero voltage between them at the non-short-circuited end. RF current, however, flows in this 'aerial', but is largely confined to the outer conductor because of skin effect. This current produces a voltage drop in the resistance of the outer conductor, and since almost no such effect occurs in the inner conductor, the voltage drop appears between inner and outer at the free end of the cable. So far as this effect is concerned, braided cables are better than lapped ones, having generally lower resistance.

Thus, by more than one mechanism, a length of microphone cable is always found to have a common-mode RF interference voltage between its inner conductors and the screening, sometimes of a few millivolts magnitude. The internal impedance with which any specific component of this interference voltage is presented is highly dependent on the length of the cable in relation to the wavelength of the interference.

If the cable inners are provided with a common-mode earth via a centre-tapped winding, as shown in broken-line in Figure 8.34,* a very substantial RF

---

*For the transformer described in Section 8.5.2, the common-mode inductance seen between the ends of the bifilar primary and its centre-tap is only 0.6 μH.

interference current may flow in this earthing lead, which, with normal constructional arrangements, would be taken to earth somewhere well inside the amplifier enclosure. An RF magnetic field is thus created which causes interference voltages to be induced into the rest of the circuitry – an effect which good designs always aim to minimize.

The general experience seems to be that it is normally better not to earth the centre-tap but to leave the primary winding floating, and this is now the nearly universal practice. Occasionally a switch has been provided so that a centre-tap may be used as an optional alternative – it is sometimes the better choice if the main interference is at high audio frequencies owing to a leaky cable screen.

With the centre-tap omitted, some interference current, usually of much smaller magnitude, will flow to earth via the transformer capacitances, and it used to be almost universal practice to include an earthed electrostatic screen between the transformer windings to prevent this current reaching the secondary winding. However, because of the low secondary impedances made possible by the advent of transistors, it is found that entirely satisfactory results can usually be obtained without involving the extra cost of a screen – many BBC microphone transformers have no screen, for example.

The low impedance allows a sizable capacitor, e.g. 1 nF, to be connected across the secondary circuit, and this capacitor can be put extremely close to the input transistor for effective VHF interference suppression. Sometimes a series inductor may be added for further filtering action. (The connection of such a capacitor directly between base and emitter of an input stage is a somewhat questionable technique, since it can cause oscillation by creating a negative input conductance.)

An overriding consideration is that the outer screen of a microphone cable should be taken solidly to earth right at the input socket and not via a lead wandering around near the amplifier circuitry.

8.5.16 Transformers and negative feedback

The benefits of negative feedback may be applied to microphone transformers, high-level line-input transformers, and output transformers.

Figure 8.35 shows two ways in which a desired value of secondary loading resistance, R_S, (see Section 8.5.4) may be produced without introducing significant degradation of the noise figure. The (a) scheme is that used in the BBC microphone amplifier circuit of Figure 8.6.

The formula given for the (b) scheme is a general one, equally relevant in other contexts where the circuit may be used to produce negative resistance. For the present application, A is made negative, say −100.

Figure 8.36 shows, slightly simplified by the omission of small capacitors across the op. amp. resistors, a circuit designed for KEF Electronics Ltd for use in a studio monitor loudspeaker with self-contained amplifiers. The aim is to provide a high-impedance floating input, suitable for bridging across any available signal line, and giving a high degree of rejection of common-mode interference present on the line.

The transformer is wound highly symmetrically on a two-section bobbin, with the half-secondaries, wound in reverse directions, at the inner radius for minimum resistance – the primary resistance is of no importance whatever, and thinner wire was in fact used for this winding. The primary and secondary are well spaced apart by low-loss insulation, giving an inter-winding capacitance of less than 30 pF.

To maintain overall symmetry of behaviour, the amplifier that follows the transformer is also balanced, both inputs being virtual earths – or virtual grounds if this term is preferred!

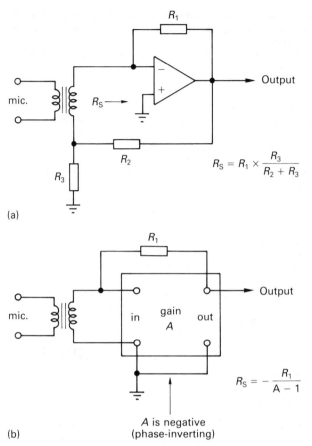

(a)

$$R_S = R_1 \times \frac{R_3}{R_2 + R_3}$$

(b)

$$R_S = -\frac{R_1}{A - 1}$$

A is negative
(phase-inverting)

Figure 8.35   *Circuits giving effective secondary loading of value R_S, with negligible sacrifice of signal-to-noise ratio.*

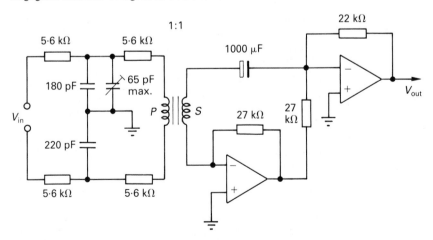

Figure 8.36   *Balanced input circuit for active studio monitoring loudspeaker.*

Feeding a transformer in this way to an almost zero impedance is enormously beneficial with regard to frequency response and distortion, since the core hardly has to develop any flux density at all.

The distortion for an input level of 0.775 V r.m.s. at 30 Hz, both as predicted using the universal curve of Figure 8.31 and as directly measured, is well under 0.001%.

The frequency response is flat within ±1 dB from 2 Hz to 100 kHz, and immaculately flat over the whole audio band. There is a peak of about 5 dB at 0.8 Hz owing to series resonance between the secondary inductance and the 1000 µF capacitor, but this may be eliminated, if desired, by replacing this capacitor by a network involving two capacitors and a resistor.

The noise level relative to 0.775 V is –108 dBA, and the unweighted hum level is also about –108 dB even with the screened input transformer about 20 cm from a large mains transformer.

With the trimmer capacitor correctly set, the common-mode rejection ratio is better than 90 dB at 1 kHz and better than 70 dB at 10 kHz.

The transformer used in the version of the scheme described above is wound on a Mumetal core of the same size as for the microphone transformer of Section 8.5.2, and has 1000 turns on both primary and secondary. A more economical version, using a ferrite transformer core, is used by Quad Electroacoustics Ltd for a professional amplifier input stage.

Reducing the shunt inductance to a fairly low value has no adverse effect on the audio frequency response, though it gives an increase in low-frequency noise, which nevertheless still remains inaudible.

A different embodiment of the same broad principle, evolved quite independently, is described in Reference 34, and employs only a single op. amp. The secondary winding feeds the inverting input of this, but the feedback resistor, instead of being taken to this input, is taken to a tertiary winding. Assuming infinite op. amp. gain, this scheme ideally reduces the flux density right down to zero, which allows an extremely small transformer to give a very good performance.

An alternative way to achieve the above result, which I tried at the time the Figure 8.36 circuit was evolved, but found it unnecessary to adopt, is to apply a little positive feedback to one of the op. amps., making it into an open-circuit-stable negative resistance (see below) of magnitude equal to the secondary copper resistance of the transformer.

Transformers are often used to provide a high-level floating line output for feeding to other equipment. Because of the use of negative feedback, the output impedance is usually very low, but in self-contained studio installations the loading imposed on such an output is normally quite light. Line output voltage levels are usually expressed in dBu, i.e. decibels relative to 0.775 V r.m.s. – see footnote near the end of Section 8.3.11. Sometimes, however, such as in outside broadcast equipment, which has to feed into a long telephone line, the requirement is for a purely resistive output impedance of specific value, such as 75 Ω.

With early transistor equipment,[17,18] push–pull PNP output stages, with a centre-tapped transformer primary, were normally used, the negative feedback being preferably derived from a tertiary winding.[35,36]

The later availability of complementary transistors, and more particularly of suitable op. amps., has enabled such transformers to be driven in a 'single-ended' manner, the simplest such arrangement being that of Figure 8.37.

The distortion in the voltage across the primary is made very low by the negative feedback, but the secondary voltage will contain a small amount of distortion owing to the flow of the highly distorted magnetizing current in the

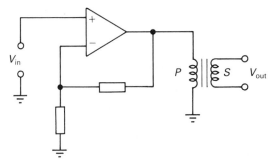

Figure 8.37   *Simplest op. amp. circuit for low-impedance driving of output line transformer. To avoid significant performance degradation caused by d.c. core polarization, the gain should preferably be limited to about ×2 and no appreciable d.c. voltage should be present at the input terminal.*

primary copper resistance (see Section 8.5.10). The distortion due to this cause is greater with small transformers than it would be with large ones, because of item (d) in Section 8.5.12, but is nevertheless likely to be much less than 1% for full output level at 30 Hz, given a reasonably good transformer.

There are two things that can be done to give a substantial reduction in the above distortion:

(a) Take the negative feedback, or some of it, from a tertiary winding. This will reduce the distortion of the flux waveform and therefore also of the output voltage.
(b) Insert negative resistance in series with the primary to annul the effect of the copper resistance.

Good results can be obtained in either of these ways, but (b) is especially attractive because it simplifies the transformer and reduces its cost, while requiring only trivial elaboration of the op. amp. drive circuit.

Negative resistance circuits are of two broad types – open-circuit stable and short-circuit stable; simple examples of these are shown in Figure 8.38. In Figure 8.38(a), with the input open-circuited, negative feedback is greater than positive, giving stability, but if the input is short-circuited, only positive feedback remains and the circuit runs into a saturated state. In circuit (b), a short-circuit on the input makes the negative feedback dominant and gives stability.

The (a) circuit, with values chosen to give a negative resistance of magnitude just slightly less than the primary copper resistance, can be inserted directly in series with the primary earth-return lead in Figure 8.37, and will give a good reduction in low-frequency distortion. However, this scheme as it stands is rather undesirable, since it is on the verge of instability, giving a large magnification of the op. amp. offset voltage, and being liable to run away with itself if resistance values change slightly.

The problem may be solved by inserting an electrolytic capacitor in series with the primary winding, the negative resistance then being set to be nominally equal to the copper resistance.

The circuit is now fully stable, but exhibits a sub-audio-frequency response peak caused by series resonance between the electrolytic capacitor and the primary shunt inductance. This can be removed by connecting resistors, ideally of value $\sqrt{(L/C)}$, across both the capacitor and the primary winding. The negative-resistance circuit then sees a constant-resistance network having a resistance

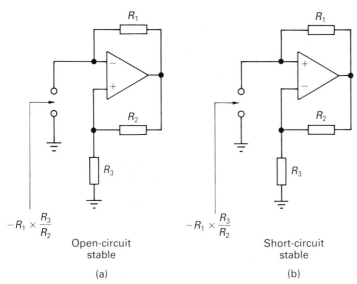

$$-R_1 \times \frac{R_3}{R_2}$$

Open-circuit
stable

(a)

$$-R_1 \times \frac{R_3}{R_2}$$

Short-circuit
stable

(b)

Figure 8.38   *Negative-resistance circuits.*

magnitude far higher than the magnitude of its own negative resistance, and the set-up is therefore thoroughly tame and non-resonant.

I have used the above scheme very effectively in a precision measurement context, with the electrolytic capacitor in series with the top end of the transformer primary. This makes available between the top end and earth, a voltage which is the true induced voltage, unadulterated by the voltage drop in the copper resistance. The ratio of this voltage to the secondary voltage is then determined with high precision purely by the turns ratio, if the windings are appropriately positioned.

Though the use of a separate op. amp. to produce the negative resistance in the above manner is a perfectly satisfactory practical scheme, an even more economical arrangement is shown in Figure 8.39. In this, stability is assured even if R_1 is increased well beyond the value that gives perfect annulling of the copper resistance, provided C_1 is not made excessively large. The effect of a very large C_1 value is to make the circuit liable to maintain a very-low-frequency relaxation oscillation if R_1 is increased a little beyond the critical value. A value of C_1 giving a response $-3\,$dB at $5\,$Hz and $-1\,$dB at $10\,$Hz is usually a sensible choice; then an increase in R_1 of even 20% above the correct value will not normally allow oscillation to be sustained. (Since the copper resistance of the winding has a positive temperature coefficient of $0.4\%/°$C, R_1 should ideally also have this coefficient.)

In the absence of C_2, and with C_1 adequately small, increasing R_1 by a large amount, say 100%, may first produce oscillation at a frequency of the order of $1\,$MHz, owing to series resonance in the transformer. Though not likely to be troublesome, the possibility of such oscillation is totally eliminated when C_2 is present, and the right value of C_2 is also effective in giving the circuit an excellent square-wave response.

A frequency-response flat to within $\pm0.1\,$dB from $20\,$Hz to $20\,$kHz is readily obtainable on no load.

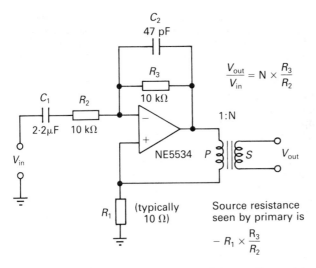

Figure 8.39 *Output circuit incorporating negative resistance to reduce distortion due to copper resistance of primary winding of line transformer.*

When a resistive load is applied, however, the presence of appreciable leakage inductance will cause the output to fall off at high frequencies, and this, at first sight, would seem to be an inherent weakness of the Figure 8.39 scheme.

The alternative tertiary-winding solution mentioned earlier can overcome this trouble, since, if sufficient care is taken, the tertiary winding can be so positioned on the bobbin that the leakage flux linking with it is equal to that linking with the secondary winding[35]. The negative feedback then effectively monitors the secondary voltage and thus ensures a flat frequency response at the output terminals. Even then, however, the primary-to-secondary leakage inductance causes the maximum output voltage obtainable on load, before voltage clipping occurs, to be less at high frequencies than at lower frequencies.

The most elegant solution is to reduce the total leakage inductance to an extremely low value by winding the primary and secondary together as a bifilar pair. This gives far lower leakage inductance than can be achieved in other ways – a ratio of main inductance to leakage inductance of well over 100 000:1 is readily obtained.

A simple bifilar winding, of course, restricts the transformer ratio to 1:1, but this limitation can be overcome as described later.

An NE5534 op. amp. on ±15 V supplies can comfortably give an output voltage swing of ±12 V, or 8.5 V r.m.s., so that, with a 1:1 transformer, this is also the output voltage and corresponds to +20.8 dBu. With a 600 Ω load on the output, the maximum obtainable output will be slightly reduced because of the winding resistances, totalling perhaps 20 Ω, but a level of +20 dBu should still be achieved. This requires a peak current of 18.3 mA, which the op. amp. is capable of supplying.

If the output resistance of the circuit is made up to 75 Ω by adding series resistors, and is fed to a 75 Ω load, the achievable output level is limited not by voltage clipping, but by the maximum output current that the op. amp. can turn on, which may be taken as 25 mA. This gives 23 mW mean to 75 Ω or approximately +14 dBm. The peak voltage at the primary is only about 4 V.

A higher output voltage, on no load or a light load, may be obtained by adopting a trifilar winding, with two sections in series.

With this 1:2 ratio, a no-load voltage level of over +26 dBu becomes available, but when a 600 Ω load is applied, the maximum output obtainable is limited to a peak voltage of 7.5 V by the maximum op. amp. current of 25 mA. This corresponds to a voltage level of +16.7 dBu, and the power to the 600 Ω load is, of course, +16.7 dBm.

On a 75 Ω load with 1:2 transformer, the maximum level obtainable is 6 dB less than with the 1:1 transformer, and is therefore only +8 dBm.

To meet all requirements when a 1:2 ratio is used, it may be necessary to increase the current capability of the drive circuit. This could be done by including complementary emitter-followers within the op. amp. feedback loop, but an alternative solution, which has appeared in Lundahl literature, is shown in Figure 8.40, and gives a doubling of the current capability. Moreover, the adoption of antiphase driving of the two primary sections has the great virtue that capacitive coupling from primary to secondary is ideally of a fully balanced nature, thus avoiding any capacitive common-mode output.

The reasons why 'multifilar' windings have come so much into their own in this application are:

(a) Simple ratios of 1:1, 1:2 or 1:0.5 are appropriate.
(b) The inherently much higher capacitances of such windings are of little consequence because of the low-impedance nature of the circuits, and because the negative feedback is taken directly from the primary rather than via transformer action from another winding. When tertiary-winding feedback is used, the behaviour of the transformer at frequencies far above the top of the audio band needs to be carefully taken into account.[35] An interesting combination of techniques, however, is to use tertiary-winding feedback, but with the normal separate tertiary winding replaced by one section of a multifilar winding.

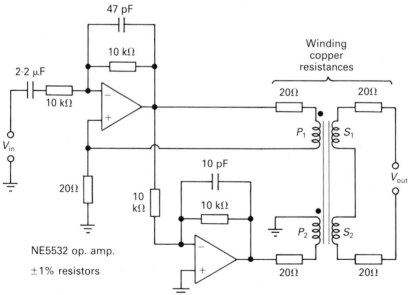

Figure 8.40 *Balanced op. amp. line output circuit employing 'multifilar' transformer, with negative-resistance compensation of primary winding resistance. The capacitively-coupled common-mode output is very low. Winding resistances are shown explicitly to aid understanding.*

(c) With multifilar windings, voltages of the same order as the input voltage exist between directly adjacent turns, but the low voltages involved eliminate any insulation problems.

Mumetal is not a satisfactory core material to use for these output transformers, since its relatively low saturation flux density would necessitate putting on so many turns that the copper resistances would be intolerably high, unless an uneconomically large size of core was used. The very high permeability of Mumetal, which is its special virtue, is not needed for hard-driven output transformers, and Radiometal, which is a nickel–iron alloy with about 50% nickel, is sometimes used. It has a saturation flux density of about 1.6 T.

A good alternative, having about the same saturation flux density, is grain orientated silicon steel (GOSS metal, invented in the mid 1930s, with extraordinary appropriateness, by Dr N. P. Goss!). This is marketed as Hipersil etc., and is used in the form of spiral-wound toroids, or as C-cores. The joining surfaces in the latter are ground flat to give intimate magnetic contact, and are held firmly together by a steel tension band. The double-saturation effect described in Section 8.5.11, and shown in Figure 8.32, is absent.

Further progress has been made, and continues to be made, in the evolution of improved transformer-core materials.

To conclude this Section, Figure 8.41 illustrates a novel technique for feeding audio signals at very low distortion through balanced lines of moderate length.

The receiving end, at the right, is basically that of Figure 8.36 but without the input resistors; it therefore presents a low impedance to the line. This low impedance appears also across the secondary of T_1, augmented somewhat by the line resistance and, at high frequencies, inductance. Thus the flux densities in both transformers are held very low, so that small Mumetal or ferrite cores may be used, which is neat and economical. A response flat to within ±0.1 dB from 20 Hz to 20 kHz is obtainable, with distortion of 0.01% or lower, and excellent signal-to-noise ratio.

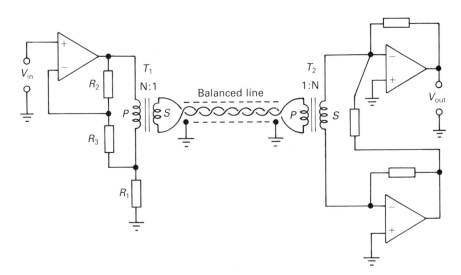

Figure 8.41   *Line operation on virtual-earth basis.*

The resistor network associated with the primary of T_1 applies a combination of current and voltage negative feedback such as to make the effective source resistance R_G feeding the primary have the value given by equation (8.38):

$$R_G = R_1 \times \frac{R_2 + R_3}{R_3} \tag{8.38}$$

R_2 and R_3 have relatively high values, giving negligible passive loading effect. This arrangement gives good stability, whereas employing only current feedback would be liable to lead to instability, and would result in the occurrence of gross overloading if the line was disconnected. A small capacitor should be shunted across R_2.

A ratio N of about 5:1 is generally appropriate, with R_G about 10 kΩ. There is no need for the transformers to have very low leakage inductance. When designing such a system it is helpful to bear in mind the typical microphone-line values $R = 0.07\,\Omega/\text{m}$ (go-and-return), $L = 0.7\,\mu\text{H}/\text{m}$ and $C = 150\,\text{pF}/\text{m}$. These give $Z_0 \approx 70\,\Omega$ and a velocity of propagation of about $10^8\,\text{m/s}$, i.e. about one third of the free-space velocity; wavelengths in the cable are therefore about three times shorter, for a given frequency, than they would be in free space.

It may be noticed that the series electrolytic capacitor of Figure 8.36 has been omitted in Figure 8.41. This is an economy that may be made, though it is liable to result in an offset at the amplifier output of a substantial fraction of a volt. However, if positive feedback is used to annul the secondary resistance of T_2, as mentioned in connection with Figure 8.36, then the capacitor must be included.

8.5.17 Higher-impedance microphone transformers

As mentioned in Section 8.5.2, it was desirable in the valve era to step the microphone impedance up to quite a high value to achieve a good noise performance and to subdue valve microphony. This, however, tends to make the frequency response much more dependent on variations in source impedance than with modern lower-impedance designs, since the resonance frequency no longer occurs several octaves above the audio band.

A very interesting example is the input transformer used in the BBC OBA/8 amplifier of Figure 8.5, in which the 300 Ω source resistance is stepped up to 300 kΩ at the valve grid. The transformer is quite large, using size 101A Mumetal laminations, which are 7.6 cm (3 in) tall. The secondary is wound in one narrow section in the middle of the bobbin, resulting in a total shunt capacitance of only about 50 pF, including the input capacitance of the top-grid valve.

The leakage inductance is intentionally made large, by having the low-profile primary well spaced away to the side of the tall secondary. A capacitor with a value of about 1.5 nF is shunted across this primary, so that the equivalent circuit, at medium and high frequencies, referred to the secondary, becomes approximately as in Figure 8.42(a), and functions as a π-section low-pass filter. The leakage inductance is such that the cut-off frequency is a little above 10 kHz, and the characteristic impedance is somewhat less than 300 kΩ; this requires a leakage inductance in the region of 7 H.

The effect of varying the source resistance over a wide range, as when series faders are used (see Section 8.2), is considerably less with such a π-filter design than when the input capacitor is omitted, and a response remaining within ±1 dB limits up to 10 kHz can be obtained.

It may be noticed, in Figure 8.42(a), that the symbols L and C have been used to denote ½-section values. This unconventional usage, much to be recommended, is due to my colleague E. F. Good, and results in the cut-off frequency for low-

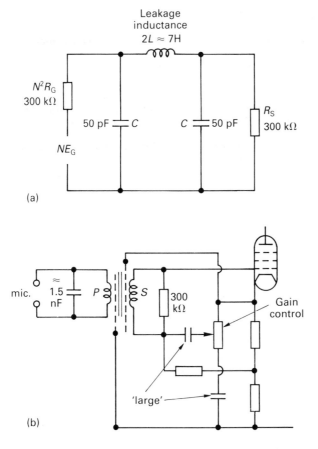

Figure 8.42   *Parameter values and input circuit relating to BBC OBA/8 amplifier illustrated in Figure 8.5.*

pass and high-pass filters, and the resonance frequency of tuned circuits, all being given by $f_0 = 1/2\pi\sqrt{(LC)}$ instead of by three different formulae. The characteristic impedance, however, is still given by $R_0 = \sqrt{(L/C)}$.

The manner in which the above transformer is incorporated in the OBA/8 circuit is shown in essence in Figure 8.42(b). The subtle use of two electrostatic screens prevents alteration in the setting of the negative-feedback gain control (which is ganged to a passive inter-stage control) from significantly affecting the value of the right-hand filter capacitance, which would cause the frequency response to vary with the gain setting.

A basic shortcoming of the OBA/8 design is that the presence of the passive 300 kΩ resistor across the transformer secondary causes a loss in signal-to-noise ratio, with a 300 Ω source, of ideally 3 dB. The possibility of virtually eliminating this loss by using negative feedback via a resistor of much higher value to terminate the filter had not been thought of in those early days.

There are many other applications in which the parameters of a transformer may be effectively exploited to obtain a wanted filter characteristic.

8.5.18 A 1000:1 test transformer

Figure 8.43 gives details of a simple but very useful test unit, which may be inserted in series with a balanced microphone line, with the microphone present, to check the frequency response and gain under normal working conditions.

It is intended for feeding from a $600\,\Omega$ oscillator. The internal $68\,\Omega$ resistor attenuates the oscillator e.m.f. by a factor of 10 (within 2%) and provides a source resistance close to $60\,\Omega$ for energizing the transformer primary. Thus an oscillator e.m.f. of, say, $1\,V$ r.m.s. injects $0.1\,mV$ r.m.s. ($-78\,dBu$) in series with the microphone.

The frequency response is flat to well within $\pm0.1\,dB$ from $20\,Hz$ to $50\,kHz$, and the total harmonic distortion for a $20\,Hz$ output level of $1\,mV$ r.m.s. ($-58\,dBu$) is less than 0.1%, being much less than this for higher frequencies and/or lower levels.

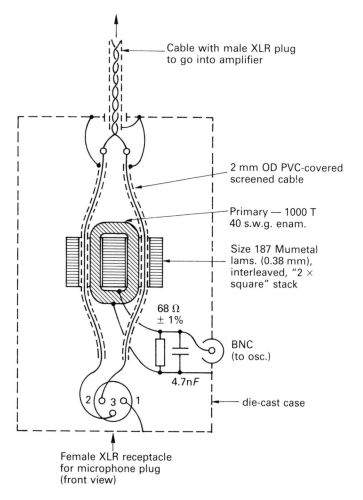

Cable with male XLR plug to go into amplifier

2 mm OD PVC-covered screened cable

Primary — 1000 T 40 s.w.g. enam.

Size 187 Mumetal lams. (0.38 mm), interleaved, "2 × square" stack

$68\,\Omega$ ± 1%

BNC (to osc.)

4.7nF

die-cast case

2 3 1

Female XLR receptacle for microphone plug (front view)

Figure 8.43 *Unit for testing microphone circuits. For clarity, the bobbin, and the top part of the core, have been omitted.*

The 4.7 nF capacitor across the input subdues the excitation of internal ringing of the 1000-turn winding, at frequencies in the region of 1 MHz, if a square-wave input is used, and an excellent non-overshooting 20 kHz square-wave output is then obtainable, with a rise-time of about 0.5 µs.

Owing to the rather curious configuration of the leakage flux field at such high frequencies, it is found that the output ringing is minimized when the effective 1-turn secondary formed by the screened leads is located midway between the two bobbin cheeks, and it is worth arranging for this to be the case.

At 20 Hz a virtually perfect square wave is obtained, the droop being only just discernible on careful inspection.

Magnetic screening of the transformer is quite unnecessary, but the effective loop area of the 1-turn circuit should be kept as small as possible, preferably by twisting the two pieces of cable together. Care should be taken, of course, to avoid allowing the cable screening to form a shorted turn – the screening of the two pieces must be earthed at one end only, as shown.

The XLR connectors and their contact numbering are arranged in accordance with the widely adopted IEC268/BS5428 Standard.[37]

8.5.19 A transformer for loudspeaker talk-back

A good loudspeaker may be operated in reverse as a very high-quality microphone.

It is shown in Reference 38 that if the loudspeaker output current is fed to an amplifier having an extremely low input impedance, and a frequency-response rising with frequency at 20 dB/decade throughout the audio spectrum, then the axial frequency response as a microphone will be identical to that obtained in normal loudspeaker use.

The Reference gives details of more elaborate circuit arrangements than that presented here, which provide lower distortion, better signal-to-noise ratio, and more accurate equalization, but the simple ferrite-cored transformer shown in Figure 8.44 constitutes a highly satisfactory practical compromise, very easy to implement.

The primary inductance is about 32 µH, which has a reactance of 2 Ω at 10 kHz, and because a ferrite core is used, the reactance is very linearly proportional to frequency. Thus, with an 8 Ω loudspeaker, a reasonable approximation to the ideal zero input impedance is provided.

Sound pressure levels up to at least 100 dB can be handled with adequately low distortion.

The 680 pF capacitor across the secondary, augmented by up to about 100 pF of cable and/or amplifier input capacitance, gives resonance at approximately 20 kHz, the resonance being well damped by the 22 kΩ resistor. The capacitor serves to suppress RF interference, and the damping, in addition to giving more closely the required frequency response, is also beneficial in reducing the risk of oscillation when this inductive source is connected to a microphone amplifier having internal negative feedback. A balanced output may be taken if preferred.

The use of a centre-tap on the bifilar primary is the best choice for this application, since unscreened loudspeaker cable is often used. It is found that this usually gives no interference problems. But it is wise to keep the loudspeaker away from television sets, whose line time-base magnetic field can cause trouble.

Provided the transformer is not much less than 1 m from mains transformers, no audible hum will be experienced even if no screening can is employed. The use of a gapless transformer core is preferable in this respect to having a gapped core with more turns, though the latter would give reduced distortion.

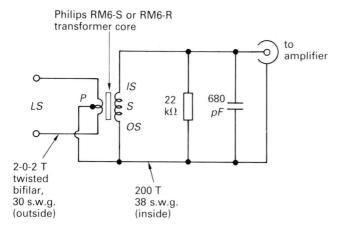

Figure 8.44 *Loudspeaker-talk-back transformer unit. An additional 1-0-1 T primary may be provided if desired, to give a more accurate frequency-response at high frequencies for measurement purposes, but at the sacrifice of 6 dB of signal-to-noise ratio.*

The transformer core employed has a minimum A_L value of 1900 nH (type RM6-S) or 2000 nH (type RM6-R). If only a gapped RM6 core is to hand – e.g. as obtainable from *R–S* Components – it may be very easily converted to the required gapless core by placing a piece of fairly fine wet-or-dry abrasive paper, used dry, on a good flat surface, and rubbing the appropriate half of the core on this for a minute or two.

Applications of the above principle to the subjective assessment of loudspeakers, to frequency response measurement, and to absolute sensitivity determination, are discussed in Reference 38.

References

1. Wente, E. C. (1917). A condenser transmitter as a uniformly sensitive instrument for the absolute measurement of sound intensity. *Phys. Rev.* **10**, 39–63.
2. West, P. E. F. A. (1972). The first five years. *BBC Engineering*, No. 92, 4–24.
3. In memoriam (1976). *JAES* **24**, 708.
4. Pawley, E. (1972). *BBC Engineering 1922–1972*, BBC Publications.
5. Microphone amplifiers (1928). *BBC Handbook*, pp. 213–215.
6. The control room amplifiers (1933). *BBC Year Book*, pp. 391–399.
7. Microphones (1933). *BBC Year Book*, pp. 371–384.
8. Recent developments in microphone design (1934). *BBC Year Book*, pp. 407–413.
9. Microphones (1932). *BBC Year Book*, pp. 339–342.
10. Barrett, A. E., Mayo, C. G. and Ellis, H. D. M. (1939). New equipment for outside broadcasts. *World Radio* (BBC), 21 July, 12–13; 28 July, 10–11.
11. Baxandall, P. J. (1980). Audio gain controls. *Wireless World*, **86**(1537), 57–62 and **86**(1538), 79–83.
12. Shorter, D. E. L. and Harwood, H. D. (1955). The design of a ribbon type pressure-gradient microphone for broadcast transmission. *BBC Eng. Mono.* No. 4.
13. Godfrey, J. W. (ed.) (1955). *Studio Engineering for Sound Broadcasting*, Iliffe, London.
14. Burrell Hadden, H. (1962). High-quality sound production and reproduction. Iliffe, London.
15. Berry, S. D. (1954). Newly developed amplifiers for the sound programme chain. *BBC Quarterly*, **9**, 111–122.

16. Berry, S. D. (1952). New equipment for outside broadcasts. *BBC Quarterly*, **7**, 120–128.
17. Berry, S. D. (1959). Transistor amplifiers for sound broadcasting. *BBC Eng. Mono*. No. 26.
18. Berry, S. D. (1963). The application of transistors to sound broadcasting. *BBC Eng. Mono*. No. 46.
19. Baxandall, P. J. (1952). Negative feedback tone control. *Wireless World*, **58**, 402–405.
20. Baxandall, P. J. (1962). Collectors upwards or positive upwards?, *Wireless World*, **68**, 23–26.
21. Baxandall, P. J. (1955). Gramophone and microphone preamplifier. *Wireless World*, **61**, 8–14 and 91–94.
22. Baxandall, P. J. (1968). Noise in transistor circuits. *Wireless World*, **74**(1397), 388–392 and **74**(1398), 454–459. Please note printer's error: in Fig. 9, middle diagram, current generator labelled $\sqrt{(2qI_cB)}/g_m$ should be $\sqrt{(2qI_cB)}/r_{b'e}g_m$.
23. Faulkner, E. A. (1968). The design of low-noise audio frequency amplifiers. *The Radio and Electronic Engineer* (JIERE), **36**(1), 17–30.
24. Fellgett, P. B. (1987). Thermal noise limits of microphones. *JIERE*, **57**(4), 161–166.
25. Cohen, G. J. (1984). Double balanced microphone amplifier. Preprint 2106, AES Australian Regional Convention, Melbourne, September. (A Philips contribution.)
26. Sowter, G. A. V. (1987). Soft magnetic materials for audio transformers: history, production and applications. *JAES*, **35**(10), 760–777.
27. Welsby, V. G. (1950). *The Theory and Design of Inductance Coils*, Macdonald, London.
28. Bush, H. D. and Tebble, R. S. (1948). The Barkhausen effect. *Proc. Phys. Soc.*, **60** Part 4 (340), 370–381.
29. Brailsford, F. (1951). *Magnetic Materials*, Methuen, London (monograph).
30. Feynman, R. (1975). *The Feynman lectures on physics*, Vol. II, Chapter 37, Addison-Wesley, London.
31. Story, J. G. (1938). The design of audio-frequency input and inter-valve transformers. *Wireless Engineer*, **15**(173), 69.
32. Gayford, M. L. (1985). The ribbon microphone. *Studio Sound*, **27**(9), 46–48; **27**(10), 98–100.
33. Ott, H. W. (1976). Noise reduction techniques in electronic systems. Wiley, Chichester.
34. Blesser, B. (1972). An ultraminiature console compression system with maximum user flexibility. *JAES*, **20**(4), 297–302.
35. Mayo, C. G., Ellis, H. D. M. and Tanner, R. H. (1939). Improvements in and relating to thermionic valve amplifiers. *Brit. Pat. Spec.* 514 729.
36. *Magnetic Circuits and Transformers* (Massachusetts Institute of Technology) (1944). Wiley, Chichester.
37. Dibble, K. (1982). Standard – what standard? *Studio Sound*, **24**(2), 54–56.
38. Baxandall, P. J. (1980). Loudspeakers as high-quality microphones. AES Preprint No. 1593 at 65th Convention, February.

9 Microphones in Stereophonic Applications

Francis Rumsey

9.1 Introduction

Modern recording and broadcasting is now firmly in the stereophonic era. Much has been written over the past century about stereo techniques, indeed it must be one of the best-documented and most thoroughly argued fields of audio engineering, and a study of this research shows that many of the principles of stereo were understood as early as 1930, although it was left to later workers to develop these principles commercially. There is good reason to go back to this original work since it reveals just how many of the principles are often ignored in modern sound recording, and how popular misconceptions have grown up concerning such matters as coincident pair techniques, binaural recording, and loudspeaker reproduction.

It is the intention of this chapter to investigate the principles of stereophony as they relate to microphone design and technique, considering both true stereo microphones as well as the use of mono microphones for stereo purposes. The discussion of 'stereo' is not limited to conventional two-loudspeaker reproduction, but considers it in the true sense of the Greek *stereo*, meaning 'solid', or 'three-dimensional'. Stereo techniques cannot be considered from a purely theoretical point-of-view, neither can the theory be ignored, the key being in a proper synthesis of theory and subjective assessment. Some techniques that have been judged subjectively to be good do not always stand up to rigorous theoretical analysis, and those that are held up as theoretically 'correct' are sometimes judged subjectively to be poorer than others. Part of the problem is that the mechanisms of directional perception are still not fully understood: a fact that is readily admitted by researchers in psycho-acoustics, although it is true to say that we are a good deal closer to understanding this subject than we were when stereo sound first appeared. Most commercial stereo reproduction uses only two loudspeakers, and thus the listening situation is already a distortion of reality (since real sonic experience involves sound arriving from all around the head), perhaps leading listeners to *prefer* distorted images because of the pleasing artefacts such as 'spaciousness' which are often side-effects, just as many listeners 'prefer' sound which is distorted in other ways.

In the following pages stereo pick-up and reproduction is considered from both a theoretical and a practical point-of-view, recognizing that theoretical rules may have to be bent or broken for operational and subjective reasons.

9.2 Directional perception

Any study of the stereophonic applications of microphones would be incomplete without first addressing the subject of directional perception in human hearing, since it is the aim of any stereo pick-up and reproduction technique to create the illusion of directionality in reproduced sound. A study of directional hearing will also help the reader considerably in appreciating the distinction between 'binaural', 'stereo', and 'ambisonic' techniques.

Directional hearing may coarsely be divided into three planes (see Figure 9.1): the lateral plane from left to right, the front–back or 'median' plane, and the vertical plane. In simple stereo reproduction it is the first of these that is the most important, although, as will be seen, it is not impossible to create illusions of the other two using transducers orientated in only the first plane. Front–back distinction becomes more important when considering microphone techniques that aim to offer some degree of 'surround-sound' information, and vertical distinction completes the picture if one is ever to attempt a complete reconstruction of the original sound field (see Section 9.6, Ambisonics).

Study of directional hearing principles extends over most of the last century and, although it can be said that we now understand the mechanisms involved fairly well, it should also be said that there remains considerable debate over

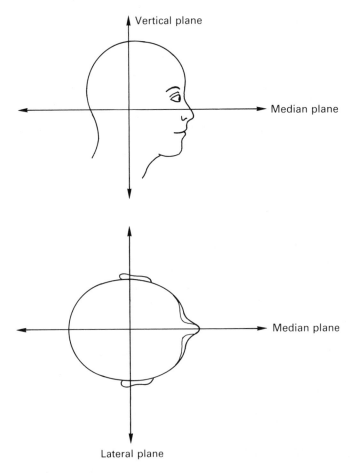

Figure 9.1 *Spatial perception works in three planes: the median (front-back), the lateral (left–right), and the vertical (up–down).*

specific theories, especially concerning the relative importance of the different mechanisms involved at different frequencies and with different types of sound. In the following section a resumé of the principal mechanisms will be given, with specific relevance to stereo sound reproduction. More comprehensive studies of directional hearing principles may be found in Moore[1] and Blauert.[2]

9.2.1 Lateral plane

In order to understand how the direction of sounds in the lateral plane are perceived (either in front of or behind the head) it is necessary to make a distinction between continuous repetitive waveforms (such as pure tones) and transient sounds or noises. The reason for this is that it is reasonable to talk about a 'phase difference' in degrees between two signals if they are continuous and have a recognizable fundamental frequency, but this concept becomes somewhat irrelevant when considering complex signals that start and stop unpredictably. In the discussion of directional hearing the existence of a phase difference between the two ears is often quoted, when what is really meant is a difference in time. When considering the mechanisms of hearing it is often convenient to use pure tones as a signal source example, but it should be remembered that one rarely listens to pure tones in real life, since real sounds are often of a more complex, noise-like nature.

One's ability to perceive directionality in sound depends almost entirely on the fact that two ears are involved, although there are more minor effects which may be perceived even if only one ear is involved (see Section (d) below). Principally, lateral distinction of direction is based on sound-level differences between the two ears, and on time-of-arrival differences between the two ears (which may be considered as phase differences in some cases). It is apparent that the brain relies on a combination of cues when assessing direction, and that the relative importance of the different cues changes with the frequency and nature of the source.

(a) Level difference

Considering for a moment the so-called 'free field' condition, in which sound energy from a source is not reflected off any boundary surface but is free to move away from the source in all directions, there will be two potential reasons for a difference in sound-pressure level (SPL) between the ears of a listener. Assuming that the sound source is at a given angle off the axis running from front to back of the head through the nose, there will be a difference in perceived level between the two ears such that the ear nearest the source will perceive a higher SPL than the distant ear. This is due firstly to the extra distance that the sound must travel between ears, and secondly to the shadowing effect of the head.

The level difference which results from the extra distance travelled is very small indeed in most real circumstances, since direct sound intensity obeys the inverse square law of decay with distance from the source. Put in practical terms inverse square law decay results in a level drop of 6 dB for every doubling of distance away from the source. Unless the source is very close to the head, the distance between the ears compared with the distance already travelled by the sound will be relatively small, and is unlikely to represent anything like a doubling of distance. To take an example (see Figure 9.2), considering a sound source 4 m from a listener, at 90° 'off centre' (i.e. resulting in the maximum path difference between the ears), and assuming that the extra distance travelled by the sound between the ears is 20 cm, the difference in intensity that would result due to inverse square law decay will be only 0.2 dB, which may be accepted as an

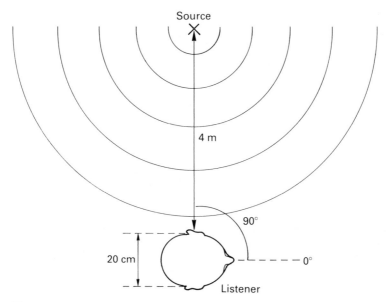

Figure 9.2 *A source at a distance of 4 m away from a listener, located at 90°
to the median plane, has already travelled quite far in relation to the path-length
distance between the ears (20 cm), and thus the drop in level due to the inverse
square law is small (~0.2 dB).*

imperceptible difference. If the source is very close to the head, then inverse
square law decay will play a larger part in the level difference between the
ears.

By far the more important reason for the existence of level difference is the
'shadowing' effect of the head on high-frequency sound. Acoustically, an object
only begins to act as an effective barrier to sound when the wavelength of the
sound is small in relation to the size of the object. Since the wavelength of audible
sound ranges from around 18 m to a few centimetres, it may be seen that at low
frequencies the head is much smaller than the wavelength, and thus will not act
as a barrier, but at high frequencies it will have a greater shadowing effect,
resulting in a loss of high frequencies on the 'dark side' (that is the side more
distant from the source). Figure 9.3 shows the resulting level difference calculated
by Steinberg and Snow,[3] based on tests with pure tones made by Sivian and
White,[4] earlier this century and it may be seen that for frequencies above about
1000 Hz the level difference between the ears becomes appreciable. This
frequency-intensity response difference has variously been referred to as a
'quality' or 'timbre' difference.

It is thus possible for the brain to base part of its estimation of direction on the
difference in spectral emphasis between the ears, the ear with the greater high-
frequency content being that closest to the source. In a reverberant room there
may be other reasons for a level difference between the ears, such as the existence
of standing wave patterns that result in pressure maxima and minima at points in
the room. In such cases the brain becomes more confused, having to rely heavily
on the transient information in the sound, using the first few milliseconds of the
arrival of the direct sound upon which to base its calculation of direction.

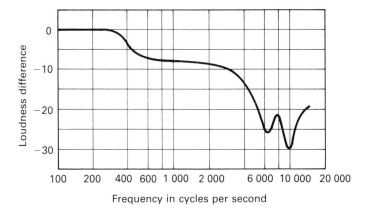

Figure 9.3 *The shadowing effect of the head results in a level difference between the ears which varies with frequency. This graph shows the level difference which results between the ears when a source of pure tone is placed to the left of a listener (after Sivian and White).*

(b) Phase difference

If one considers an imaginary sound source emitting a continuous repetitive waveform such as a sine wave, it is possible to talk in terms of a phase difference between the two ears of a listener for a given frequency and offset of the source. In such a case it should be noted that for any given position of the source the phase difference will be a function of frequency, becoming greater as the frequency increases up to a point where it becomes unclear which ear is lagging and which leading, since at the frequency where the distance between the ears is equal to half the wavelength there will appear to be a 180° phase difference between the ears, making it difficult to tell which ear is lagging and which leading, and also at multiples of this frequency (see Figure 9.4). Above the frequency where $d = \lambda/2$ (around 800 Hz) the phase difference cue becomes unreliable since the same phase difference may arise at a number of frequencies.

It has been shown that the phenomenon of 'phase locking' exists between the peaks of the audio waveform and the pattern of nervous discharge from the auditory fibres of the inner ear,[5] resulting in 'spikes' of electrical activity at peaks of the waveform. This provides a means for the brain to estimate the relative phase difference between the ears, since it can time the interval between these discharge spikes. Above about 700 Hz this phase-locked discharge becomes less reliable, with each fibre able to respond less often since each has a finite recovery time, although the neural discharge does not become completely asynchronous until around 4000 Hz. Between 700 and 4000 Hz the pattern of synchronous 'spikes' is maintained in the combination of outputs from many hundreds of fibres, all of which are stimulated, but not all of which 'fire' on each cycle of the waveform, the principle of which is shown diagrammatically in Figure 9.5. Above 4000 Hz the brain can no longer rely on the timing of nervous discharge from the ear, since the pattern becomes disorganized and somewhat random, but it is interesting to note that above this frequency a marked increase is noticed in the *level* difference between the ears due to the shadowing effect of the head. There

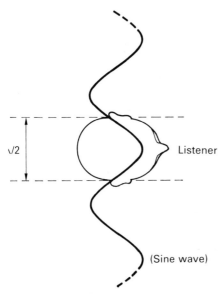

Figure 9.4 *At a certain frequency (~800 Hz) the half-wavelength of sound in air is approximately equal to the distance between the ears. Above this frequency confusion will result over which ear is lagging and which is leading in phase.*

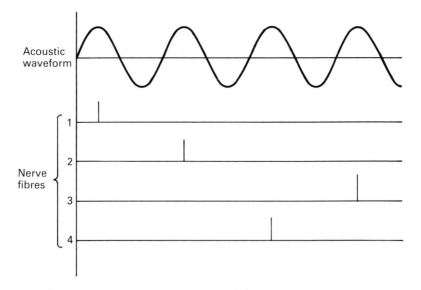

Figure 9.5 *Between about 700 Hz and 4000 Hz it is not possible for nerve fibres within the inner ear to discharge synchronously once per cycle of the audio waveform, but the periodic frequency of the wave is signalled to the brain in the combination of outputs from different fibres.*

is thus a 'cross-fade' area in the middle frequency range over which the brain begins to rely less on the relative timing of synchronous discharge, and more on the relative sound levels between the ears. This explains the increased errors noted by researchers in the location of pure tones by test subjects in this middle frequency range.

It must be noted that all of the above discussion refers principally to continuous repetitive waveforms, and that there is room for argument over the relevance of 'phase' as a parameter for measurement by the brain. As previously stated, for a given position of sound source the phase difference between the ears is a function of frequency, and so it is not possible to say that a source positioned at, say, 40° off-centre will result in a given number of degrees phase difference, since the number of degrees depends on the frequency. Similarly, if the frequency is held constant and the source is moved, the phase difference changes with position. It is clear therefore that for every position of sound source and for every frequency there is a different phase difference, and this would require that the brain maintained an infinitely long look-up table of frequency versus phase difference versus position, which seems unlikely.

Perhaps therefore it is less ambiguous, and more realistic when discussing real sounds in real surroundings, to talk in terms of a *time-of-arrival* difference between the ears, since this is not a function of frequency and can be stated for any position of the source whatever its frequency. Pure tones are generally more difficult to locate than transient sounds, especially in the middle-frequency region, where the phase cue becomes confusing and there is not enough level difference between the ears to be helpful. Most experiments that have shown accuracy by subjects in locating pure tones have been conducted in near free-field conditions, thus avoiding the confusion that arises out of reflections and pressure differences in the reverberant field. This is not to say that one should treat the phase cue as an irrelevance, but that its value should be regarded warily.

It should be noted that the actual response of the basilar membrane within the inner ear to a transient click is in fact similar to a decaying sinusoidal waveform, and Moore[1] suggests that it may be that the brain is able to compare the phases of these decaying oscillations and use this as an aid to localization. He points to research by Yost *et al.*,[6] which showed that tests using clicks that had been high-pass filtered so as to include only energy above 1500 Hz resulted in a deterioration of localization ability, but that the same clicks low-pass filtered so as to include only energy *up to* 1500 Hz resulted in little change in localization ability, suggesting that localization ability is largely related to the LF content of transients.

(c) Time-of-arrival difference

The ears in an average head are spaced apart by about 15 cm, resulting in a maximum path-length difference between them of around 20–23 cm due to the extra distance that must be travelled by a sound wave front to the more distant ear. The maximum path-length difference results when the source is at 90° to the so-called 'median plane' (running from front to back through the nose), and at this angle, assuming that the speed of sound is 330 m/s, the time-of-arrival (TOA) difference is around 600 μs, or 0.6 ms. At angles less than this, the TOA difference is proportional to the sine of the angle θ (see Figure 9.6).

Although a single ear is incapable of distinguishing between the arrivals of two sounds within such a short period,[7] binaural time differences between arrivals of single clicks as small as 10–20 μs, corresponding to angles of only a few degrees, are perceptible in terms of a change in apparent direction of the source. This assumes a single point source being perceived by two separate ears, and the

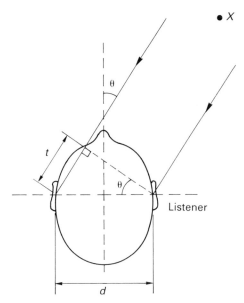

Figure 9.6 *A sound source at position X, offset from the median plane by angle θ, gives rise to a time-of-arrival difference t between the two ears of a listener, such that* $t = \dfrac{d \sin \theta}{330}$.

situation can be verified by feeding clicks on headphones to a listener with an adjustable delay in the feed to one ear. The angle of perceived offset is related directly to the number of microseconds delay between the ears, being always towards the advanced ear, up to the limit of around 0.6 ms delay, after which a 'phasey' image or indistinct position begins to result.

So far only the single point source has been considered, resulting in delays of up to 0.6 ms between the ears (the so-called 'binaural delay'), but we must also consider the situation in which more than one source exists, since this is one of the keys to understanding stereo sound reproduction. In such a situation, such as that which might exist when listening to the same sound emitted from two sources, Haas,[8] studying the effects of echoes on the perception of direction of a source, showed that the perceived direction of the sound tended towards the advanced (in time) source, and that the delays over which the phenomenon was noticed extended up to around 50 ms, a much greater delay than that over which the binaural effect applies. In the case of two sources both ears would hear both sources, and thus the situation would not be the same as that with a single source where one wave front arrived at each ear at slightly different times. Haas showed that for delays up to about 50 ms the sounds from the two sources would be 'fused' together by the brain, appearing as one source with the perceived location towards that of the first arrival. Beyond 50 ms the brain begins to perceive the sounds as distinct, and the second appears as an 'echo' of the first. With single clicks, the effect breaks down more quickly than with complex sounds, allowing delays of only up to around 5 ms before the 'fusing' effect disappears. Wallach *et al.*[7] also studied the precedence effect in sound localization, and their results bear detailed examination.

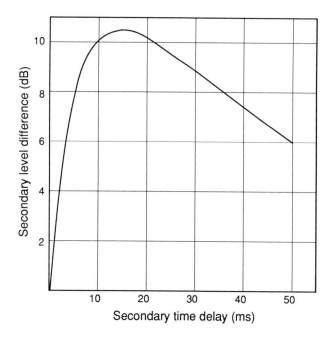

Figure 9.7 *The so-called 'Haas effect' shows the combination of time delay and level difference between primary and secondary sound sources for apparent equal loudness.*

The so-called 'Haas effect curve' (see Figure 9.7), shows that for the delayed source to appear equally as loud as the undelayed source it must be a certain number of decibels higher in level to compensate for the precedence effect of the first arrival, the peak of the effect being at a delay of around 15 ms, where the delayed source must be 11 dB louder than the undelayed source in order to be perceived as equally loud. We begin to see, therefore, the beginnings of a possibility for trading level difference against time difference to achieve the same directional effect.

An interesting recent discovery by Hafter *et al.*[9] has shown that there is evidence to suggest that the binaural system 'locks onto' the onset of a sound (Hafter used a train of clicks), adapting its perception of direction to that suggested by the first click (and the resulting delay between the ears), largely ignoring subsequent clicks, although suitable 'triggers' may release the perception mechanism from this adaptation. In his experiments such 'release triggers' took the form of a low-intensity burst of noise, a tone burst, and a pause in the train of clicks. It is as if the brain is prepared to accept a direction suggested by the onset of a sound which can thereafter be cancelled by a suitable trigger which resets the mechanism, such triggers being related to a change in the sound. This is of considerable importance to our understanding of stereo sound.

(d) The rôle of the pinna

In addition to the primary effects noted above there are more subtle effects that also have a bearing on the perception of directionality in the lateral plane. These are principally related to the rôle of the *pinna* (the visible skin and bone structure of the outer ear), which has recently been reconsidered. The effect of the pinna has not historically been considered very important in human sound localization, and does not feature greatly in older papers on stereo reproduction, but there is evidence to show that even though each person's ears have a different shape, the person relies considerably on information derived from reflections off different parts of it, resulting in slightly delayed components of the sound which combine additively and subtractively with the direct arrivals, thus resulting in a comb filter-like pattern of 'notches' in the perceived frequency spectrum, the position and depth of which depends on the angle of the source. Furthermore, resonances may result in the outer ear, which will have a similar effect on frequency response. By learning the patterns of spectral emphasis resulting from sounds in different locations the brain gains an extra tool for localization.

In experiments where people have had their pinnae filled in with a mouldable material, leaving simply a passage to the auditory canal, the ability to detect direction has been noticeably impaired. This will be discussed further when considering 'binaural' microphone techniques and sound reproduction.

(e) Combining the cues

The ability to detect the direction of sounds in the lateral plane therefore depends on a combination of frequency-dependent level difference (due principally to the shadowing effect of the head) at mid-to-high frequencies, on frequency-dependent phase differences for continuous low-frequency sounds, on time-of-arrival differences for complex transient sounds, and on the subtle spectral shaping effects of the pinna. There appears to be a cross-over zone for continuous sounds between about 700 and 4000 Hz in which confusion may arise due to the breaking-down of phase cues and the gradual increase in level difference, above which level difference becomes the most important cue. There is a need to distinguish between the 'binaural effect' in which the two ears are independently stimulated by the same wave front (resulting in delays from 0 to 600 μs) and the 'precedence' effect in which the two ears are both stimulated by two or more sources arriving at different times (which operates over delays up to about 50 ms, depending on the nature of the source). There is some evidence to suggest that the binaural hearing mechanism adapts itself to the direction suggested by the onset of a sound, ignoring subsequent information to some extent, unless that information succeeds in re-triggering the localization process. It is possibly true, therefore, that *changes* in the sound have a more persuasive influence on localization than continuous features.

9.2.2 Front–back or median plane

Less time will be spent on the discussion of the mechanisms for front–back and vertical perception, although they are still important for some stereophonic techniques.

Initially it is hard to see what distinguishes a point source at a given number of degrees off-centre in front of the head from the same source at the same angle behind the head, since the TOA difference will be the same and so, to a large extent, will be the shadowing effect of the head. Even so, it is clear that the freedom to move the head plays an important part in localization in this plane, since people whose heads have been held perfectly still have been found to display greater difficulty in front–back distinction. The reason for this is that even a small change in the rotation angle of the head will alter the TOA difference between the ears, and for a given direction of rotation this difference will get either smaller or greater, depending on whether the source is in front of or behind the head. Considering a source at 30° to the right of the median plane in front of the head, we see that turning the head to the right would lessen the TOA difference, whilst for the same source at 30° to the right behind the head we see that turning the head to the right would *increase* the TOA difference. Thus the brain may make small movements of the head and note the change in TOA that results.

A further factor is the rôle that sight plays in this distinction, since we use our eyes much more to determine the location of a source in front, whereas we must use our ears to determine the location of a source behind. If a source cannot be seen then it must be behind (ignoring the vertical for the moment)! This factor must not be over-played, since it is still possible to determine front–back location with the eyes closed, although there is some evidence to show that blind people have better aural localization to the front than sighted people.

A final factor is the effect of the pinna on sounds from the rear, since its size is such as to act as a partial barrier to very high-frequency sounds from the rear, changing the spectral emphasis of a sound from the rear when compared with the same sound from the front.

9.2.3 Vertical plane

Localization in the vertical plane is concerned partly with the effect of reflections from the ground and from the shoulders, in addition to pinna effects, since sounds at different angles of elevation will reach the ears directly and will also be reflected, reaching the ear slightly later, via indirect paths. It must be noted that there is a basic difference between the indoor and outdoor situations, since indoors there is a ceiling to consider, whereas outdoors there are likely to be no reflections from above. Furthermore, the only reliable reflection path distance to be accounted for is that from the shoulders, which for a given person remains fixed within limits of the movement of the head. For sounds from above, the difference in path length between the sound reflected from the shoulders and the floor compared with the direct path will result in cancellation and addition at different frequencies (in the same way as discussed above with reflections from the pinna, although reflections from the shoulders and floor result in notches and peaks which are more widely spaced, and which extend to a lower frequency, due to the distances involved). It is partly in the comparison of these frequency spectra with stored templates that the brain will locate sounds in the vertical plane.

The memory of learned situations, and also the expectation that particular sounds will emanate from certain directions, is also important for vertical localization, since there are not many sounds that come from below, as a general rule. The experience of standing on a mountain top whilst an aeroplane flies along a valley below is a most unusual one, since it is not the expected position of an aeroplane. Similarly, listeners played a high volume recording of a jet aeroplane taking off over the head will tend to duck, even though there is no obvious height information in the replay system.

9.3 Principles of directional sound reproduction

Now that the mechanisms by which direction is perceived in sound have been introduced, the principles of directional sound reproduction in audio systems will be discussed as a precursor to the discussion of stereo signals and microphone configurations. 'Stereophonic sound' will be considered in a broader sense than is often the case, since there is now a marked rekindling of interest in unconventional forms of directional sound reproduction such as three-channel stereo, wide-image stereo and ambisonics, due in part to the work being done on wide-screen and high definition television systems around the world.

9.3.1 Historical development

We have become used to stereo sound as a two-channel format, although a review of the last century shows that two channels really only became the norm through economic and domestic necessity, and through the practical considerations of encoding directional sound easily for gramophone records and radio. A two-loudspeaker arrangement is practical in the domestic environment, is reasonably cheap to implement, and provides good phantom images for a central listening position.

Early work on directional reproduction undertaken at Bell Labs in the 1930s[10] involved attempts to recreate the 'sound wavefront' that would result from an infinite number of microphone/loudspeaker channels (see Figure 9.8) by using a smaller number of channels (see Figure 9.9). In all cases, spaced pressure response (omnidirectional) microphones were used, each connected via a single amplifier to the appropriate loudspeaker in the listening room. Steinberg and Snow[3] found that when reducing the number of channels from three to two, central sources appeared to recede towards the rear of the sound stage and that the width of the reproduced sound stage appeared to be increased. They attempted to make some calculated, rather than measured, deductions about the way that loudness differences between the channels affected directional perception, choosing to ignore the effects of time or phase difference between channels. Some twenty years later Snow[11] made comment on those early results, reconsidering the effects of time difference in a system with a small number of channels, since, as he pointed out, there was in fact a marked difference between the multiple-point source configuration shown in Figure 9.8 and the small number of channels shown in Figure 9.9.

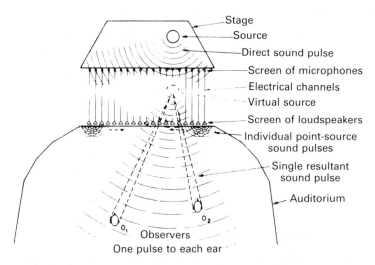

Figure 9.8   *A good virtual sound stage could be constructed by an infinite number of microphone-loudspeaker channels, giving rise to correct binaural differences for listeners in any position (after Steinberg and Snow).*

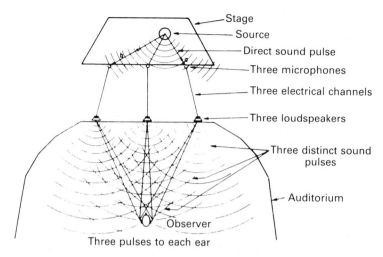

Figure 9.9 *The reduction in number of channels to three gives rise to three distinct arrivals of sound from the three loudspeakers, dependent on the microphone spacing. The differences between the ears are not the same as in natural listening, and the precedence effect takes over (after Steinberg and Snow).*

It was suggested that in fact the 'ideal' multi-source system recreated the original wave front very accurately, allowing the ears to use exactly the same *binaural* perception mechanisms as used in the real-life sound field. It may be appreciated that the 'wall' of multiple loudspeakers acted as a source of Huygens' wavelets, recreating a new plane wave with its virtual source in the same relative place as the original source, thus resulting in a TOA difference between the listener's ears in the range 0–600 μs, depending on position. In the two- or three-channel system, far from this simply being a sparse approximation to the 'wavefront' system, the ears are subjected to two or three discrete arrivals of sound, the delays between which are likely to be considerably in excess of those normally experienced in binaural listening, being a number of milliseconds depending on the original spacing of the microphones. In this case, the effect of directionality relies much more on the precedence effect described in the previous section, and also on the relative levels of the channels. Snow therefore begs us to remember the fundamental difference between 'binaural' situations, and what he calls 'stereophonic' situations (see Section 9.2.1(e)).

This difference was also recognized by Alan Blumlein, whose now famous patent specification of 1931 (accepted 1933)[12] allows for the conversion of signals from a binaural format suitable for spaced pressure microphones to a format suitable for reproduction on loudspeakers, and for other formats of pick-up that result in an approximation of the original time and phase differences at the ears when reproduced on loudspeakers. This will be discussed in more detail later on, but it is interesting historically to note how much writing on stereo reproduction, even in the early 1950s, appears unaware of Blumlein's most valuable work, which appears to have been ignored for some time, perhaps due to war and the financially depressed situation in the 1930s.

A British paper presented by Clark *et al.* (of EMI) in 1958[13] revives the Blumlein theories, and shows in more rigorous mathematical detail than Blumlein's patent specification how a two-loudspeaker system may be used to create an accurate correlation between the original angle of offset of a sound source and the perceived angle of offset on reproduction by controlling only the relative signal amplitudes of the two loudspeakers (derived in this case from a pair of coincident velocity-sensitive microphones). The system also adapts Blumlein's concept of sum-and-difference processing (see Section 9.4.1(a)) to increase the relative 'width' at low frequencies compared with that at high frequencies, since tests showed that increased channel level difference at LF in relation to that at HF was needed for accurate localization of phantom images. The authors discuss the three-channel spaced microphone system of Bell Laboratories, and suggest that although it produces convincing results in many listening situations it is uneconomical for domestic use, and that the two-channel simplification (again using spaced microphones, at about 3 m apart) has a tendency to result in the 'hole-in-the-middle' effect with which many modern users of spaced micro-phones may be familiar (the sound appearing to come either from the left or the right but not anywhere in between). They concede, in a discussion, that the Blumlein method adapted by them does not take advantage of all the mechanisms of binaural hearing, especially the precedence effect, but that they have endeavoured to take advantage of, and recreate, a few of the directional cues that exist in the real-life situation.

There is therefore a historical basis for both the spaced microphone arrange-ment which makes use of the precedence effect (with only moderate level differences between channels), as well as the coincident microphone technique or any other technique that results in only level differences between channels, with some evidence to show that the spaced technique is more effective with three channels than with only two. Later, we shall see that spaced techniques have a fundamental theoretical flaw from the point-of-view of 'correct' imaging of continuous sounds, which has not always been appreciated, although such techniques may result in subjectively acceptable sounds.

Interestingly, three front channels are the norm in cinema sound reproduction, since the central channel has the effect of stabilizing the important central image for off-centre listeners, having been used ever since the Disney film *Fantasia* in 1939. Gerzon[14] points out that people have perhaps misunderstood the intentions of Bell Laboratories in the 1930s, since it is not generally realized that they were working on a system suitable for auditorium reproduction with wide-screen pictures, as opposed to a domestic system. He suggests that three-channel stereo will become more important again, but this time in the home with wide-screen TV, requiring careful optimization of the signals for the new environment.

Alternatives to three-channel stereo exist for the coverage of a wider listening area, these involving loudspeaker arrangements that are directional in such a way as to compensate off-centre listeners with a higher audio level from the more distant speaker to make up for the increased precedence effect of the closer speaker. One interesting example of this is the recent Canon 'wide-imaging stereo' speaker system[15] which makes use of shaped conical reflectors mounted above mid- and high-frequency drivers pointing upwards, in order to give the loudspeaker a polar response that favours listeners on the opposite side of the listening area, compensating for the increased precedence effect of the closer speaker.

9.3.2 Two channels on loudspeakers

The principles of stereo reproduction using two channels on loudspeakers (as opposed to headphones) will now be discussed in more detail, since this is the most common listening arrangement in use today.

From the above discussions it will now be clear that in most cases stereo reproduction from two loudspeakers can only hope to achieve a modest illusion of the original sound field, since reproduction is from the front quadrant only (although see Section 9.3.5(b), Transaural stereo). It might be acceptable to state that the best illusion will be created when the sound signals present at the two ears are as similar as possible to those perceived in natural listening, even if one cannot entirely rationalize the psycho-acoustic mechanisms involved. It is possible to create this illusion using either time differences between the speaker outputs or using level differences between them. It is also possible to use a combination of the two, although there is a problem with 'time-difference' stereo in that contradictions may arise between transient and continuous sounds (see Section 9.5.3). The main point to be considered with loudspeaker reproduction is that both ears receive the signals from both speakers, whereas in headphone listening each ear receives only one signal channel. The result of this is that the loudspeaker listener seated in a centre seat (see Figure 9.10) receives at his left ear the signal from the left speaker first, followed by that from the right speaker, and at his right ear the signal from the right speaker first, followed by that from the left speaker, the time t being the time taken for the sound to travel the extra distance from the more distant speaker.

It can be shown that, (at least for low frequencies up to around 700 Hz),[13] if the outputs of the two speakers differ only in level and not in phase (time) then the vector summation of the signals from the two speakers at each ear results in two signals which, for a given frequency, differ in phase angle proportional to the

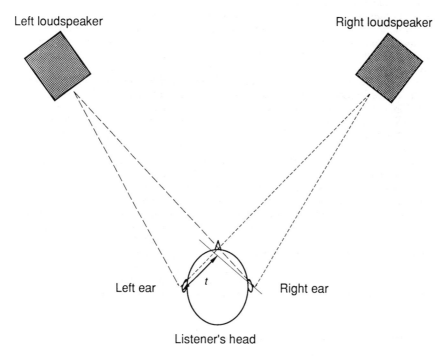

Figure 9.10 *When listening to sound on two loudspeakers, both ears receive sound from both loudspeakers, the signal from the more distant loudspeaker arriving a time t later than that from the nearer speaker for each ear.*

relative amplitudes of the two signals (the level difference between the ears being negligible at LF). For a given level difference between the speakers, the phase angle changes approximately linearly with frequency, which is the case when listening to a real point source. At higher frequencies the phase difference cue becomes largely meaningless (see Section 9.2.1 (b)) but the shadowing effect of the head results in level differences between the ears. If the amplitudes of the two channels are correctly controlled it is possible to produce resultant phase and amplitude differences for continuous sounds that are very close to those experienced with natural sources, thus giving the impression of virtual images anywhere between the left and right loudspeakers. This is the basis of Blumlein's 1931 stereophonic system 'invention'[12] although the mathematics is quoted by Clark *et al.* in 1957,[13] and further analysed by Bauer,[16] Bennett *et al.*,[17] and Lipshitz.[18] The result of the mathematical phasor analysis is a simple formula that can be used to determine, for any angle subtended by the loudspeakers at the listener, what the apparent angle of the virtual image will be for a given difference between left and right levels. Firstly, referring to Figure 9.11, it can be shown that

$$\sin \alpha = ((L - R)/(L + R)) \sin \theta_0$$

where α is the *apparent* angle of offset from the centre of the virtual image, and θ_0 is the angle subtended by the speaker at the listener. Secondly, it can be shown that

$$(L - R)/(L + R) = \tan \theta_t$$

where θ_t is the *true* angle of offset of a real source from the centre–front of a coincident pair of figure-eight velocity microphones. $(L - R)$ and $(L + R)$ are the well-known difference (S) and sum (M) signals of a stereo pair, defined in Section 9.4.1(a).

It is a useful result since it shows that it is possible to use positioning techniques such as 'pan-potting' which rely on the splitting of a mono signal source into two components, with adjustment of the relative proportion fed to the left and right channels without affecting their relative timing. It also makes possible the combining of the two channels into mono without cancellations due to phase difference.

Experiments with wideband speech signals replayed on a standard loud-speaker arrangement (see Section 9.3.6) have shown that a level difference of approximately 18 dB between channels is necessary to give the impression that a sound comes from either 'fully left' or 'fully right' in the image (see Figure 9.12), and that there is a measure of disagreement between listeners as to the positions of 'half left' and 'half right', which might be expected. If a time difference also exists between the channels, then transient sounds will be 'pulled' towards the advanced speaker because of the precedence effect, the perceived position depending to some extent on the time delay. If the left speaker is advanced in time relative to the right speaker (or more correctly, the right speaker is delayed!) then the sound appears to come more from the left speaker, although this can be corrected by increasing the level to the right speaker. There is a trade-off between time and level difference, and it can be shown that if the left channel is, say, 2 ms earlier than the right, then the right must be made approximately 5 dB louder in order to compensate and bring the signal back to the centre. A time difference of between 2 and 4 ms (depending on the nature of the signal) appears to be required for a sound to appear either fully left or fully right, although this holds true only for transient signals and may be contradicted by the resulting LF phase differences for continuous sounds, as argued by Lipshitz.[18] Stereo microphone techniques operate using combinations of level and time difference between the channels.

It may be shown[13] that a coincident arrangement of velocity (figure-eight) microphones at 90° to one another produce outputs that differ in amplitude with varying angle over the frontal quadrant by an amount that gives a very close

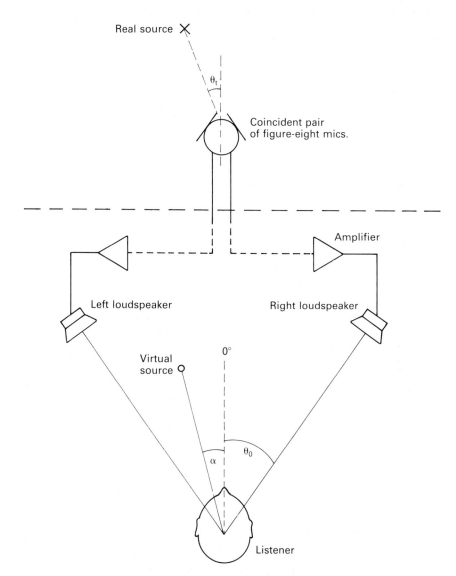

Figure 9.11   *A sound source is offset by angle θ_t from the centre–front of a pair of coincident figure-eight microphones, and is reproduced on loudspeakers each subtending an angle of θ_0 to the listener. The virtual source is reproduced at angle α (see text).*

correlation between the true angle of offset of the original source from the centre line and the apparent angle on reproduction from loudspeakers that subtend an angle of 120° (*sic*) to the listening position. At lesser angles the change in apparent angle is roughly proportionate as a fraction of total loudspeaker spacing, maintaining a correctly proportioned 'sound stage'. It may also be shown that the difference in level between channels should be smaller at HF than at LF in order to preserve constant correlation between actual and apparent angle, and this may be achieved by using an equalizer in the difference channel (see Section 9.4.1(a)) which attenuates the difference channel by a few decibels for frequencies above 700 Hz (this being the subject of a British patent by Vanderlyn[19]). Gerzon has

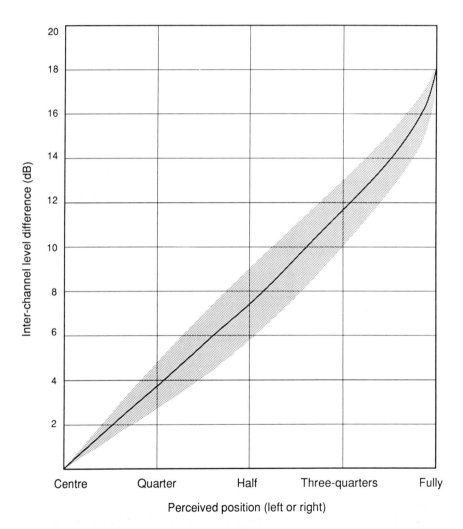

Figure 9.12 *Speech sources appear at different positions on reproduction, depending on the level difference between the channels. The shaded areas shows measured disagreement between listeners.*

suggested a figure between 4 and 8 dB[20] depending on programme material and spectral content, calling this 'spatial equalization'.

The ability of a system based only on level differences between channels to reproduce the correct timing of *transient* information at the ears has been questioned, not least in the discussion following the original paper presentation by Clark *et al.*, but these questions have been tackled to some extent by Vanderlyn in a much later paper of 1979,[21] in which he attempts to show how such a system can indeed result in timing differences between the neural discharges between the ears, taking into account the integrating effect of the hearing mechanism in the case of transient sounds. He quotes experimental evidence to support his hypothesis which is convincing. If a system based only on level differences did not cope accurately with transients, then one would expect transients to be poorly localized in subjective tests, and yet this is not the case, with transients being very clearly located in coincident-pair recordings.

9.3.3 More than two channels on loudspeakers

Sound for pictures has long made use of more than two loudspeaker channels to cover a wide listening area, to 'fix' the central image where dialogue tends to reside, and to offer some additional spatial information such as rear images or 'surround sound'. Multi-channel reproduction for domestic usage has had a chequered and not altogether successful history. Quadraphonic reproduction systems appeared in the 1970s to a number of different standards, but they have not found continued favour and thus will not be discussed further here. A system that appears to have survived the test of time, although still not widely used, is the so-called 'ambisonic' system, developed under the auspices of the NRDC in the UK during the 1970s and based mainly on work by Gerzon, Fellgett and Barton. This system is based on well-researched psycho-acoustic principles and aims to reproduce the original sound field as accurately as possible in a wide variety of possible configurations. Its principal child in commercial terms is the 'sound field' microphone, which is discussed in Section 9.6.4.

(a) Cinema-style surround sound

The most popular multi-channel sound format for the cinema at the present time is based on four channels: left (L), centre (C), right (R) and 'surround' (S), with sound signals panned into positions around the listener during post-production by varying the relative levels between the channels. The typical positions of loudspeakers are as shown in Figure 9.13. All dialogue is normally sent to the centre channel, with music and effects using the space of the other three channels. The *LCRS* format for sound signals is inherent in the 'Dolby stereo' format for cinema sound tracks, and Dolby Laboratories manufacture an encoder and decoder matrix which allows for the four channels to be combined into two channels for recording onto film with reasonable compatibility in the case of two-channel listening.

(b) Ambisonics (introduction)

The ambisonic system is really a complete and unified approach to directional sound pick-up, storage and reproduction, allowing for anything from mono up to full 'periphonic' sound in three dimensions. Ambisonic signals in the 'B-format' consist of a pressure component (W) and three velocity components in quadrature for directional encoding (X, Y and Z), out of which may be derived signals to be fed to loudspeakers around the listener, by suitable matrixing. This system is covered in greater detail in Section 9.6.

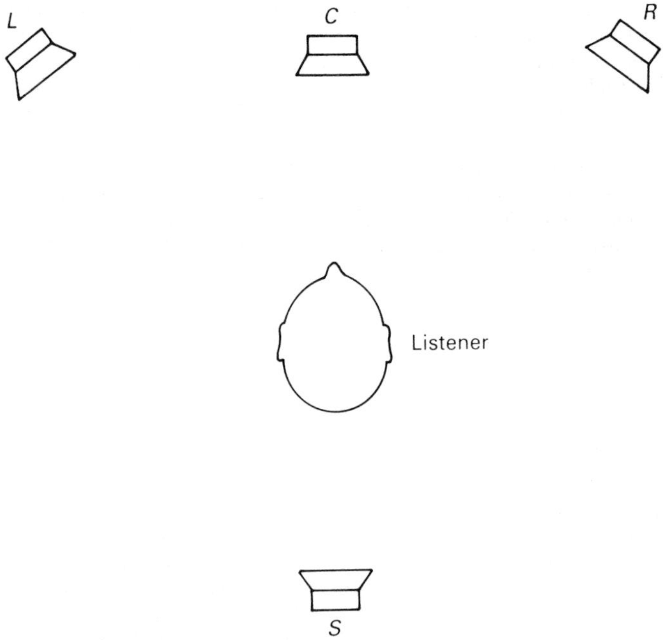

L

C

R

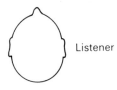

Listener

S

Figure 9.13   *In cinema surround-sound, loudspeakers are placed left (L), centre (C), right (R) and surround (S). There may be a number of surround loudspeakers arrayed behind the listener.*

(c) 3/2 surround formats

Although it has been the goal for many years, a universal surround loudspeaker format has proved elusive to date, although now there may be some light at the end of the tunnel. Working groups of the SMPTE,[22] the EBU,[23] and the CCIR[24] have all made draft recommendations for a 3/2-channel surround reproduction system – that is a system with three front channels (*L, C, R*) and two rear channels (stereo surround). It is also intended optionally to include a '0.2' channel carrying only low-frequency non-directional information for the bass 'boom'. It is intended that although the surround format would be used for high-definition television (HDTV), there would also be audio-only applications.

In the 3/2 standard it is expected that the front channels will carry directional information, and the rear channels will carry stereo ambience and effects signals that are not critically located, so that there could be considerable flexibility in the placement of the surround loudspeakers. Indeed, it is suggested that the surround loudspeakers will often be placed nearer the sides, rather than at the rear of domestic rooms. Theile[25] says that 'the addition of side/rear loudspeakers does not enlarge the listening angle (by delivering genuine surround localization of phantom sources); rather it adds an acoustic environment to the frontal stereophonic presentation of directional sound'.

This channel configuration does not preclude the use of the 3/2 loudspeaker layout for attempts at three-dimensional spatial localization, and the working groups have been clever in specifying principally the reproduction format, rather than saying how material should be produced for that format. That way they have left the way clear for audio-only applications of the 3/2 format which may wish to use the rear channels for all-round localization. The CCIR has also produced a draft hierarchy of alternative lower-level formats for smaller numbers of channels, allowing configurations such as 3/1, 3/0, 2/2, 2/0 (conventional stereo) and 1/0 (mono), with matrix equations to suit each.

Although proposals have been made for many loudspeaker arrangements and required numbers of channels for HDTV surround sound, the 3/2 format is a pragmatic compromise, which stands a good chance of being implementable, and which is capable of convincing reproduction. It is suggested that it will also prove acceptable for audio-only applications, such as surround broadcasts of music in future DAB (digital audio broadcasting) systems.

(d) Holophonics

A 'holophonic' system was described by Zucarelli and used on a number of projects in the early 1980s to remarkably good effect. The technique appeared to be a modified and highly optimized binaural technique,[26] the results of which were particularly stunning on headphones, being one of the only binaural techniques capable of producing accurate headphone localization in front of the listener. Recordings were also not unconvincing when reproduced on loud-speakers. Unfortunately a certain amount of mystery surrounded the technique, and Zuccarelli's attempts at a theoretical explanation for the success of his system[27] were ridiculed in subsequent letters. Since unmodified binaural techniques do not work well on loudspeakers this system appears to have sunk into the background, like most binaural techniques, despite the wide use of 'personal' cassette players with headphones.

9.3.4 Two channels on headphones

Headphone reproduction is different from loudspeaker reproduction since, as already stated, each ear is fed only with one channel's signal. This is therefore an example of the binaural situation and allows for the ears to be fed with signals which differ in time by up to the binaural delay (600 μs), and also differ in amplitude by amounts similar to those differences which result from the shadowing effects of the head. This suggests the need for a microphone technique using microphones spaced apart by the binaural distance, and baffled by an object similar to the human head, in order to produce signals with the correct differences (see Section 9.5.2(a)(iv)).

Bauer[28] pointed out that if stereo signals designed for reproduction on loudspeakers were fed to headphones there would be a too-great level difference between the ears compared with the real-life situation, and that the correct interaural delays would not exist. This results in an unnatural stereo image that lacks the expected sense of space. He therefore proposed a network which introduced a measure of delayed cross-talk between the channels to simulate the correct interaural level differences at different frequencies, as well as simulating the interaural time delays resulting from loudspeaker signals incident at 45° to the listener. He based the characteristics on research done by Weiner[29] which produced graphs for the effects of diffraction around the human head for different angles of incidence. The characteristics of Bauer's circuit are shown in Figure 9.14 (with Weiner's results shown dotted). It may be seen that Bauer chooses to reduce the delay at HF, partly because the circuit design would have been too complicated, and partly because localization relies more on amplitude difference at HF anyway.

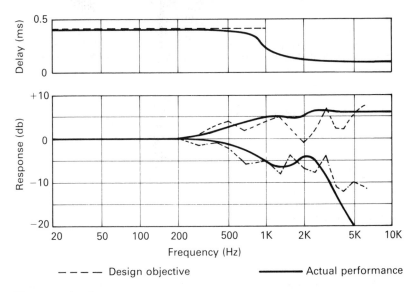

Figure 9.14 *Bauer designed a circuit that introduced binaural delay and cross-talk into signals designed for loudspeaker reproduction so that they could be reproduced satisfactorily on headphones. The characteristics of his circuit are shown here. The upper graph shows the amount of delay introduced in the cross-feed between channels, whilst in the lower graph L_g and R_g are left and right channel gains to imitate the shadowing effects of the head.*

Interestingly Bauer's example of the stereophonic versus binaural problem chooses spaced pressure microphones as the means of pick-up, showing that the output from the right microphone for signals at the left of the image will be near zero. This is likely only if the microphones are very close to the source (as in a multi-microphone balance), whereas in many spaced omnidirectional arrays there will be a considerable output from the right microphone for sounds at the left of the sound stage, thus there will also be a time delay equivalent to the path-length difference between the source and the two microphones which will add to that introduced by his network. In fact what Bauer is suggesting is really a variation on Blumlein's 'shuffler' network[12] which would work best with non-spaced microphones (i.e: a directional coincident pair). Blumlein's shuffler converted the phase differences between two binaurally spaced microphones into amplitude variations to be reproduced correctly on loudspeakers, whereas Bauer is trying to insert a phase difference between two signals which differ only in level (as well as constructing a filter to simulate diffraction effects).

Bauer also suggests the reverse process (turning binaural signals into stereo signals for loudspeakers), pointing out that cross-talk must be *removed* between binaural channels for correct loudspeaker reproduction, since the cross-feed between the channels will otherwise occur twice (once between the pair of binaurally spaced microphones, and again at the ears of the listener), resulting in poor separation and a narrow image. He suggests that this may be achieved using the subtraction of an anti-phase component of each channel from the other channel signal, although he does not discuss how the time difference between the binaural channels may be removed.

Further work on a circuit for improving the stereo headphone sound image was done by Thomas [30] in 1977, quoting listening tests which showed that all of his listening panel preferred stereo signals on headphones which had been subjected to the 'crossfeed with delay' processing. Such processes form the basis for 'transaural stereo', introduced below.

9.3.5 Processing of binaural signals

(a) Binaural synthesis

Recent work involving digital signal processing[31] has resulted in a system which can be used for processing audio signals so as to introduce the correct 'head-related transfer function' (HRTF) for every possible angle of incidence of a monophonic sound source. In practice, the system employs a series of digital filters, delays and mixers which simulate the effect of the pinnae on sound signals as well as simulating the delay and shadowing effects of the head. This has resulted in a system capable of acting as a 'binaural mixer', with the facility for positioning mono sounds at any elevation or offset angle around the listener. It is also possible to use the system as a 'room simulator' in order to test the effects of placing sources in virtual rooms of different shapes, sizes and surface character-istics, allowing a listener to place himself anywhere in this virtual room and receive the correct binaural signals. This could be of considerable benefit to acoustic designers.

(b) Transaural stereo

A recent development in the field of stereo recording and reproduction exists in the form of 'transaural stereo', a technique partly based on the ideas of Bauer (cited above) in relation to the cancellation of cross-talk between signals designed for binaural monitoring on headphones, so as to make them suitable for loudspeaker reproduction. The basis of transaural stereo, as described by Cooper and Bauck[32] is in the pre-conditioning of loudspeaker signals with a component representing the *inverse* of the cross-talk which occurs between the two ears when listening to loudspeakers, thus allowing the ears to become independent once more, and allowing them to be supplied with the correct binaural signal relationships. He suggests the use of binaural recording techniques, or binaural synthesis, to originate the signals which are then pre-conditioned to create a transaural master, suitable for loudspeaker reproduction.

Experiments by Atal and Schroeder in 1962 (cited by Cooper and Bauck) showed that it was possible to introduce a cross-talk-cancelling component into the signals fed to two loudspeakers which would compensate for the interaural cross-talk, although the system was so position-critical on the part of the listener (he had to be within 75 mm of the 'correct' position), as well as requiring that the listening environment be anechoic, that the concept did not proceed into general acceptance. Nonetheless, the results, for listeners in the correct environment and position, were said to be stunning, with directional information present in all dimensions.

The breakthrough described by Cooper lies in a modification of this early work, resulting in improvements in ordinary stereo reproduction derived from binaural recordings, which are less dependent on listener position, may be achieved with relatively simple electronics, and do not require anechoic listening conditions. He also suggests that the system is capable of offering full surround-sound effects from only two loudspeakers, and that it allows for the creation of 'virtual loudspeakers' in positions other than those of the real loudspeakers. The work is too lengthy even to be summarized here, for reasons of space, and the listener is referred to Cooper's excellent paper. Further recent studies of similar transaural techniques may be found in Griesinger,[33] Moller,[34] and Gierlich and Genuit,[35] all of which are concerned to a greater or lesser extent with the processing of binaural signals for loudspeaker reproduction.

9.3.6 Room acoustics and loudspeaker placement

The detailed effects of room acoustics and loudspeaker placement or performance on stereo reproduction will not be discussed further here. The reader is referred to the excellent study of the subject which exists in Davis.[36]

9.4 Stereo signals

9.4.1 Terminology

In the following section the nature of stereo audio signals will be discussed, together with definitions of the terms used to describe the various formats of stereo sound, since stereo microphones may operate in a number of possible ways. Also included is a discussion of the effects of misalignment on stereo signals.

(a) A, B, M and S signals

It is conventional in broadcasting terminology to refer to the left (L) channel of a stereo pair as the *A* signal and the right (*R*) channel as the *B* signal. In the case of some stereo microphones or systems these are called respectively the *X* and the *Y* signals. In a two-channel stereo system the *A* signal feeds the left loudspeaker and the *B* signal feeds the right loudspeaker. In colour-coding terms (for meters, cables, etc.), the *A* signal is coloured red and the *B* signal is coloured green. This may be confusing when compared with some domestic hi-fi wiring conventions that use red for the right channel, but it is the same as the convention used for port and starboard on ships. Furthermore there is a German convention that uses yellow for the left channel and red for the right.

It is sometimes convenient to work with stereo signals in the so-called 'sum-and-difference' format, since it allows for the control of image width and ambient signal balance. The sum or main signal is denoted *M*, and is based on the addition of *L* and *R* signals, whilst the difference or side signal is denoted *S*, and is based on the subtraction of *R* from *L* to obtain a signal which represents the difference between the two channels (see Section 9.4.2). The *M* signal is that which would be heard by someone listening to a stereo programme in mono, and thus it is important in situations where the mono listener must be considered, such as in broadcasting. Colour-coding convention holds that *M* is coloured white, whilst *S* is coloured yellow, but it is sometimes difficult to distinguish between these two colours on certain meter types.

(b) W, X, Y and Z signals in ambisonic systems

Ambisonic encoding involves the representation of a sound field by a number of signal components. There are various formats for ambisonic signals, but the so-called *B* format is that designated for professional use in the studio since it allows for great flexibility in processing, and is relatively robust in the face of system alignment errors. There are some similarities between *B* format signals and the *MS* format described above, but the *B* format allows for three-dimensional soundfield representation.

The *B* format consists of four components: *W*, which is an omnidirectional pressure component, and *X*, *Y*, and *Z*, which are directional velocity components (equivalent to the outputs of figure-eight velocity microphones whose output varies with the cosine of the angle of incidence). *X* represents the forward component (roughly equivalent to the *M* signal defined above), *Y* represents a sideways component (roughly equivalent to the *S* signal defined above), and *Z* represents a vertical component (which has no parallel in conventional two-channel stereo.) *W*, *X* and *Y* are sufficient to represent a sound field in the horizontal plane, whilst *Z* may be added to give height information. The maximum directional gains of the *X*, *Y* and *Z* components are 3 dB above that of the *W* component in order that all signals will have roughly equal energy for typical sound fields.

As will be seen in Section 9.6, it is possible to combine these pressure and vector components in different proportions and configurations to achieve virtually any directional pattern of pick-up, pointing in any direction. For example, the combination of the *W* (pressure) and *X* (cosine) components will give the equivalent of a forward-facing cardioid pick-up pattern ($1 + \cos \theta$).

L, *C*, *R* and *S* in cinema surround systems

Cinema surround sound makes use of four directional signals: left (*L*), centre (*C*), right (*R*) and surround (*S*). The *L*, *C* and *R* signals are distributed to three loudspeakers spread across the screen width, and the mono surround signal is divided between a number of speakers behind the listening area. This four-channel sound signal is often encoded into two channels for recording onto film, by means of a Dolby stereo encoder (see Section 9.4.4(a)).

9.4.2 Derivation of stereo signals

LR, *XY*, or *AB* stereo signals (they are all the same) may be derived by many means. Most simply, they may be derived from a pair of coincident directional microphones orientated at a fixed angle to each other (see Section 9.5.1(a)). Alternatively they may be derived from a pair of spaced microphones, either directional or non-directional, with an optional third microphone bridged between the left and right channels. Finally they may be derived by the splitting of one or more mono signals into two, by means of a 'pan-pot', which is really a dual-ganged variable resistor which controls the relative proportion of the mono signal being fed to the two legs of the stereo pair, such that as the level to the left side is increased that to the right side is decreased. Panpot laws are covered in further detail in Section 9.5.2(b)(ii).

MS or 'sum and difference' format signals may be derived by conversion from the *AB* format using a suitable matrix (see Section 9.4.3) or by direct pick-up in that format. A coincident pair of microphones may be used for direct pick-up, such that the *S* signal is derived from a sideways-facing figure-eight, and the *M* signal is derived from one of a variety of polar patterns, depending on the desired final balance (see Section 9.5.1(a)). For every *AB* stereo pair of signals it is possible to derive an *MS* equivalent, since *M* is the sum of *A* and *B*, whilst *S* is the difference between *A* and *B*. Likewise, signals may be converted from *MS* to *AB* formats using the reverse process.

9.4.3 Conversion between signal formats

In order to convert an *AB* signal into *MS* format it is necessary to follow some simple rules. Firstly, the *M* signal is not usually a simple sum of *A* and *B*, as this will result in over-modulation of the *M* channel in the case where a maximum level signal exists on both *A* and *B* (representing a central image); therefore a correction factor is applied, ranging between − 3 dB and − 6 dB (equivalent to a division of the voltage by between $\sqrt{2}$ and 2 respectively). For example,

$$M = (A + B) - 3\,\mathrm{dB} \quad \text{or} \quad M = (A + B) - 6\,\mathrm{dB}$$

The correction factor depends on the nature of the two signals to be combined. If identical signals exist on the *A* and *B* channels (representing 'double mono' in effect), then the level of the uncorrected sum channel (*M*) will be two times (6 dB) higher than the levels of either of *A* or *B*, requiring a correction of − 6 dB in the *M* channel in order for the maximum level of the *M* signal to be reduced to a comparable level. In a television production where most of the sound sources are panned centrally (e.g. central dialogue) the 6 dB correction factor would be more appropriate, since there is a strong likelihood that a large number of operational situations will result in identical *A* and *B* signals close to maximum level. If the *A* and *B* signals are non-coherent (random phase relationship), then only a 3 dB rise in the level of *M* will result when *A* and *B* are summed, requiring the − 3 dB correction factor to be applied. This is more likely with stereo music signals. As most stereo material has a degree of coherence between the channels, the actual rise in level of *M* compared with *A* and *B* is likely to be somewhere between the two limits for real programme material.

The *S* signal results from the subtraction of *B* from *A*, and is subject to the same correction factor. For example,

$$S = (A - B) - 3\,\mathrm{dB} \quad \text{or} \quad S = (A - B) - 6\,\mathrm{dB}$$

S can be used to reconstruct *A* and *B* when matrixed in the correct way with the *M* signal (see below), since (*M* + *S*) = 2*A* and (*M* − *S*) = 2*B*. It may therefore be appreciated that it is possible at any time to convert a stereo signal from one format to the other and back again.

If either format is not provided at the output of microphones or mixing equipment it is a relatively simple matter to derive one from the other electrically. Figure 9.15 shows two possible methods. Figure 9.15(a) shows the use of transformers, where *A* and *B* signals are fed into the primaries, and *M* is derived by wiring the two secondaries in series and in phase, while *S* is derived by wiring two secondaries in series and out of phase. Figure 9.15(b) shows the use of summing amplifiers, whereby *A* and *B* are summed in phase to derive *M*, and summed with a phase inversion (− 180°) in the *B* leg to derive *S*. Both matrixes have the advantage that they will also convert back to *AB* from *MS*, in that *M* and *S* may be connected to the inputs and they will be converted back to *A* and *B*.

9.4.4 Encoding of multi-channel stereo into two channels

(a) Dolby stereo

The Dolby stereo encoding process takes LCRS format signals (see Section 9.4.1(c)) and matrixes them together into a two-channel format which may subsequently be decoded back to four channels. The aim of any matrixing system

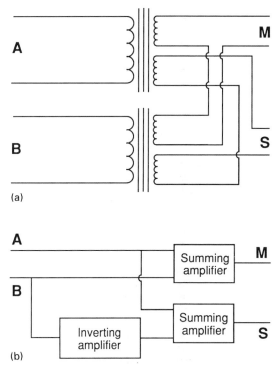

Figure 9.15 *Stereo signals may be converted from the AB format to MS format either by using transformers wired with secondaries in-phase to derive M and out-of-phase to derive S, as shown in (a), or by using summing amplifiers as shown in (b).*

like this is to achieve the encoding with minimum side-effects for the person who listens to the two channel version without decoding.

The Dolby 'Prologic' decoder matrix block diagram is shown in Figure 9.16. Left, centre and right signals are distributed in phase and in correct proportion between the left and right channels of the two-channel encoder output, whereas the surround information is added equally to both channels but out of phase. On decoding the equal in-phase information is sent to the centre–front, and the difference information is sent to the surround channel, being subjected also to a delay (about 30 ms) and low-pass filter, as well as noise reduction. The delay is added to ensure that no confusion exists between centre–front and rear signals, and takes advantage of the Haas effect (see Section 9.2.1(c)) to ensure that listeners towards the rear of the cinema hear sounds from the front first. The filtering ensures that HF directional information, which might pull sibilant central sounds towards the rear, is reduced in the surround channel, as well as making the surround channel less 'obvious'.

(b) UHJ encoding

The principles of UHJ encoding of ambisonic signals are discussed in detail in Section 9.6.3.

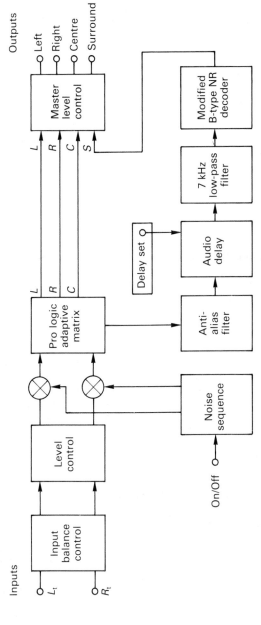

Figure 9.16 *Dolby Surround Pro Logic decoder block diagram (courtesy of Dolby Labs).*

9.4.5 Effects of misalignment on stereo signals

Differences in level, frequency response and phase may arise between signals of a stereo pair, perhaps due to losses in cables, misalignment, and performance limitations of equipment. It is important that these are kept to a minimum for stereo work, as inter-channel anomalies result in various audible side-effects. Differences also result in poor mono compatibility. These differences and their effects are discussed below.

(a) Frequency response and level

A difference in level or frequency response between A and B channels results in a stereo image biased towards the channel with the higher overall level or that with the better HF response. Also, an A channel with excessive HF response compared with that of the B channel will result in the apparent movement of sibilant sounds towards the A loudspeaker. Level and response misalignment on MS signals results in increased cross-talk between the equivalent A and B channels, such that if the S level is too low at any frequency the AB signal will become more monophonic (width narrower), and if it is too high the apparent stereo width will be increased.

(b) Phase

Inter-channel phase anomalies will affect one's perception of the positioning of sound source, and it will also affect mono compatibility. Phase differences between A and B channels result in 'comb-filtering' effects in the derived M signal due to cancellation and addition of the two signals at certain frequencies where the signals are either out of or in phase.

(c) Cross-talk

It was stated earlier that an inter-channel level difference of only 18 dB was required to give the impression of a signal being either fully left or fully right. Cross talk between A and B signals is not therefore usually a major problem, since the performance of most audio equipment is far in excess of these requirements. Excessive cross-talk between A and B signals results in a narrower stereo image, whilst excessive cross-talk between M and S signals results in a stereo image increasingly biased towards one side.

9.5 Stereo microphone configurations

9.5.1 The true stereo microphone

It is difficult to define the term 'stereo microphone', but one might attempt to define it as any microphone which is capable of providing outputs which between them represent directional information about the sound field that exists at the microphone in two or more planes. A stereo microphone typically comprises two or more directional capsules, and as such a stereo microphone is really at least two mono microphones in one housing. It is difficult to make a stereo microphone which operates as a spaced pair, unless the spacing is very small, for obvious physical reasons, and so stereo microphones tend to be based on the coincident-pair principle. Spaced pair stereo is usually achieved using separate mono microphones, unless a 'dummy head' is used for binaural recording, in which case two small microphones may be buried in the ear canals of the head (see Section 9.5.2(a)(iv)

(a) Coincident pair principles

The coincident-pair stereo microphone incorporates two directional capsules which may be angled over a range of settings to allow for different configurations and operational requirements. The pair may be operated in either the *AB* or *MS* modes, and a matrixing unit is sometimes supplied with microphones that are intended to operate in the *MS* mode in order to convert the signal to *AB* format for recording. The directional patterns (polar diagrams) of the two microphones need not necessarily be figure-eight, although if the microphone is used in the *MS* mode the *S* capsule must be figure-eight (see below). Directional information is encoded solely in the level differences between the capsule outputs, since the two capsules are mounted physically as close as possible, and therefore there are no phase differences between the outputs except at the highest frequencies, where inter-capsule spacing may become appreciable in relation to the wavelength of sound.

(i) *AB* pairs

AB stereo microphones (sometimes known as *XY* microphones), such as the one shown in Figure 9.17, may be mounted vertically in relation to the sound source,

Figure 9.17 *A typical coincident stereo microphone (courtesy of Beyer).*

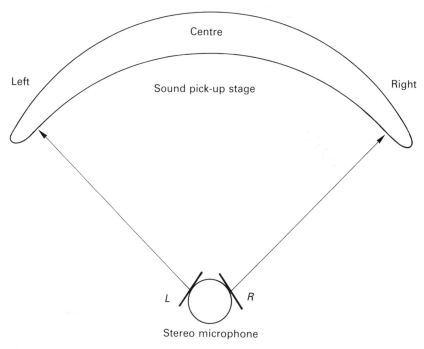

Centre

Left

Sound pick-up stage

Right

L   R

Stereo microphone

Figure 9.18 *In the AB mode a coincident stereo microphone's capsules are oriented so as to point left and right of the centre of the sound stage.*

such that the two capsules are angled to point left and right (see Figure 9.18). The choice of angle depends on the polar response of the capsules used, since most stereo microphones allow for either capsule to be switched through a number of pick-up patterns between figure-eight and omni. As shown theoretically in the papers described in Section 9.3, a coincident pair of figure-eight microphones at 90° gives a very accurate correlation between the actual angle of the source and the apparent position of the virtual image when reproduced on loudspeakers, but there are also operational disadvantages to the figure-eight pattern in many cases. We shall proceed to investigate coincident pairs that use a number of different polar patterns, bearing in mind that the theoretical arguments about correct correlation between real and phantom images can only be taken so far when there are conflicting operational considerations.

Figure 9.19 shows the polar pattern of a stereo microphone that is set to figure-eight pattern, and it serves as a useful example for the discussion of a number of phenomena which are not so easy to visualize with other pairs. Firstly, it may be seen that the fully left position corresponds to the null point of the right capsule's pick-up. This is the point at which there is maximum level difference between the two capsules, and it is a 'red herring' that the fully left position also corresponds to the maximum pick-up of the left capsule, since it does not in other stereo pairs. As a sound moves across the sound stage from left to right it results in a gradually decreasing output from the left microphone, and an increasing output from the right microphone. Since the microphones have cosine responses, the output at 45° off axis is $\sqrt{2}$ times the maximum output, or 3 dB down in level, thus the takeover between left and right microphones is smooth for music signals.

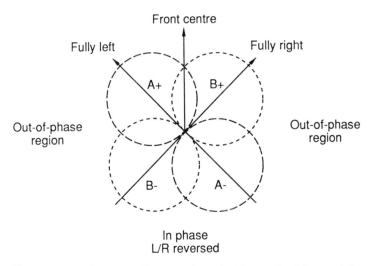

Figure 9.19   *The polar diagram of a coincident pair of figure-eights shows that there is a large out-of-phase region in the side quadrants, and a left–right reversal at the rear.*

The second point to consider with this pair is that the rear quadrant of pick-up suffers a left–right reversal, since the rear lobes of each capsule point in the opposite direction. This is important when considering the use of such a microphone in situations where confusion may arise between sounds picked up on the rear and in front of the microphone, such as in television sound where the viewer can also *see* the positions of sources. The third point is that pick-up in both side quadrants results in out-of-phase signals between the channels, since a source further round than 'fully left' results in pick-up by both the negative lobe of the right capsule and the positive lobe of the left capsule. There is thus a large region around a crossed pair of figure-eights that results in out-of-phase information, this information often being ambient or reverberant sound. Any sound picked up in this region will suffer cancellation if the channels are summed to mono, with maximum cancellation occurring at 90° and 270°, assuming 0° as the centre-front.

The operational advantages of the figure-eight pair are the crisp and accurate positioning of sources, together with a natural blend of ambient sound from the rear. Some cancellation of ambience may occur, especially in mono, if there is a lot of reverberant sound picked up by the side quadrants. Disadvantages lie in the large out-of-phase region, and in the size of the rear pick-up which is not desirable in all cases and is left–right reversed.[37] Stereo pairs made up of capsules having less rear pick-up may be preferred in cases where a 'drier' or less-reverberant balance is required, and where frontal sources are to be favoured over rearward sources. In such cases the capsule responses may be changed to be nearer the cardioid pattern, and this requires an increased angle between the capsules to maintain good correlation between actual and perceived angle of sources.

The cardioid crossed pair shown in Figure 9.20 is angled at approximately 131°, although angles of between 90° and 180° may be used to good effect depending on the 'width' of the sound stage to be covered. At an angle of 131° a centre source

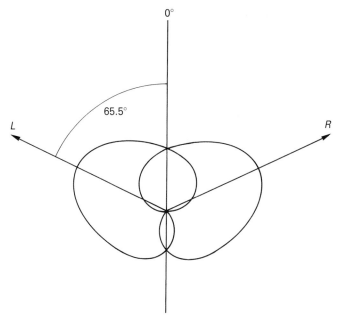

Figure 9.20 *A coincident pair of cardioid microphones should theoretically be angled at 131°, but deviations either side of this may be acceptable in practice.*

is 65.5° off-axis from each capsule, resulting in a 3 dB drop in level compared with the maximum on-axis output (the cardioid microphone response is equivalent to $0.5(1 + \cos\theta)$, where θ is the angle off-axis of the source, and thus the output at 65.5° is $\sqrt{2}$ times that at 0°). A departure from the theoretically correct angle is often necessary in practical situations, and it must be remembered that the listener will not necessarily be aware of the 'correct' location of each source; neither may it matter that the true and perceived positions are different. A pair of 'back-to-back' cardioids has often been used to good effect (see Figure 9.21), since it has a simple MS equivalent (see Section 9.5.1(a)(ii)) of an omni and a figure-eight, and has no out-of-phase region. Although theoretically 'fully left' is at 90° off-centre, it may be appreciated that there will in fact be a satisfactory level difference between the capsules to make fully left appear at a somewhat smaller angle.

With any *AB* crossed pair, fully left or fully right corresponds to the null point of pick-up of the opposite channel's microphone, as already stated, and, as will be seen below, this corresponds also to the point where the *M* signal equals the *S* signal (where the sum of the channels is the same as the difference between them). It is very important to grasp the difference between the angle subtended by the two capsules and the angle between fully left and fully right. As the angle between the capsules is made larger, the angle between the null points will become smaller (see Figure 9.22). Many people become confused over the effect that this will have on the width of the stereo image, and this requires some discussion.

Operationally, if one wishes to 'widen' the stereo image one will widen the angle between the microphones, which is intuitively the right thing to do. This results in a narrowing of the angle between fully left and fully right, which is also

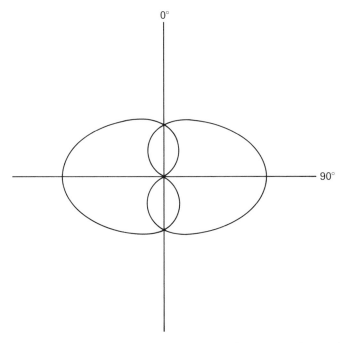

Figure 9.21   *Back-to-back cardioids have been proved to work well in practice, and have no out-of-phase region.*

correct, since sources which had been, say, half left in the original image will now be further towards the left. A narrow angle between fully left and fully right results in a very wide sound stage, since sources have only to move a small distance to result in large changes in reproduced position. This corresponds to a wide angle between the capsules.

Further coincident pairs are possible using any polar pattern between figure-eight and omni, although the closer that one gets to omni the greater the required angle to achieve adequate separation between the channels. The hypercardioid pattern is often chosen for its smaller rear lobes than the figure-eight, allowing a more distant placement from the source for a given direct-to-reverberant ratio. Since the hypercardioid pattern lies between figure-eight and cardioid, the angle required between the capsules is correctly around 110°.

The theoretical foundation (see Section 9.3.2) should not be forgotten in that it suggests the need for an electrical 'narrowing' of the image at high frequencies in order to preserve the correct angular relationships between low and high-frequency signals, although this is rarely implemented in practice with coincident pair recording. A further consideration to do with the theoretical versus the practical is that although microphones tend to be referred to as having a particular polar pattern, this pattern is unlikely to be consistent across the frequency range and this will have an effect on the stereo image. Cardioid crossed pairs should theoretically exhibit no out-of-phase region since there should be no negative rear lobes, but in practice most cardioid capsules becomes more omni at LF and more hypercardioid at HF, and thus some out-of-phase components may be noticed in the HF range whilst the width may appear too narrow at LF.

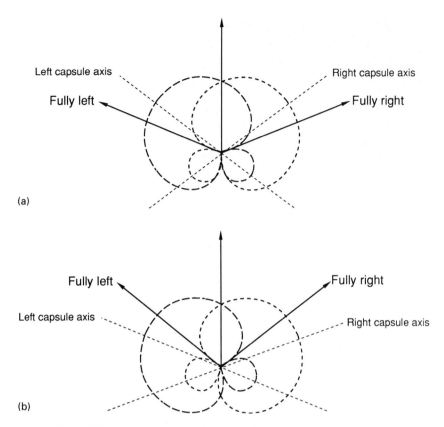

Left capsule axis

Right capsule axis

Fully left

Fully right

(a)

Fully left

Fully right

Left capsule axis

Right capsule axis

(b)

Figure 9.22 *There is a difference between the acceptance angle of a stereo pair (the angle between the pick-up null points) and the angle between the capsules. As the angle between the capsules is increased (as shown in (b)) the acceptance angle decreases, thus widening the stereo image.*

Attempts have been made to compensate for this in stereo microphone design, one example of which has been used by Sanken in an MS microphone[38] depicted in Figure 9.23, by adding high- and low-frequency acoustic compensation mechanisms to maintain relatively constant directional response over the frequency range.

AB pairs in general have the possible disadvantage that central sounds are off-axis to both microphones, perhaps considerably so in the case of crossed cardioids. This may result in a central signal with a poor frequency response and possibly an unstable image if the microphone's polar response is erratic. Whether or not this is important depends on the importance of the central image in relation to that of offset images, and will be most important in cases where the main source is central (such as in television, with dialogue). In such cases the *MS* technique described in the next section is likely to be more appropriate, since central sources will be on-axis to the *M* microphone. For music recording it would be hard to say whether central sounds are any more important than offset sources, and thus either technique may be acceptable.

Figure 9.23 *The Sanken CMS-2 is a small MS microphone with fixed directivity pattern of the MS capsules (courtesy of Stirling Audio).*

In commercial 'side-fire' *AB* microphones, it is common for one of the capsules to be fixed and for the other to rotate in order to set the angle between the capsules. It is sometimes difficult to see where the centre–front position is with some microphones, and this may have to be discovered through trial and error by getting an assistant to walk around the microphone making a suitable noise until equal level is obtained from both capsules. Alternatively, some stereo microphones have a small LED which moves with each capsule to identify the direction in which each is pointing.

(ii) *MS* pairs

Although some stereo microphones are built specifically to operate in the *MS* mode, it is possible to take any coincident pair with the capability that at least one capsule can be switched to figure-eight, and orientate it so that it will produce suitable signals. The *S* component (being the difference between left and right signals) is always a sideways-facing figure-eight with its positive lobe facing left. The *M* or mono component may be any polar pattern facing to the centre–front, although the choice of *M* pattern depends on the desired equivalent *AB* pair (see below), and will be the signal which a mono listener would hear. True *MS* microphones usually come equipped with a control box which matrixes the *MS* signals to *AB* format if required, using a technique similar to that shown in Section 9.4.3. A control for varying *S* gain is often provided as a means of varying the effective acceptance angle between the equivalent *AB* pair (see below).

It should be remembered that *MS* signals may not be monitored directly: they are sum and difference components and must be converted to *AB* format at a convenient point in the production chain. The advantages of keeping a signal in the *MS* format until it needs to be converted will be discussed below, but the major advantage of pick-up in the *MS* format is that central signals will be on-axis to the *M* capsule, resulting in the best frequency response. Furthermore, it is possible to operate an *MS* microphone in a similar way to a mono microphone (so long as precautions are taken when moving the microphone, as detailed in Section 9.5.1(a)vii)) which may be useful in television operations where the MS microphone is replacing a mono microphone on a pole or in a boom. Hibbing,[39] amongst others, points to the reduced audible effects of variations in microphone polar pattern with frequency when using the MS pick-up technique.

To see how *MS* and *AB* pairs relate to each other, and to draw some useful conclusions about stereo width control, it is informative to consider again a coincident pair of figure-eight microphones. For each *MS* pair there is an *AB* equivalent. This stands to reason, since *M* is simply the sum of *A* and *B* signals, whilst *S* is simply the difference between them. The polar pattern of the *AB* equivalent to any *MS* pair may be derived by plotting the level of $(M + S)/2$ and $(M - S)/2$ for every angle around the pair ($M + S = 2A$ and $M - S = 2B$). Taking the *MS* pair of figure-eight microphones shown in Figure 9.24, it may be seen that the *AB* equivalent is simply another pair of figure-eights, but rotated through 45°. Thus the correct *MS* arrangement to give an equivalent *AB* signal where both 'capsules' are orientated at 45° to the centre–front (the normal arrangement) is for the *M* capsule to face forwards and the *S* capsule to face sideways.

A number of interesting points arise from a study of the *AB/MS* equivalence of these two pairs, and these points apply to all equivalent pairs. Firstly, fully left or right in the resulting stereo image occurs at the point where $S = M$ (in this case at 45° off centre). This is easy to explain, since the fully left point is the point at which the output from the right capsule (*B*) is zero. Therefore $M = A + 0$, and $S = A - 0$, both of which equal *A*. Secondly, at angles of incidence greater than 45° off centre in either direction the two channels become out of phase, as was seen above, and this corresponds to the region in which *S* is greater than *M*. Thirdly, in the rear quadrant where the signals are in phase again, but left–right reversed, the *M* signal is greater than *S* again. The relationship between *S* and *M* levels, therefore, is an excellent guide to the phase relationship between the equivalent *AB* signals. If *S* is lower than *M*, then the *AB* signals will be in phase. If $S = M$, then the source is either fully left or right, and if *S* is greater than *M*, then the *AB* signals will be out of phase.

To show that this applies in all cases, and not just that of the figure-eight pair, look at the MS pair in Figure 9.25 together with its *AB* equivalent. This MS pair is made up of a forward-facing cardioid and a sideways-facing figure-eight (a popular arrangement). Its *AB* equivalent is a crossed pair of hypercardioids, and again the extremes of the image (corresponding to the null points of the *AB* hypercardioids) are the points at which *S* equals *M*. Similarly, the signals go out of phase in the region where *S* is greater than *M*, and come back into phase again for a tiny angle round the back, due to the rear lobes of the resulting hypercardioids. Thus the angle of acceptance (between fully left and fully right) is really the frontal angle between the two points on the *MS* diagram, where *M* equals *S*.

Now, consider what would happen if the gain of the *S* signal were raised (imagine expanding the lobes of the *S* figure-eight in Figure 9.26). The result of this would be that the points where *S* equalled *M* would move inwards, making the acceptance angle smaller. As explained earlier, this results in a wider stereo image, since off-centre sounds will become closer to the extremes of the image, and is equivalent to increasing the angle between the equivalent *AB* capsules. Conversely, if the *S* gain is reduced, the points at which *S* equals *M* will move further out from the centre, resulting in a narrower stereo image, equivalent to decreasing the angle between the equivalent *AB* capsules. This is neatly exemplified in a commercial example, the Neumann RSM 191i, (depicted in Figure 9.26) which is an *MS* microphone in which the *M* capsule is a forward-facing short shotgun microphone with a polar pattern rather like a hypercardioid. The polar pattern of the equivalent *AB* pair is shown in Figure 9.27 for three possible gains of the *S* signal with relation to *M* (– 6 dB, 0 dB and + 6 dB). It will

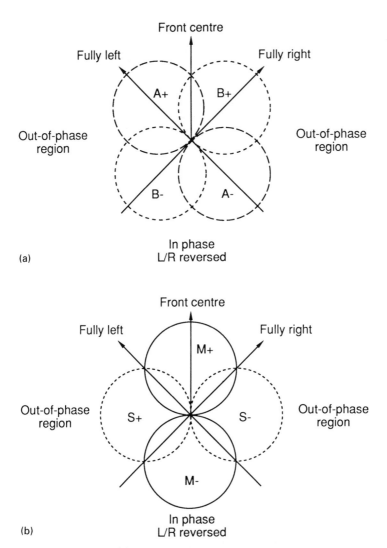

Figure 9.24   *Every AB pair has an MS equivalent. An AB pair of figure-eights is shown in (a) and its MS equivalent in (b).*

be seen that the acceptance angle (φ) changes from being large (narrow image) at – 6 dB, to small (wide image) at + 6 dB.

The other thing to be noted about the effect of changing S gain is the size of the rear lobes of the AB equivalent. It may be seen that the higher the S gain the larger the rear lobes. Therefore, not only does S gain change stereo width, it also affects rear pick-up, and thus the ratio of direct to reverberant sound.

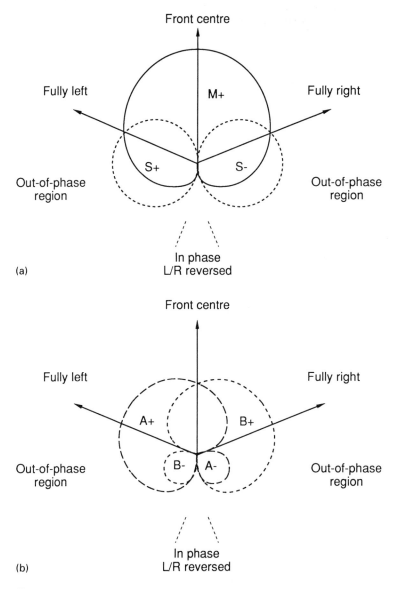

Figure 9.25 *The MS equivalent of a forward-facing cardioid and sideways figure-eight, as shown in (a), is a pair of hypercardioids whose effective angle depends on S gain, as shown in (b).*

Figure 9.26 *The Neumann RSM191i is an end-fire MS microphone whose control box allows control of S gain (courtesy of FWO Bauch).*

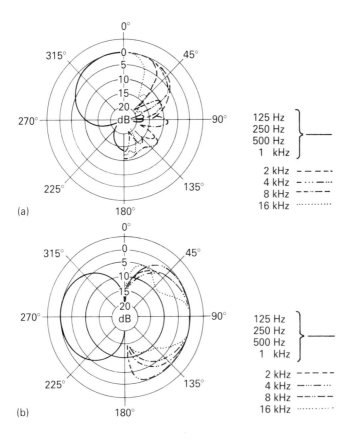

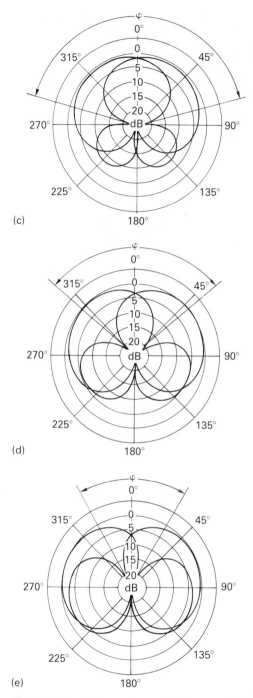

Figure 9.27 *Polar patterns of Neumann RSM191i microphone: (a) M capsule,*
(b) S capsule, (c) AB equivalent with – 6 dB S gain, (d) O dB S gain, (e) + 6 dB
S gain (courtesy FWO Bauch).

MS microphones, therefore, are useful as a means of deriving signals in the sum and difference format, for subsequent matrixing back to *A* and *B* at a later stage. With microphones in the *MS* format, the proportions of *M* and *S* may be varied in order to affect the resulting stereo width and direct-to-reverberant balance. Any *AB* stereo microphone may be operated in the *MS* configuration, simply by orientating the capsules in the appropriate directions and switching them to an appropriate polar pattern, but certain microphones are dedicated to *MS* operation simply by the physical layout of the capsules. Such microphones normally include a matrix for conversion to *AB* format, although this conversion may be achieved in a mixing console, as shown in Section 9.5.1(a)(v). There is not space here to show all the possible *MS* pairs and their *AB* equivalents, but a comprehensive review may be found in Dooley and Streicher.[40]

(iii) 'End-fire' versus 'side-fire' configurations

Physically, there are two principal ways of mounting the capsules in a coincident stereo microphone, be it *MS* or *AB* format: either in the 'end-fire' configuration where the *M* capsule 'looks out' of the end of the microphone such that the microphone may be pointed at the source (see Figure 9.28), or in the 'side-fire' configuration where the capsules look out of the sides of the microphone housing.

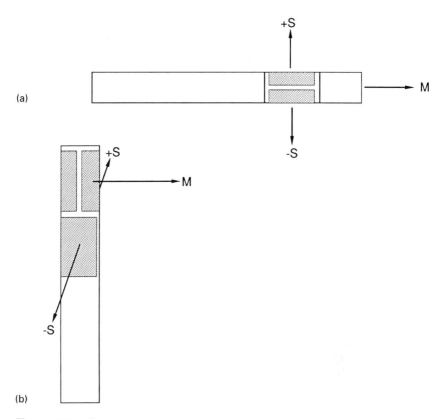

Figure 9.28 *Stereo microphones may be orientated in either (a) the end-fire mode or (b) the side-fire mode. An MS microphone is shown here.*

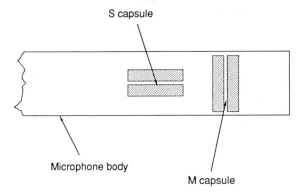

Figure 9.29   *Typical end-fire capsule configuration.*

It is less easy to see the direction in which the capsules are pointing in a side-fire microphone, but such a microphone makes it possible to align the capsules vertically above each other so as to be time-coincident in the horizontal plane, as well as allowing for the rotation of one capsule with relation to the other. An end-fire configuration is more suitable for the MS capsule arrangement (see Figure 9.29), since the *S* capsule may be mounted sideways behind the *M* capsule, and no rotation of the capsules is required, although there is a commercial example of an *AB* end-fire microphone for television ENG use which houses two fixed cardioids side-by-side in an enlarged head (see Figure 9.30)

Figure 9.30   *The AKG C522 is an end-fire AB-format microphone using crossed cardioids (courtesy AKG).*

(iv) Some commercial implementations

Although some commercial examples have already been introduced in the preceding sections, it would be worth while examining the approaches to stereo microphone design taken by a selection of manufacturers.

A good example of the standard side-fire microphone is the Neumann USM 69i, which is identical acoustically to the earlier SM 69 but incorporates switching of the polar pattern of both capsules into the microphone body, as opposed to putting this in a separate control box. As can be seen from the photograph in Figure 9.31, the polar pattern may be switched from omni, through cardioid, to figure-eight in six steps. The upper capsule may be rotated so as to be at any angle between 0° and 180°. The frequency response and polar characteristics of this microphone are shown in Figure 9.32

The Sanken CMS-2, on the other hand, shown in Figure 9.23, is claimed to be the world's smallest *MS* stereo condenser microphone, with an extremely low equivalent noise level of 16 dBA. It is a side-fire *MS* microphone, as opposed to the end-fire Neumann RSM 191 depicted earlier in Figure 9.26, and it is made up

Figure 9.31 *The Neumann USM69i is a side-fire AB microphone with variable directivity patterns (courtesy FWO Bauch).*

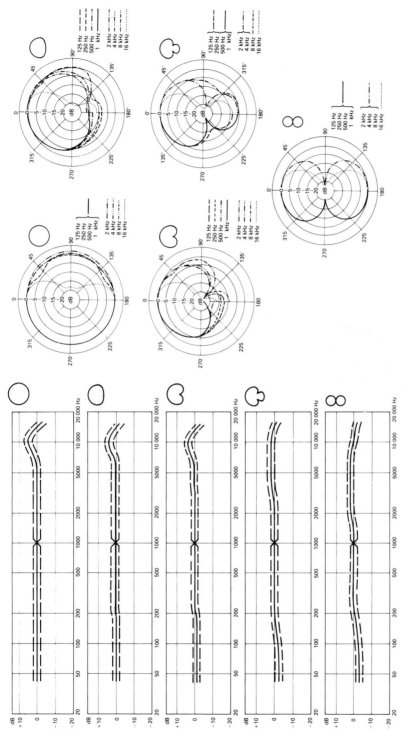

Figure 9.32 *Frequency responses and polar patterns of the USM 69i in different modes.*

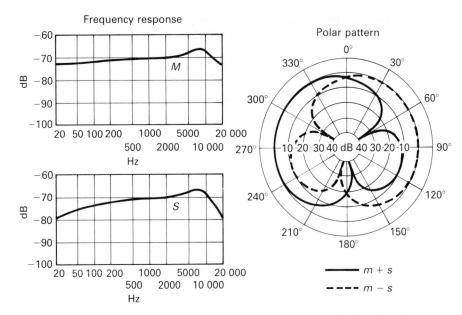

Figure 9.33   *Frequency responses and polar pattern of the Sanken CMS-2 (courtesy Stirling Audio).*

of a fixed cardioid and figure-eight. Its frequency response and *AB* equivalent polar pattern are shown in Figure 9.33. In normal use the relative gains of *M* and *S* are set such that the acceptance angle equals 127°, but this may be altered by the connection of an optional *MS* matrix controller.

The AKG C34 MS microphone is a modified version of the original C34 *AB* microphone, in which one capsule has been fixed in the figure eight mode (to make *S*), but where the *M* capsule has been turned to face out of the end of the housing and is left as a fully variable polar pattern. This is one of the few end-fire MS microphones with a variable-pattern *M* capsule, and has proved popular in television companies because, when mounted on a boom, the boom operator can alter the 'tightness' of the *M* pattern using the associated control box to suit the current scene or pick-up situation. If someone is 'off microphone' the *M* pattern can be made a softer cardioid to pick up over a wider angle, and if tight coverage of a single source is required the *M* pattern can be made more narrowly hypercardioid.

The AMS ST250 *MS* microphone (shown in Figure 9.34) is based on 'sound field' principles, developed for the AMS/Calrec Soundfield microphone (see Section 9.6), and as such has a number of novel features such as the facility for operating the microphone in either the end- or side-fire modes simply by pressing a switch which re-configures the capsules electrically. Furthermore, if the microphone is used upside-down (usually resulting in a swapping-over of left and right positions) a switch can be pressed to flip over the channels. As will be seen in the next section, this is simply a matter of phase-reversing the *S* signal.

A similar feature is found on the control box of the Neumann RSM 191, although it is labelled in a confusing manner. A four-position switch on the control box is labelled – *XY*, *XY*, *MS*, – *MS*, and has the function of allowing the microphone to be used upside-down in either the *XY* (*AB*) mode or the *MS* mode.

Figure 9.34 *The AMS ST250 stereo microphone is based on Soundfield principles.*

In the *MS* mode (when the control box does no matrixing), the – *MS* position phase-reverses the *S* signal, with the effect of flipping over left and right in any resulting stereo image, and in the *XY* mode (when the box is converting the output of the microphone to *XY* format) the – *XY* position simply swaps over left and right outputs. The box also has a six-position switch to control *S* gain, labelled both in decibels of *S* relative to *M* (from – 9 dB to + 6 dB), and in acceptance angle (fully left to fully right). The acceptance angle varies from 170° at – 9 dB *S* gain, to 60° at + 6 dB *S* gain.

(v) Operational considerations with stereo pairs

The control of *S* gain is an important tool in determining the degree of width of a stereo sound stage, and for this reason the *MS* output from a microphone might be brought (unmatrixed to *AB*) into a mixing console, so that the engineer may have control over the width. This in itself can be a good reason for keeping a signal in *MS* form during the recording process, although *M* and *S* can easily be derived at any stage using a conversion matrix. If the number of channels can be afforded on the mixing console, then it is possible to have, say, one fader for the *M* signal, one for the *S* signal, and one for a phase-reversed *S* signal.

The diagram in Figure 9.35 shows how it is possible to derive an *AB* mix with variable width from an *MS* microphone using three channels on a mixer. It should be noted that no *MS* matrices are required. *M* and *S* outputs from the microphone are fed in-phase through two mixer channels and faders, and a post-fader feed of *S* is taken to a third channel line input, being phase-reversed on this channel. The *M* signal is routed to both left and right mix buses (panned centrally), whilst the *S* signal is routed to the left mix bus ($M + S = 2A$) and the – *S* signal (the phase-reversed version) is routed to the right mix bus ($M - S = 2B$). It is important that the gain of the – *S* channel is matched very closely with that of the *S* channel. A means of deriving *M* and *S* from an *AB* format input is to mix *A* and *B* together with the *B* signal phase-reversed to get *S*, and without the phase reverse to get *M*. In order to convert back to *A* and *B* a similar method to the above could be employed, but this is wasteful of mixer channels and a matrix might be more convenient.

From a handling point of view a number of points should be made. Outdoors, stereo microphones are more susceptible to wind noise and rumble than most mono microphones, as they incorporate velocity-sensitive capsules which always give more problems in this respect than omnis. Most of the interference would reside in the *S* channel, since this has always a figure-eight pattern, and thus

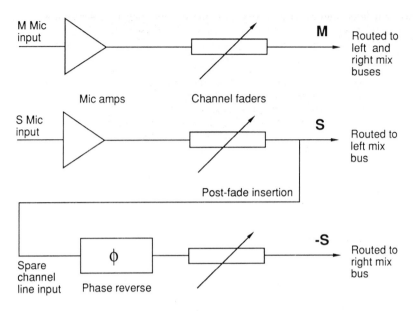

Figure 9.35   *An AB mix with variable width can be derived from an MS microphone connected to three channels on a mixer (see text).*

would not be a problem to the mono listener. Similarly, physical handling of the stereo microphone, or vibration picked up through a stand, will be much more noticeable audibly than with a mono pressure microphone. Stereo microphones should not be used close to people speaking, as small movements of their heads can cause large changes in the angle of incidence, and thus considerable movement in their apparent position in the sound stage. This is especially true in cases where the *S* gain is high (narrow angle of *AB* acceptance), as very small head or microphone movements result in large swings of the sound image.

9.5.2 Near-coincident microphone configurations

It is generally admitted that the Blumlein/Clark *et al.* theory of vector summation of sounds from loudspeakers having only level differences between them holds true only for continuous low-frequency sounds up to around 700 Hz, where head-shadowing effects begin to take over (see Section 9.3.2). Indeed at the time of presentation of the Clark *et al.* paper (1957) questions were asked about the ability of a system based only on level differences to cope with the important localization cues provided by short transients. The authors replied that they had decided to ignore precedence effects, because of the dependence on seating position, and that they believed that very few real sounds consisted of short pulses, many short sounds in fact being capable of being analysed into a series of sinusoidal components (that is to say, Fourier analysis). This is interesting, since in the earlier section on directional perception (Section 9.2.1 (c)) we saw that the response of the basilar membrane to transient signals resembled a decaying sinusoidal wave-form, possibly allowing for phase difference cues to be extracted. Although at the time the authors believed that their system did not exploit time difference cues,

Vanderlyn later suggested a theory to explain how such a system copes with transient information, as described at the end of Section 9.3.2.

There appears to be some basis, though, for the adoption of 'near coincident' pairs of directional microphones, to introduce small additional timing differences which may help in the localization of transient sounds and increase the 'spaciousness' of a recording, whilst at the same time remaining nominally coincident at low frequencies, adequately to maintain the level differences between channels required by Blumlein *et al.* for the reconstruction of low-frequency phase differences with continuous sounds, although resulting in an upper frequency limit for this effect which is lower than that for truly coincident pairs. Subjective evaluations bear this principle out, with listeners appearing to be considerably impressed by recordings made on near-coincident pairs. One comprehensive subjective assessment of stereo microphone arrangements, performed at the University of Iowa,[41] consistently resulted in the near-coincident pairs scoring amongst the top few performers for their sense of 'space' and realism. Lipshitz[18] attributes these effects to 'phasiness' at high frequencies (which some people may like, nonetheless), and argues that truly coincident pairs are preferable. Gerzon[2] suggests that a very small spacing of crossed cardioid microphones (about 5 cm) actually compensates for the phase differences introduced when 'spatial equalization' is used (the technique described earlier of increasing *LF* width relative to *HF* width by introducing equalization into the *S* channel).

A number of examples of near-coincident pairs exist as 'named' arrangements, although, as Williams has shown,[42] there is a whole family of possible near-coincident arrangements using combinations of spacing and angle. The so-called 'ORTF pair' is an arrangement of two cardioid microphones, deriving its name from the organization which first adopted it (the *Office de Radiodiffusion-Télévision Française*). The two microphones are spaced apart by 170 mm, and angled at 110°. The 'NOS' pair (*Nederlande Omroep Stichting*, the Dutch Broadcasting Company), uses cardioid mics spaced apart by 300 mm and angled at 90°. Figure 9.36 illustrates these two pairs, along with a third pair of figure-eight microphones spaced apart by 200 mm, which has been called a 'Faulkner' pair, after the British recording engineer who first adopted it. This latter pair can be found to offer good 'focusing' on a small-to-moderate-sized central ensemble with the microphones placed further back than would normally be expected.

9.5.3 Spaced arrays

(a) Principles

Although spaced arrays have a historical precedent for their usage, since they were the first to be documented (in the work of Clement Ader at the Paris Exhibition in 1881)[43] and have been widely used, especially in the United States, since the 1930s, they are perhaps less 'correct' theoretically. Spaced arrays rely principally on the precedence effect, discussed in Section 9.2, in that the delays which result between the channels tend to be of the order of a number of milliseconds (as opposed to the binaural delay of up to 600 μs). Wallach *et al.*[7] discussed the relevance of the precedence effect to stereo reproduction, and, as already stated, it is possible to combine time and level difference to obtain a particular phantom image position, and it is known that one parameter may be traded off against the other to some extent.

With spaced arrays the level and time difference resulting from a source at a particular left-right position on the sound stage will depend on how far the source

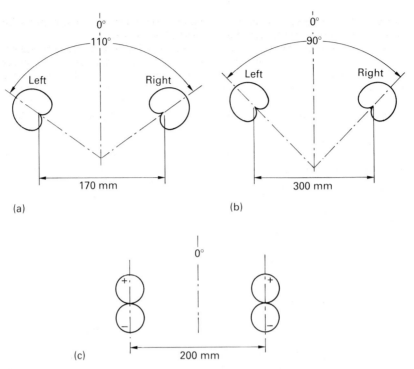

(a)

(b)

(c)   200 mm

Figure 9.36 (a)   *ORTF near coincident pair, (b) NOS pair, (c) Faulkner pair.*

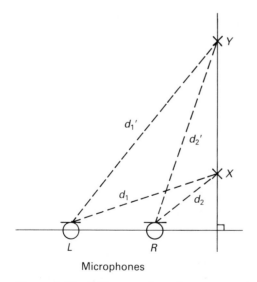

Microphones

Figure 9.37   *With spaced omnis a source at position X results in path lengths d_1 and d_2 to each microphone respectively, whilst a source in the same LR position but at a greater distance (source Y) the path length difference is smaller, resulting in a smaller time difference than for X.*

is from the microphones (see Figure 9.37), with a more distant source resulting in a much smaller delay and level difference. In order to calculate the time and level differences that will result from a particular spacing it is possible to use the following two formulae:

$$\Delta t = (d_1 - d_2)/c, \qquad \Delta L = 20 \log_{10}(d_1/d_2)$$

where Δt is the time difference and ΔL the pressure-level difference which results from a source whose distance is d_1 and d_2 respectively from the two microphones, and c is the speed of sound (340 m/s).

When a source is very close to a spaced pair there may be a considerable level difference between the microphones, but this will become small once the source is more than a few metres distant. The positioning of spaced microphones in relation to a source is thus a matter of achieving a compromise between closeness (to achieve satisfactory level and time differences between channels), and distance (to achieve adequate reverberant information relative to direct sound). When the source is large and deep, such as a large orchestra, it will be difficult to place the microphones so as to suit all sources, and it may be found necessary to raise the microphones somewhat so as to reduce the differences in path length between sources at the front and rear of the orchestra.

Spaced microphone arrays do not stand up well to theoretical analysis when considering the imaging of continuous sounds, the precedence effect being related principally to impulsive or transient sounds. Because of the phase differences between signals at the two loudspeakers created by the microphone spacing, interference effects at the ears at low frequencies may in fact result in a contradiction between level and time cues at the ears.[18] It is possible in fact that the ear on the side of the 'earlier' signal may *not* experience the higher level, thus producing a confusing difference between the cues provided by impulsive sounds and those provided by continuous sounds. The lack of phase coherence in spaced-array stereo is further exemplified by phase-inverting one of the channels on reproduction, an action that does not always appear to affect the image particularly, as it would with coincident stereo, showing just how uncorrelated the signals are.

It may thus be found that stereo perspective and accuracy of positioning are less realistic with spaced arrays, although many convincing recordings have resulted from their use. Lipshitz suggests that the impression of 'space' that results from the use of spaced arrays is in fact simply the result of 'phasiness' and comb-filtering effects. Cooper and Bauck[32] suggest that there is a place for both the 'spaciousness' that results from spaced techniques and the imaging that results from coincident techniques, since the highly decorrelated signals resulting from spaced techniques are also a feature of concert-hall acoustics. This therefore supports to some extent the arguments for near-coincident techniques, which appear to combine the merits of both coincident and spaced microphones. Many recording engineers prefer spaced arrays because the omni microphones often used in such arrays tend to have a flatter and more extended frequency response than their directional counterparts, although it should be noted that spaced arrays do not *have* to be made up of omni microphones (see below). Mono compatibility of spaced pairs is variable, although not always as poor in practice as might be expected. Coincident pairs, from a superficial point of view, appear to promise better mono compatibility due to the lack of phase differences between the capsules, but in reverberant surroundings there is often a lot of pick-up in the 180° out-of-phase region of such a pair (see Figure 9.19), which suffers cancellation in mono.

(b) 'Decca tree'

The so-called 'Decca tree' is a popular arrangement of three spaced mono microphones with an omnidirectional (pressure) pick-up pattern. The name derives from the usage of this technique by the Decca Record Company, although even that company would admit to not sticking rigidly to this arrangement in all circumstances. A similar arrangement is described by Grignon[44] as early as 1949.

A 'tree' of three omnis is configured according to the diagram in Figure 9.38, with the centre microphone spaced so as to be slightly forward of the two outriggers, although it is possible to vary the spacing to some extent depending on the size of the source stage to be covered. The reason for the centre microphone and its spacing is to stabilize the central image which tends otherwise to be rather imprecise, although the existence of the centre microphone will also complicate the phase relationships between the channels, thus exacerbating the comb-filtering effects that may arise with spaced pairs. The advance in time experienced by the forward microphone will tend to 'solidify' the central image, due to the precedence effect, avoiding the 'hole-in-the-middle' often resulting from spaced pairs. The outriggers are angled outwards slightly, so that the axes of best HF response favour sources towards the edges of the stage whilst central sounds are on-axis to the central microphone.

(c) Other spaced pairs

Spaced microphones with either omnidirectional or cardioid patterns may be used in configurations other than the Decca tree described above, although the 'tree' has certainly proved to be the more successful arrangement in practice. The precedence effect begins to break down for delays greater than around 40 ms, since the brain begins to perceive the two arrivals of sound as being discrete rather than integrated, thus beginning to sound like an echo or 'phasy' effect. It is therefore reasonable to assume that spacings between microphones that exceed this delay between channels should be avoided. This maximum delay, though, corresponds to a microphone spacing of around 17 m, and such extremes have not proved to work well in practice due to the great distance of central sources from either microphone compared with the closeness of sources at the extremes, resulting in a considerable level drop for central sounds and thus a 'hole in the

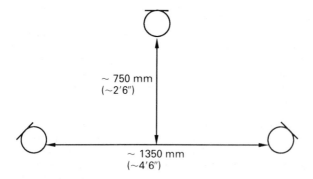

~ 750 mm
(~2'6")

~ 1350 mm
(~4'6")

Figure 9.38 *The classic 'Decca tree' involves three omnis, with the centre microphone spaced slightly forward of the outriggers.*

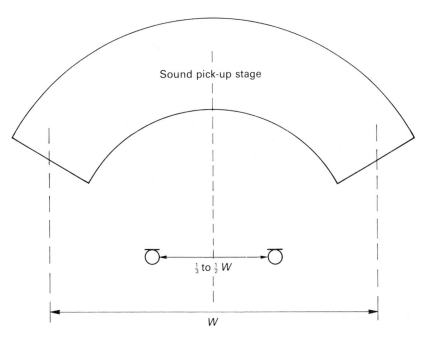

Sound pick-up stage

$\frac{1}{3}$ to $\frac{1}{2}$ *W*

W

Figure 9.39 *One source suggests that spaced omni microphones may be distanced at between one-third to one-half the width of the sound stage.*

middle'. Furthermore, the Haas effect for longer time delays than about 10 ms begins to result in the sound being firmly in the leading speaker.

Work by Dooley and Streicher has shown that good results may be achieved using spacings of between one-third and one-half of the width of the total sound stage to be covered (see Figure 9.39), although closer spacings have also been used to good effect. Brüel and Kjaer manufacture matched stereo pairs of omni microphones together with a bar that allows variable spacing, as shown in Figure 9.40, and suggest that the spacing used is somewhat smaller than one-third of the stage width (they suggest between 5 cm and 60 cm, depending on stage width), their principal rule being that the distance between the microphones should be small compared with the distance from microphones to source. Informal experiments by the author suggest that interchannel delays of between 2 and 4 ms are adequate to give the impression that a transient sound is either fully left or right in the resulting image, depending on the nature of the signal. This agrees well with Lipshitz,[18] who also points out that for a signal to appear half-left or half-right, only about one-quarter of the delay is needed compared with that needed for fully left or right, further reinforcing the likelihood of a hole-in-the-middle with spaced pairs.

Omni microphones are often used in addition to coincident pairs, as 'outriggers' to cover the extremes of an orchestra or choir, and to add some 'body' to the sound of the string section (see Figure 9.41). This is hard to justify theoretically, and is beginning to move towards the realms of multi-microphone pick-up, which is also hard to justify theoretically but which may be used for convenience and in order to produce a 'commercially acceptable' sound. Once more than around three microphones are used to cover a sound stage one has to

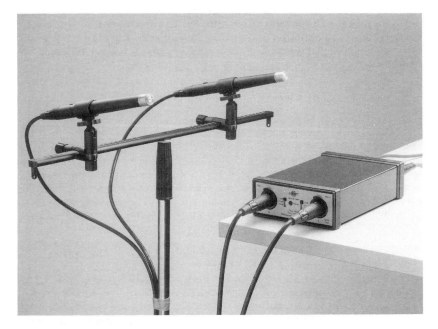

Figure 9.40   *B&K omni microphones mounted on a stereo bar that allows variable spacing*

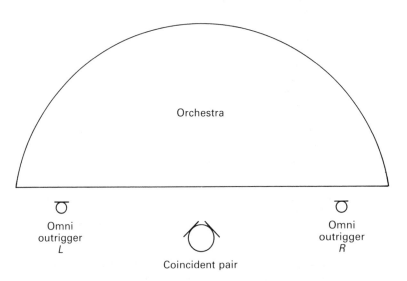

Figure 9.41   *Omni outriggers may be used in addition to a coincident pair for wide sources*

consider a combination of theories, possibly suggesting conflicting sonic information between the outputs of the different microphones. In such cases the sound balance will be optimized on a mixing console and theory can be relegated to a back seat, since what is now being created is a synthetic balance subject to the creative control of the recording engineer (see Section 9.54).

(d) Binaural pairs and 'dummy head' techniques

The binaural hearing and reproduction process has already been thoroughly explained, and it is now intended to look into the means of sound pick-up for binaural reproduction on headphones. Binaural recordings can have an uncanny realism when monitored on headphones, although they tend not to work well on loudspeakers for reasons already stated (see Section 9.3.4), although transaural processing (see Section 9.3.5(b)) involving tailored cross-talk cancellation may be the future solution to this problem.

Binaural microphone techniques depend on the spacing of pressure microphones at the distance apart of a normal listener's ears, so as to produce two 'ear signals' which are as close as possible to a representation of the sound pressures at each ear. When these are reproduced at the same two ears on headphones, the result is a close approximation to the originally recorded sound field, and thus a very life-like listening experience. Factors that must be considered are the effects of the pinna and the ear canal on the sound to be picked up and reproduced, as well as the similarity (or lack of it) between different peoples' ears. The use of artificial 'heads' and baffles to simulate the shadowing effect of the head should also be considered.

Small pressure microphones (such as those used as 'tie-clip' microphones, e.g. Sony ECM-50) may be mounted either flush with the side of a real head on a headphone-like headband, this being useful if a person wishes to make a recording using his own ear spacing, perhaps whilst walking around with a portable tape recorder. The shadowing effects of his own head would then be introduced between the microphones, although the effect of the pinna would be limited by the intrusion of the microphone and headband. Alternatively, a 'dummy head', such as the Neumann KU81i, may be used, in which the microphones are mounted inside the ear canal, so that the response of the microphone is modified by the resonant and reflective properties of the outer ear, as well as by the resonant cavities of the head in some cases, and also reflections off the torso in the case of a 'head and torso simulator' such as the B&K Type 5930 (shown in Figure 9.42(a)). Dummy heads vary in their degree of 'reality', some having imitation skin as well as the correct resonant and absorptive properties as a real head, in order that the result is as 'life-like' as possible.

It is important to consider at which stage in the recording and reproduction process the pinna and ear canal effects will have been added: whether during recording or during reproduction. On replay, the existence of a headphone over the ear would largely supress any pinna effects (although open-backed headphones may allow some to remain). Headphones particularly suppress the all-important conch response, the conch being the main resonant cavity of the pinna, disturbances of which have been shown to have a large effect on the perception of a sound as being 'at the ear'. To be accurate, the sound should only have been 'encoded' by the pinna and ear canal characteristics once, and thus it might be suggested that the most effective solution would be to mount microphones such that they are just at the entrance to the ear canal, in order that they are subject to pinna effects during recording, but not ear canal resonances (which are supplied on replay by the listener's own ears). For measurement purposes a slightly different approach applies, since it may be that the ear canal,

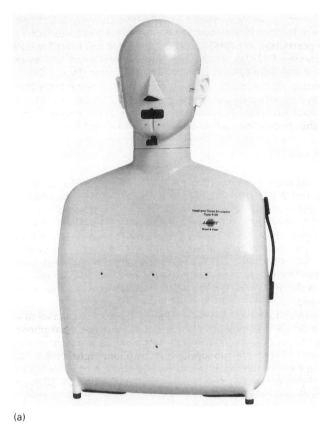

(a)

Figure 9.42(a) *B&K head and torso simulator for binaural recording*

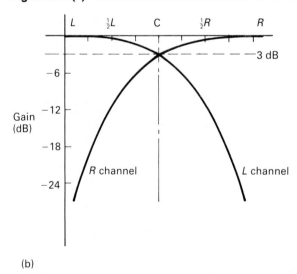

(b)

Figure 9.42(b) *Panpot law for sound source placement by a mixer.*

pinna, and perhaps even inner ear responses are required to affect the microphone output, since the aim is often to measure noise levels in a situation as close to that of human perception as possible, without the intention to replay that noise again on headphones. For this reason B&K makes two 'head and torso simulators', one for binaural recording and one for measurement (Type 4128).

Although not simulating all the head effects, a convincing binaural recording may be made using two microphones spaced apart by the ear distance (15–20 cm) and separated by a wooden or perspex baffle with a diameter of around 25 cm. This approximates to the shadowing and delaying effects of a real head.

9.5.4 Multi-microphone arrangements

We have so far considered the use of a small number of microphones to cover the complete sound stage, but it is also possible to make use of a large number of microphones, each covering a small area of the sound stage and intended to be sonically 'independent' of the others. In the ideal world, each microphone in such an arrangement would pick up sound only from the desired sources, but in reality there is considerable 'spill' from one to another. It is not the intention in this chapter to provide a full resumé of studio microphone technique, and thus discussion will be limited to an overview of the principles of multi-microphone pick-up as distinct from the more simple techniques described above (sometimes known as 'purist' techniques).

In multi-microphone recording each microphone feeds a separate channel of a mixing console, where levels are individually controlled and the microphone signal is 'panned' to a virtual position somewhere between left and right in the sound stage. The pan control takes the monophonic microphone signal and splits it two ways, controlling the proportion of the signal fed to each of the left and right mix buses. Typical pan control laws follow a curve which gives a 3 dB drop in the level sent to each channel at the centre, resulting in no perceived change in level as a source is moved from left to right (see Figure 9.42(b)) due to the summation of sound intensities from left and right loudspeakers. The –3 dB panpot law is not correct if the stereo signal is combined electrically to mono, since the summation of two equal signal voltages would result in a 6 dB rise in level for signals panned centrally, and thus a – 6 dB law is more appropriate for mixers whose outputs will be summed to mono (e.g. radio and TV operations) as well as stereo, although this will then result in a drop in level in the centre for stereo signals. A compromise law of – 4.5 dB is often adopted by manufacturers for this reason.

Multi-microphone balances therefore rely on channel level differences, separately controlled for each microphone, to represent the positions of virtual sources in a synthesized sound stage, with relative level between channels used to adjust the 'strength' and thus perhaps the perceived distance of a source in a mix, although distance is more a function of the associated reflections and reverberant balance than of level. Artificial reverberation may be added to restore a sense of space to a multi-microphone balance. It is common in 'classical' music recording to use close microphones in addition to a coincident pair or spaced pair in order to reinforce sources which appear to be weak in the main pick-up, these close microphones being panned as closely as possible to match the true position of the source. The results of this are variable and can have the effect of 'flattening' the perspective, removing any depth which the image might have had, and thus the use of close microphones must be handled with subtlety.

The recent development of cheaper digital signal processing (DSP) has made possible the use of delay lines, sometimes as an integral feature of digital mixer channels, to adjust the relative timing of spot microphones in relation to the main pair so as to prevent this distortion of depth, and to equalize the arrival times of distant microphones in order that they do not exert a precedence 'pull' over the output of the main pair. As described in Section 3.5, it is possible to process the outputs of multiple microphones to simulate binaural delays and head-related effects in order to create the effect of sounds at any position around the head, when the result is monitored on headphones.

9.6 Ambisonics

9.6.1 Principles

The ambisonic system of directional sound pick-up and reproduction is discussed here because of its relative thoroughness as a unified system, based on some key principles of psycho-acoustics, and because of its relevance to the Calrec Soundfield microphone, which is unique as a commercial product. It has its theoretical basis in work by Gerzon, Barton and Fellgett, good summaries of which may be found in Gerzon[45] and Gerzon and Barton.[46]

Ambisonics aims to offer a complete hierarchical approach to directional sound pick-up, storage or transmission and reproduction, which is equally applicable to mono, stereo, horizontal surround-sound, or full 'periphonic' reproduction including height information. Depending on the number of channels employed it is possible to represent a smaller or greater number of dimensions in the reproduced sound. A number of formats exist for signals in the ambisonic system, and these are as follows: the A-format for microphone pick-up, the B-format for studio equipment and processing, the C-format for transmission, and the D-format for decoding and reproduction. A format known as UHJ ('Universal HJ', 'HJ' simply being the letters denoting two earlier surround sound systems), described originally by Gaskell of the BBC Research Department,[47] is also used for encoding surround-sound ambisonic information into two or three channels, whilst retaining good mono and stereo compatibility for 'non-surround' listeners.

Ambisonic sound should be distinguished from quadraphonic sound, since quadraphonics explicitly requires the use of four loudspeaker channels, and cannot be adapted to the wide variety of pick-up and listening situations that may be encountered. Quadraphonics generally works by creating conventional stereo phantom images between each pair of speakers, and, as Gerzon states, conventional stereo does not perform well when the listener is off-centre or when the loudspeakers subtend an angle larger than 60°. Since in quadraphonc reproduction the loudspeakers are angled at roughly 90° there is a tendency towards a hole in the middle, as well as there being the problem that conventional stereo theories do not apply correctly for speaker pairs to the side of the listener. Ambisonics, however, correctly encodes sounds from all directions in terms of pressure and velocity components, and correctly decodes these signals to a number of loudspeakers, with psycho-acoustically optimized shelf filtering above 700 Hz to correct for the shadowing effects of the head, and an amplitude matrix which determines the correct levels to be fed to each speaker for the layout chosen. Ambisonics is thus the theoretical successor to coincident stereo on two loudspeakers, since it is the logical extension of Blumlein principles to surround sound (an area in which such principles may in fact be more appropriate than in two-channel reproduction, since the sense of 'space', which is often said to be sacrificed for the sake of imaging in two-channel Blumlein recordings, can now be correctly reconstructed in the shape of reverberant signals from all angles, whilst retaining imaging).

The source of an ambisonic signal may be an ambisonic microphone such as the Calrec Soundfield, described in Section 9.6.4, or it may be an artificially panned mono signal, split into the correct B-format components (see below) and placed in a position around the listener by adjusting the ratios between the signals. Since we are principally concerned with microphones in this book, it is intended not to cover the principles of ambisonic mixing. Good introductions to the subject of mixing may be found in Daubney[48] and Elen.[49]

It is often the case that theoretical 'correctness' in a system does not automatically lead to widespread commercial adoption, and despite considerable coverage of ambisonic techniques the system is still only used rarely in commercial recording and broadcasting. It is true that the Soundfield microphone is used widely, but this is principally because of its unusual capacity for being steered electrically so as to allow the microphone to be 'pointed' in virtually any direction without physically moving it, and set to any polar pattern between omni and figure eight, simply by turning knobs on a control box. It is used in this respect as an extremely versatile stereo microphone for two-channel recording and reproduction.

Because of its unified nature and its elegance it seems likely that ambisonics will not die as the various quadraphonic systems died, and it is possible that the system may find new life in the developing area of high-definition television with surround-sound, if only it can be integrated in a usable way. It is quite possible that ambisonics *appears* to be too complicated to the majority of users, based as it is on psycho-acoustics and mathematics, and designed by academics, leading operators to prefer the relative simplicity of two-channel stereo, or simple LCRS surround sound with pictures (see Section 9.4.1(c)). Domestically, more than two loudspeakers are always problematical, for purely practical reasons of space and living-room logistics. It would be a pity, though, to conclude any investigation into stereo sound without coverage of this intriguing system.

9.6.2 Signal formats

As indicated above there are four basic signal formats for ambisonic sound: *A*, *B*, *C* and *D*. The *A*-format consists of the four signals from a microphone with four sub-cardioid capsules orientated as shown in Figure 9.43. These are capsules mounted on the four faces of a tetrahedron, and correspond to left–front (*LF*), right–front (*RF*), left–back (*LB*) and right–back (*RB*), although two of the capsules point upwards and two point downwards. Such signals should be equalized so as to represent the sound field at the centre of the tetrahedron, since the capsules will not be perfectly coincident. The *A*-format is covered further in the discussion of the Soundfield microphone.

The *B*-format consists of four signals which between them represent the pressure and velocity components of the sound field in any direction, as shown

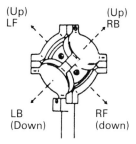

(Up)
LF

(Up)
RB

LB
(Down)

RF
(down)

Figure 9.43 *A-format capsule directions in an ambisonic microphone.*

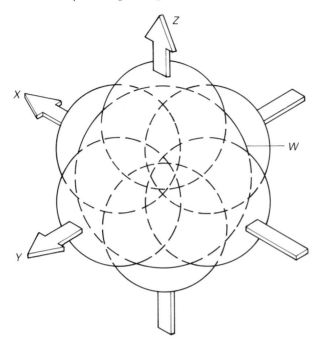

Figure 9.44   *B-format component polar patterns in ambisonics.*

in Figure 9.44. It can be seen that there is a similarity with the sum and difference format of two-channel stereo, described earlier in Section 9.5.1(a)(ii), since the B-format is made up of three figure-eight components in quadrature (X, Y and Z), and an omni component (W). All directions in the horizontal plane may be represented by scalar and vector combinations of W, X and Y, whilst Z is required for height information. X is equivalent to a forward-facing figure-eight (equivalent to M in MS stereo), Y is equivalent to a sideways-facing figure-eight (equivalent to S in MS stereo). The X, Y and Z components are subject to a gain of − 3 dB with relation to the W signal (0 dB) in order to achieve roughly similar energy responses for sources in different positions.

A B-format signal may be derived from an A-format microphone as described in Section 9.6.4(b), or it may be created directly by arranging capsules or individual microphones in the B-format mode. The Z component is not necessary for horizontal information. B-format signals may be recorded directly, allowing for subsequent manipulation of the sound field, and are robust in the face of interchannel errors.

The C-format consists of four signals L, R, T and Q, which conform to the *UHJ* hierarchy, and are the signals used for mono or stereo-compatible transmission or recording. The C-format is, in effect, the consumer format. L is the left channel, R is the right channel, T is a third channel which allows more accurate horizontal decoding, and Q is a fourth channel containing height information. The proportions of B-format signals that are combined to make up a C-format signal have been carefully optimized for the best compatibility with conventional stereo and mono reproduction. If L + R is defined as Σ (roughly equivalent to M in MS stereo) and L − R is defined as Δ (roughly equivalent to S in MS stereo), taking θ as the angle of incidence in the horizontal plane (the azimuth), and η as the angle of elevation above the horizontal, then in the B-format,

$$W = 1$$

$$X = \sqrt{2} \cos \theta \cos \eta$$

$$Y = \sqrt{2} \sin \theta \cos \eta$$

$$Z = \sqrt{2} \sin \eta$$

and in the C-format,

$$\Sigma = 0.9397W + 0.1856X$$

$$\Delta = j(- 0.3420W + 0.5099X) + 0.6555Y$$

$$T = j(- 0.1432W + 0.6512X) - 0.7071Y$$

$$Q = 0.9772Z$$

where j (or $\sqrt{(-1)}$) represents a phase advance of 90°.

Two, three, or four channels of the C-format signal may be used depending on the degree of directional resolution required, with a two-and-a-half channel option available where the third channel (T) is of limited bandwidth. For stereo compatibility only L and R are used (L and R being respectively $0.5(\Sigma + \Delta)$ and $0.5(\Sigma - \Delta)$. The UHJ or C-format hierarchy is depicted graphically in Figure 9.45.

D-format signals are those distributed to loudspeakers for reproduction, and are adjusted depending on the selected loudspeaker layout. They may be derived from either B- or C-format signals using an appropriate decoder, and the number of speakers is not limited in theory, nor is the layout constrained to a square. Four speakers gives adequate surround sound; six provide better immunity against the drawing of transient and sibilant signals towards a particular speaker; eight may be used for full periphony with height. The decoding of B- and C-format components into loudspeaker signals is too complicated and lengthy a matter to go into here, and is the subject of several patents granted to the NRDC (the UK National Research and Development Council), but it is sufficient to say that the principle of decoding involves the passing of two or more UHJ signals via a phase-amplitude matrix, resulting in B-format signals which are subjected to shelf filters (in order to correct the levels for the head-related transfer functions such as shadowing and diffraction), which are passed through an amplitude matrix which feeds the loudspeakers (see Figure 9.46). A layout control is used to vary the level sent to each speaker, depending on the physical arrangement of speakers. Formulae relating to a number of decoding options may be found in Gerzon.[45]

9.6.3 Ambisonic encoding and decoding

Commercial devices exist for converting signals to and from the various formats of the ambisonic and UHJ hierarchies. There also exist units that pan and rotate mono and stereo signals around the sound field. It is necessary for any interested designer to obtain a licence from the NRDC to use the technology involved, since it is subject to patent laws. Audio & Design have manufactured a number of

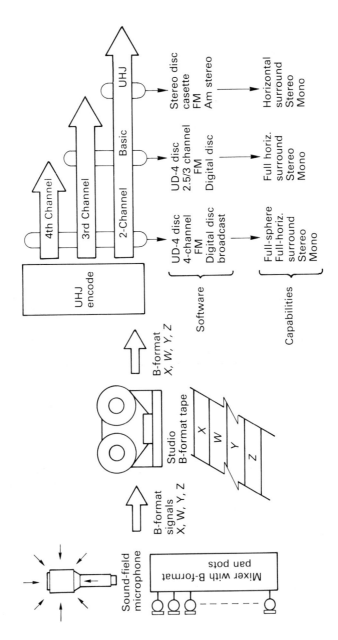

Figure 9.45 *C-format (UHJ) signal hierarchy in an ambisonic system.*

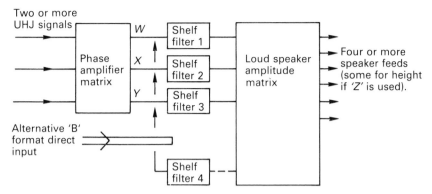

Figure 9.46 *Decoding of ambisonic signals to D-format for loudspeaker reproduction.*

useful processor units to handle ambisonic signals, and some of these are described below.

The so-called 'transcoder' is a useful device, since it takes either a *B*-format input and converts it to a two- or three-channel *UHJ* output, or takes two stereo pairs (four channels) and converts them to *UHJ*. The significance of the latter is that the two stereo pairs can be the conventional group outputs of a four-group mixer in which sounds have been panned 'quadraphonically', i.e. *LF, RF, LB* and *RB*. Facilities are provided for varying the stage widths of both front and back pairs, and thus the unit provides a good interface between conventional mixing and ambisonic systems.

A device known as a 'pan rotate' unit takes a number of mono inputs and allows them to be panned ambisonically to any position in the horizontal plane, both front-back and side-to-side. Once panned to their positions the whole sound field may be rotated using another control. The 'convertor' unit allows conventional console panpots and groups to be used ambisonically, by taking in the four console group outputs and making panning between odd and even groups give positions across a particular quadrant.

David Malham of York University designed a 'digitally programmable soundfield controller' which aimed to manipulate sound fields by digital processing of *B*-format signals. The prototype unit was capable of rotation of sound fields by given increments, but it was suggested that further work could produce a device capable of comprehensive sound-field manipulation.

9.6.4 The soundfield microphone

The AMS/Calrec Soundfield microphone, pictured in Figure 9.47, is designed for picking up full periphonic sound in the *A*-format (see Section 9.6.2), and is coupled with a control box designed for converting the microphone output into both the *B*-format and the *D*-format. *UHJ* encoders are available for *C*-format encoding. The Soundfield microphone is capable of being steered electrically by using the control box (see Figure 9.48), in terms of any one of azimuth, elevation, tilt or dominance, and as such it is also a particularly useful stereo microphone for two-channel work. The microphone correctly encodes directional information in all planes, including the pressure and velocity components of indirect and reverberant sounds, making it possible for these sounds to be reproduced correctly at the ears of the listener on replay.

Figure 9.47 *AMS Calrec Soundfield microphone (enclosed) (Courtesy AMS).*

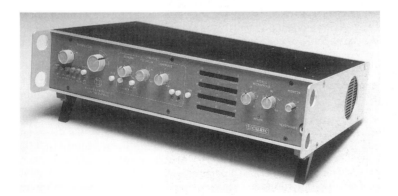

Figure 9.48 *AMS Calrec Soundfield microphone (control box) (Courtesy AMS).*

(a) Capsule arrangement

The photograph in Figure 9.49 shows the physical capsule arrangement of the microphone, this having been shown diagramatically in Figure 9.43. Four capsules with sub-cardioid polar patterns (between cardioid and omni, with a response equal to $2 + \cos v$) are mounted so as to face in the *A*-format directions, with electronic equalization to compensate for the inter-capsule spacing, such that the output of the microphone truly represents the sound field at a point (true coincidence is maintained up to about 10 kHz). The capsules are matched very closely and each contributes an equal amount to the *B*-format signal, thus resulting in cancellation between variations in inherent capsule responses.

The capsules are paired so as to produce figure-eight patterns, by subtracting $(LF - RB)$, and $(RF - LB)$, this resulting in crossed figure-eights at 90°, comparable with an *AB* pair in conventional stereo (see Figure 9.50(a)). Similarly different pairings are used to produce a vertical crossed pair of figure-eights, using $(LF - LB)$ and $(RB - RF)$, as shown in Figure 9.50(b).

(b) Derivation of B-format signals

In order to derive *B*-format signals from these capsule-pair signals it is a simple matter of using sum and difference technique, since it has already been shown

Figure 9.49   *AMS Calrec Soundfield microphone (showing capsules) (Courtesy AMS).*

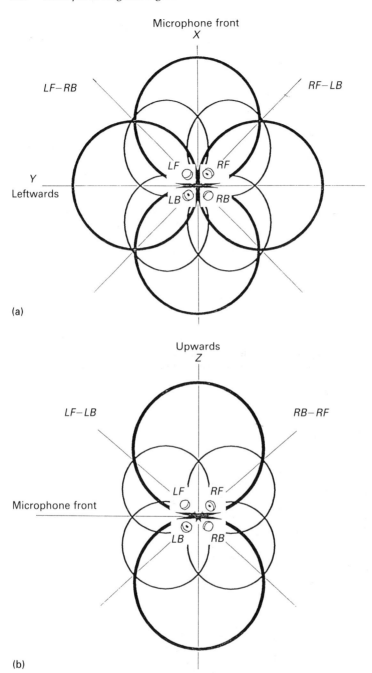

Figure 9.50 (a) *Figure-eight components resulting from horizontal subtraction of A-format signals in the Soundfield microphone. (b) Figure-eight components resulting from vertical subtraction of A-format signals in the Soundfield microphone (Courtesy of Ken Farrar).*

that the sum and difference equivalent of a crossed pair of figure-eights is simply
another pair of figure-eights rotated through 45° (see Section 9.5.1(a)(ii)). Thus

$$X = (LF - LB) + (RF - LB)$$

$$Y = (LF - RB) - (RF - LB)$$

$$Z = (LF - LB) + (RB - RF)$$

W, being an omni pressure component, is simply derived by adding the outputs
of the four capsules in phase, thus:

$$W = LF + LB + RF + RB$$

W, X, Y and Z are corrected electrically for the differences in level between them,
so as to compensate for the differences between pressure and velocity
components. For example, W is boosted at very low frequencies, since it is derived
from velocity capsules which do not have the traditionally extended bass
response of omnis.

(c) Synthesis of different polar patterns

When used as a stereo microphone, the combination of B-format signals in
various proportions can be used to derive virtually any polar pattern in a
coincident pair configuration, using a simple circuit as shown in Figure 9.51.
Crossed figure-eights are the most obvious and simple stereo pair to synthesize,

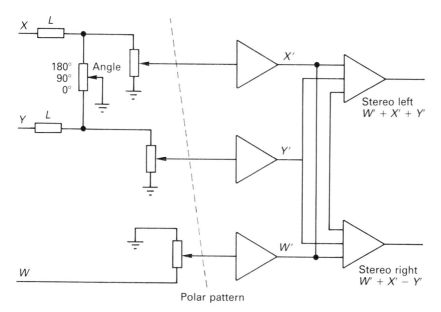

Figure 9.51 *Circuit used for controlling stereo angle and polar pattern in
Soundfield microphone (Courtesy of Ken Farrar).*

since this requires the sum-and-difference of X and Y, whilst a pattern such as crossed cardioids requires that the omni component be used also, such that

Left $= W + (X/2) + (Y/2)$

Right $= W + (X/2) - (Y/2)$

From the circuit it will be seen that a control also exists for adjusting the effective angle between the synthesized 'pair' of microphones, and that this works by varying the ratio between X and Y in a sine/cosine relationship.

(d) Control of azimuth, elevation, dominance and inversion

The microphone may be controlled, without physical re-orientation, so as to 'point' in virtually any direction (see Figure 9.52). It may also be electrically inverted, so that it may be used upside-down.

Azimuth is controlled by taking X and Y components and passing them through twin-ganged sine/cosine potentiometers, as shown in Figure 9.53, and processing them such that two new X and Y components are produced (X' and Y') which are respectively

$X' = X \cos \theta + Y \sin \theta$

$Y' = Y \cos \theta - X \sin \theta$

Elevation (over a range of $\pm 45°$) is controlled by acting on X and Z to produce X' and Z', using the circuit shown in Figure 9.54. Firstly, a circuit produces sum and difference signals equivalent to rotations through $45°$ up and down in the vertical plane, and then proceeds to combine these rotated components in appropriate proportions corresponding to varying angles between $\pm 45°$.

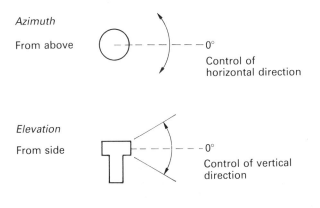

Azimuth

From above

0°

Control of
horizontal direction

Elevation

From side

0°

Control of vertical
direction

Dominance

From above

Control of front-back
precedence

Figure 9.52 *Azimuth, elevation, and dominance in Soundfield microphone.*

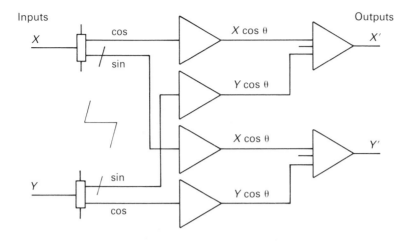

Figure 9.53   *Circuit used for azimuth control (Courtesy of Ken Farrar).*

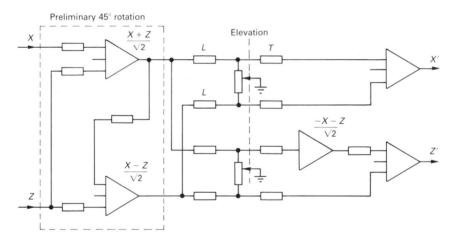

Figure 9.54   *Circuit used for elevation control (Courtesy of Ken Farrar).*

Dominance is controlled by varying the polar diagram of the W component, such that it ceases to be an omni, and becomes more cardioid, favouring sounds either from the front or from the rear. It has been described by the designer as a microphone 'zoom' control, and may be used to move the microphone 'closer' to a source by rejecting a greater proportion of rear pick-up, or further away by making W more cardioid in the reverse direction. This is achieved by adding or subtracting amounts of X to and from W, using the circuit shown in Figure 9.55.

Inversion of the microphone is made possible by providing a switch that reverses the phase of Y and Z components. W and X may remain unchanged since their directions do not change if the microphone is used upside-down.

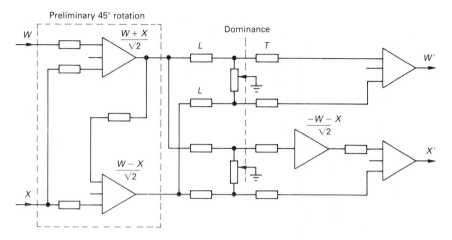

Figure 9.55 *Circuit used for dominance control (Courtesy of Ken Farrar).*

9.7 Conclusion

In this chapter it has been shown how direction is perceived in human hearing, and how phantom images, corresponding closely to the original positions of sources, may be created between a pair of loudspeakers. The creation of such phantom images requires that only a level difference exists between the loudspeakers, although the loudspeaker situation is not the same as that which exists for headphone listening, since headphone listening is closer to true binaural hearing which requires, in addition, a time delay between the two channels, having a maximum value corresponding to the time taken for a sound to travel the distance between the ears. Stereo microphone techniques optimized for loudspeaker reproduction do not produce signals that work well on headphones, and vice versa, requiring modifications to the delay and cross-talk between the signals for compatibility to exist. It is therefore important to understand how the inter*channel* signal differences relate to the inter*aural* differences which result from these, the latter depending mainly on the method of reproduction.

Conventional two-loudspeaker stereo reproduction (that is without suitable transaural processing) projects sounds which had existed in three dimensions onto a sound stage which is mainly between the two loudspeakers, thus can only hope to create an approximate illusion of reality. Both direct and reflected sound are reproduced from the front quadrant. For this reason, theoretical arguments about accurate imaging cannot be taken too far. Coincident pair microphone configurations appear to provide a close correlation between the actual and reproduced positions of sound sources when recordings are reproduced on loudspeakers, although they may achieve this at the expense of a sense of 'space' in two-channel reproduction. Near-coincident pairs have some merit in this respect, since they appear to combine a sense of 'space' with most of the imaging capabilities of coincident pairs. Widely spaced microphone techniques appear to be fundamentally flawed in theory, but are popular as a means of stereo pick-up. They do not produce a consistent image over the audio frequency range, and the transient information which results from their use may contradict continuous low-frequency information.

Microphones spaced apart by the binaural distance, and baffled to simulate the human head effects, may be used to produce outputs which can be monitored on headphones to very good effect, and which may be monitored on loudspeakers with the incorporation of suitable transaural processing. The representation of pinna effects is very important for good binaural reproduction. Multi-microphone techniques depend on the mixing and panning of a number of monophonic signals, allowing phantom images to be placed between a pair of loudspeakers at positions corresponding to the level difference between the channels. Without assistance from more-distant microphones or artificial reverberation, close multi-microphone balances sound flat and 'dead'. For best results, the outputs from each microphone should be time-equalized to represent the correct distances of each source from the position of a nominal 'main pair', in order that comb-filtering effects do not arise.

Ambisonic techniques are the logical extension of coincident Blumlein stereo to surround-sound pick-up and reproduction, although the widespread commercial use of ambisonics has yet to occur.

It is possible that the adoption of '3/2' surround formats for the cinema and high-definition TV may result in audio-only applications due to the installation of side/rear loudspeakers in many listening areas. It is likely, though, that two-channel stereo in its various forms will remain the reproduction format of choice in most domestic situations for purely practical reasons.

References

1. Moore, B.C.J. (1989). *An Introduction to the Psychology of Hearing* (3rd edn). Academic Press, London and San Diego.
2. Blauert, J. (1983). *Spatial hearing: the psychophysics of human sound localisation.* (Transl. J.S. Allen). MIT Press, Cambridge, MA.
3. Steinberg, and Snow, (1934) Auditory perspectives – physical factors. *Electrical Engineering*, **53** (1), 12–15.
4. Sivian L.J. and White, S.D. (1931) Minimum audible sound fields. *J. Acoustical Society*, July.
5. Galombos, and Davies, (1943). The response of single auditory fibres to acoustic stimulation. *J. Neurophysiology*, **6**.
6. Yost, W.A., Wightman, F.L., and Green, D.M. (1971). Lateralisation of filtered clicks. *JASA*, **50**, 1526–1531.
7. Wallach, Newman, and Rosenzweig, (1949). The precedence effect in sound localisation. *Am. J. Psych*, **62** (3), 315.
8. Haas, H. (1951). Uber den Einfluss eines Einfachechos und die Hörsamkeit von Sprache, *Acustica*, No. 2, 49–58.
9. Hafter, E.R., Buell, T.N. and Richards, V.M. (1988). Onset coding in lateralisation: Its form site and function. In Edelmann, Gall, and Cowan (eds.), *Auditory Function*, Wiley, New York.
10. Fletcher, H. (1934). Auditory perspective. *Bell System Technical Journal*, **13**, 239.
11. Snow, W. (1953). Basic principles of stereophonic sound. *JSMPTE*, **61**, 567–589.
12. Blumlein, A.D. (1958). British Patent Specification 394, 325, *JAES*, **6** (2), 91.
13. Clark, Dutton, and Vanderlyn, (1958). The stereosonic recording and reproduction system: A two channel system for domestic tape records. *JAES*, **6** (2), 102–117.
14. Gerzon, M. (1990). Three channels, the future of stereo?, *Studio Sound*, June.
15. Negishi, H. (1988). The wide imaging stereo. *Proc. Inst. Acoustics*, **10** (7), 11.
16. Bauer, B. (1961). Phasor analysis of some stereophonic phenomena. *JASA* **33** (11), 1536–1539.
17. Bennett, J.C., Barker, K. and Edeko, F.O. (1985). A new approach to the assessment of stereophonic sound system performance. *JAES*, **33** (5), 314–321.
18. Lipshitz, S. (1986). Stereo microphone techniques: are the purists wrong? *JAES*, **34** (9), 716–735.

19. Vanderlyn, P.B. (1957). British Patent 781 186. Granted 14 August.
20. Gerzon, M. (1986). Stereo shuffling: new approach, old technique. *Studio Sound*, July.
21. Vanderlyn, P. (1979). Auditory cues in stereophony. *Wireless World*, September, 55–60.
22. SMPTE (1991). *Proposed Recommended Practice: Loudspeaker Placements for Audio Monitoring in High Definition Electronic Production.* Document N15.047152, December.
23. EBU (1992). *A Possible World Standard for Multi-channel Sound.* V3/HTS Draft Report, Document GT V3. 1303, January.
24. CCIR (1991). *Multichannel Stereophonic Sound System with and without Accompanying Picture.* Draft recommendation (10–1/Temp 1), Document 10/11, December.
25. Theile, G. (1993). The new sound format '3/2 stereo'. Presented at the *94th AES Convention*, Berlin.
26. Elen, R. (1984). Surround sound. *Studio Sound*, February.
27. Zuccarelli, H., "Holophony", New Scientist, Nov 10 1983, p 483.
28. Bower, B.B. (1961). Stereophonic earphones and binaural loudspeakers. *JAES* **9** (2), 148–151.
29. Weiner, F. (1947). On the diffraction of a progressive sound wave by the human head. *JASA*, **19**, 143–146.
30. Thomas, M.V. (1977). Improving the stereo headphone image. *JAES*, **25** (7/8), 474–478.
31. Richter, F. and Persterer, A. (1989). Design and applications of a creative audio processor. *AES Convention Preprint 2782* (U-4). Presented at 86th Convention, March, Hamburg.
32. Cooper, D.H. and Bauck, J.L. (1989). Prospects for transaural recording. *JAES*, **37** (1/2), 3–19.
33. Griesinger, D. (1989). Equalisation and spatial equilisation of dummy head recordings for loudspeaker reproduction. *JAES*, **37** (1/2), 20–29.
34. Moller, H. (1989). Reproduction of artificial head recordings through loudspeakers. *JAES*, **37** (1/2), 30–33.
35. Gierlich, H.W. and Genuit, K. (1989). Processing artificial head recordings. *JAES*, **37** (1/2), 34–39.
36. Davis, D. and Davis, C. (1987). *Sound System Engineering.*
37. Streicher, R. and Dooley, W. (1985). Basic stereo microphone perspectives, a review. *JAES* **33** (7/8), 548–556.
38. Mizoguchi, A. and Ichikawa, M. (1977). *Miniature Stereomicrophone.* Note No. 213, NHK Research Laboratories, July.
39. Hibbing, M. (1989). XY and MS microphone techniques in comparison. Presented at the *86th AES Convention*, Hamburg, Preprint 2811 (A-5).
40. Dooley, W.L. and Streicher, R. (1982). MS stereo: a powerful technique for working in stereo. *JAES*, **30** (10), 707–718.
41. Cross, L. (1985). Performance assessment of studio microphones. *Recording Engineer and Producer*, February.
42. Williams, M. (1987). Unified theory of microphone systems for stereophonic sound recording. Presented at *82nd AES Convention*, London, preprint 2466 (H-6).
43. Reprinted (1981). *JAES* **29** (5), 368–372, from *Scientific American* (1881), 3 December.
44. Grignon, L.D. (1949). Experiment in stereophonic sound. *JSMPTE*, **52,** 280.
45. Gerzon, M.A. (1983). Ambisonics in multichannel broadcasting and video. Presented at *74th AES Convention*, New York, 1983 October, Preprint No 2034 (J-1).
46. Gerzon, M.A. and Barton, G.J. (1984). Ambisonic surround sound mixing for multitrack studios. Presented at *AES Convention*, Anaheim, LA, May.
47. Gaskell, P.S. (1979). System UHJ: A hierarchy of surround sound transmission systems. *The Radio and Electronics Engineer*, **49**, 449–459.
48. Daubney, C. (1982). Ambisonics – an operational insight. *Studio Sound*, August, 52–58.
49. Elen, R. (1983). Ambisonic mixing – an introduction. *Studio Sound*, September, 40–46.

10 International, regional and national standards

John Woodgate

10.1 Introduction

A standard is fundamentally an agreement, entered into by a group of people, to adopt a defined practice. The key words are 'agreement', 'defined practice' and 'adopt'. All the people concerned must agree freely, rather than under any form of duress, such as commercial or political pressure, even (perhaps especially) when conformance to the standard is backed by legal sanctions. This 'backing' may come into existence when reference is made to the standard in contemporaneous or subsequent legislation (a procedure which is much to be preferred to the inclusion of standardization texts in legislation, because it is very difficult to amend or update such texts). The standard must define the practice clearly, completely and unambiguously. The 'practice' may be the use of a definition, an abbreviation or symbol, a method of measurement, a performance requirement (or related series of any of these) or a code of practice. Finally, the standard must be 'adopted'; if it is not used in the real world it remains a useless curiosity, and the time and effort that went into its compilation is wasted. Violation of any of these conditions is likely, sooner or later, to give rise to troublesome and expensive problems.

The status of the body which publishes or endorses a standard establishes only the initial authority of that standard. It is the widespread acceptance and use of the standard that determines its on-going authority and usefulness, and these in turn have direct economic significance in terms of the number of copies of the published standard that are sold and hence of the recovery of the costs of production. It is for this reason that standards-making bodies now require strong assurances on the need for a standard on a given subject, before permission is given for drafting to begin.

10.2 Origins of standardization

It should be understood that this classification by origin does not give any indication of the breadth of acceptance and usage of a standard: an international standard may be of significance only to a few metrology laboratories, while an industry standard may be of world-wide acceptance and influence the design of millions of products.

10.2.1 Industry standards

In the past, these standards have tended to grow up by practical need and experience, often first in one country and then, as the appropriateness of the standard is realized, either spreading to other countries or, regrettably, resulting in the generation of other, conflicting local standards. An example of the first result is the ⅜ in 27 t.p.i. thread for microphone stands, derived from usage on gun barrels! Many of these standards are not documented. Those that are are mostly drawn up under the aegis of an industry association or a learned society, such as the Audio Engineering Society.

In recent years, another form of industry standard has appeared. Here, after an initial agreement by a few interested parties, a company is set up to exploit the intellectual property rights of the standard. A very successful example of this is the MIDI (Musical Instrument Digital Interface) standard. A disadvantage of this procedure is that, while there is still a desire to preserve a commercial element, such standards cannot be adopted as national or international standards.

Industry standards valid in the UK and USA for microphones and associated equipment include screw threads for stands etc.:

(1) ¼ in 20 t.p.i. Whitworth (camera tripod thread; mostly found on sound-level meters)
(2) ⅜ in 16 t.p.i. Whitworth
(3) ½ in 26 t.p.i. Whitworth form
(4) ⅝ in 27 t.p.i. Whitworth form.

These standards are not entirely undocumented, but the subject appears to be dealt with in only one published standard on microphones, the EIA (USA) standard SE105, dated August 1949. Currently, threads (2) and (4) are favoured for new designs. In spite of being based on inch sizes and the Whitworth thread form, these standards are also widely used in Europe and Japan. Further industry standards still current in the USA, despite being published many years ago, are EIA-215 Basic requirements for broadcast microphone cables (November 1958, adopted for Department of Defense use in October 1981), and ANSI/EIA-221-A Polarity or phase of microphones for broadcasting, recording and sound reinforcement (October 1979). This latter is now a national, rather than an industry, standard, because it has been adopted by the American National Standards Institute (ANSI).

10.2.2 National standards

Most industrial countries have a national standards body which, in the past, has served to establish requirements for components and products, for purposes of interchangeability and to provide means for the control of imports, as and when desired. With the growth of international trade, and of multinational manufacturers, the need has been seen for unnecessary differences between national standards to be eliminated. This is in principle a good thing, because reducing the number of product variants improves the efficiency of manufacture, and thus tends to reduce end-user prices. Such a process is termed 'harmonization'. The relationship between national and international standards for microphones is discussed in Section 10.2.4.

10.2.3 Regional standards

Although the precise origin is not clear, the concept of regional (i.e. supra-national) standards was probably first exploited in Eastern Europe. This was based largely on the uniform and complete acceptance of international standards as written, and as such, had almost exclusively beneficial effects on standardization as a whole. Regional standardization within the European Union (EU), however, has had, and will continue to have, a less clearly beneficial effect. There is an inherent paradox in that the harmonization of national standards of EU member countries facilitates trade between members, but the same process also facilitates the penetration of the European market by manufacturers based outside the Union. This is not contrary to the trade policy of the Union, but nevertheless has potentially important economic consequences. In an attempt to resolve this paradox, there is a very unwelcome trend to reintroduce, into established international standards, special requirements for the EU with little or no technical justification. Naturally, this has not gone unnoticed in other countries, and has led to a considerable worsening of international trade relations, which, at the time of writing, is not entirely resolved.

Regional standardization for electronic products in Europe is mainly under the control of three bodies: CEN (which deals with acoustics standards, among a wide range of other subjects), CENELEC (which deals with electrical, electronic and electroacoustic standards) and ETSI (European Telecommunications Standards Institute), which deals with telecommunications standards. The radio transmission aspects of radio (wireless) microphones are dealt with by ETSI. CEN and CENELEC are organs of the European Union, but ETSI is a private organization whose membership is open to national telecommunications administrations, operating companies and equipment manufacturers. These bodies are committed, in principle, to adopt international standards, where they exist, but in some areas of technology this principle is not being strictly observed, and regional deviations, which are of doubtful justification, are being proposed. Apathy and lack of appreciation by many manufacturers of the importance of standards to modern industry tend to allow these reactionary trends to enjoy some success.

Proposals introduced by the European Commission late in 1990 envisaged the reorganization of standardization work within the Union under a European Standards Council. Parts of these proposals are highly controversial: in particular, large multinational companies could obtain a completely overriding authority under the new structure, and the authority of national standards organizations would be greatly diminished, in spite of this being frequently denied in the text of the proposals. It is not possible to predict the outcome of the discussions on these proposals at present.

10.2.4 International standards

International standards are produced by ISO (International Organization for Standardization), which deals with acoustics as well as many other subjects, and IEC (International Electrotechnical Commission), which deals with electronics and electroacoustics. Responsibility for standards for microphones used as measuring instruments rests with IEC Technical Committee No. 29 Electroacoustics (TC29), while standards for microphones *per se* are dealt with by IEC Technical Committee No. 84 Equipment and systems in the field of audio, video and audiovisual engineering (TC84).

The international standards specifically concerning microphones are, at the time of writing:

- IEC268–4 (1972): Sound system equipment. Part 4 Microphones. Defines the characteristics to be included in specifications, and the corresponding methods of measurement. This standard is currently under revision in IEC TC84. Corresponding British Standard BS6840 Part 4(1987).
- IEC327 (1971): Precision method for pressure calibration of one-inch standard condenser microphones by the reciprocity technique. This standard is still current, despite the publication of IEC1094–2 (see below). Corresponding British Standard BS5677 1979(1988) (Identical).

- IEC402 (1972) Simplified method for pressure calibration of one-inch condenser microphones by the reciprocity technique. No corresponding British Standard.
- IEC486 (1974) Precision method for free-field calibration of one-inch standard condenser microphones by the reciprocity technique. This standard is still current. Corresponding British standard BS5679 1979(1988) (Identical).
- IEC581–5 (1981) High-fidelity household equipment and systems; Minimum performance requirements. Part 5 – Microphones. Corresponding British Standard BS5942 Part 5 1983(1989) (Identical).
- IEC655 (1979) Values for the difference between free-field and pressure sensitivity levels for one-inch standard condenser microphones. Corresponding British Standard BS5941 1980(1988) (Identical).
- IEC1094–1 (1992) Specifications for laboratory standard microphones. No corresponding British standard at present.
- IEC1094–2 (1992) Primary method for pressure calibration of laboratory standard microphones by the reciprocity technique. This standard can also be applied to microphones which do not conform to IEC1094–1 but have the same critical dimensions. No corresponding British Standard at present.

Further Parts of IEC1094 are in preparation. National standards of many countries are consistent with these international standards.

In addition to those standards that specifically concern microphones, the following international standards include information and requirements that apply to microphones and microphone systems, as well as to other equipment:

- IEC50 (801) (1984) International Electrotechnical Vocabulary: Chapter 801: Acoustics and Electroacoustics. Corresponding British Standard BS4727 Part 3 Group 08: 1985 (Identical, except that the BS does not include all the languages given in the international standard).
- IEC268–11 (1987) Sound system equipment. Part 11 Application of connectors for the interconnection of sound system equipment. Deals with circular (IC130–9, 'DIN audio') and concentric (jack and plug, etc.) connector applications. Corresponding British Standard BS6840 Part 11 1988 (Identical), which also implements CENELEC HD 483.11 S2.
- IEC268–12 (1987) Sound system equipment. Part 12 Applications of connectors for broadcast and similar use. Deals with jack and circular ('XLR') connector applications. Corresponding British Standard BS6840 Part 12 1987 (Identical). An amendment to the IEC standard was issued in 1991: this has now been adopted by CENELEC as HD483.12 S2 and the BS will therefore be amended in turn.
- IEC268–15 (1987) Sound system equipment. Part 15 Preferred matching values for the interconnection of sound system components. Deals with preferred values for output voltages, amplifier sensitivities and headroom, and source and load impedances. Preferred arrangements and values for phantom power systems are also included. This standard is under revision and consolidation with IEC574–4 (see below) in WG7 of IEC TC84, and will become a new Part of a comprehensive standard for 'Audio, video and audiovisual systems – Interconnections and matching values' in due course. Corresponding British Standard BS6840 Part 15 (1988), which also implements CENELEC HD 483.15 S3.
- IEC574–3 (1983) Audiovisual, video and television equipment and systems. Part 3 Connectors for the interconnection of equipment in audiovisual systems. Corresponding British Standard BS5817 Part 3 1989 (Identical). Also harmonized within the EC as HD369.3.

- IEC574–4 (1982) Audiovisual, video and television equipment and systems. Part 4 Preferred matching values for the interconnection of equipment in a system. This standard is under revision (see under IEC268–15 above). Corresponding British Standard BS5817 Part 4 1989 (Identical). Also harmonized within the EC as HD369.4. An amendment to the IEC standard was issued in 1991: this has now been adopted by CENELEC as HD369.4 S2 and the BS will therefore be amended in turn.
- IEC764 (1983) Sound transmission using infra-red radiation. This standard refers to what are usually regarded as cordless headphone systems, but may equally be regarded as microphone systems. This standard is under revision in IEC TC84. Corresponding British Standard BS6418 1983(1989).

National standards of many countries are consistent with these international standards.

10.3 Safety standards

Passive microphones do not represent an electrical safety hazard and are therefore not subject to such standards. The installation of microphones, especially in public auditoria, is usually subject to general safety regulations for buildings, which are often of purely local origin and may in any case be rigidly enforced. Increasingly, microphones are marketed in the form of 'microphone systems', incorporating power supplies and/or preamplifiers or control equipment. Mains-powered equipment for professional use is now subject (since November 1989) to the requirements of the international standard IEC65 'Safety requirements for mains operated electronic and related apparatus for household and similar general use', unless it is intended for use in a special environment, such as an explosive atmosphere. The current edition of IEC65 is the fifth (1985), on which the present harmonization of safety standards within the European Community is based, with some additional texts and some 'common modifications', which are detailed in CENELEC HD195 S6. The corresponding British Standard is BS415 1990. It is expected that a new European standard, EN60065, will be approved for introduction in 1994. A sixth edition of IEC65 is in preparation by the IEC committee, formerly SC12B but now elevated to full Technical Committee status as TC92. It is the intention of this committee to consider any special safety requirements for equipment for professional use, after the work on the sixth edition is completed. It seems unlikely that many such special requirements will be identified: one that is at present known is the need for at least a clarification of the requirements for the terminations of cables in connectors carrying voltages above 34 V peak (which could include 48 V phantom power), but not conductively connected to the mains supply.

Some microphone systems employ only battery power supplies, but some supplies contain considerable amounts of stored energy (nickel–cadmium batteries) and/or high concentrations of energy (lithium batteries). It is therefore advisable to apply the principles of IEC65 to such equipment, particularly if 48 V phantom power is provided, even though that standard is not formally applicable. Since it applies to equipment which can be powered either from mains or batteries, it does include relevant requirements, and some further requirements, applicable at present only in some EC countries, are included in HD195 S6. Safety standards in the USA and Canada differ from the international standard. In the USA, ANSI/UL1270 is restricted in scope to equipment for household use, but the Canadian standard CAN/CSA C22.2 No. 1 M-90 applies to equipment which is not necessarily under the 'jurisdiction' (a word of obscure significance in this context) of a skilled person, and this could be said in many instances of the use of microphones.

10.4 Electromagnetic compatibility (EMC)

10.4.1 General

The subject of electromagnetic compatibility divides into two: the emission of energy that could interfere with the operation of other equipment, and the immunity from interference due to energy emitted by other equipment. 'Emission' may be in the form of 'conduction' – currents injected into the mains supply or into wiring to other equipment, static electric and/or magnetic fields, dynamic electric and/or magnetic fields ('induction fields', which seem to be rather neglected at present) and 'radiation' – electromagnetic radiation which may propagate to a considerable distance from the source. The increasing use of electronics in safety-of-life systems whose malfunction could be disastrous, the increasing use of digital techniques which can produce copious amounts of potentially interfering stray signals, the introduction of microelectronic devices with rather limited resistance to damage by excessive voltages (such as those due to electrostatic discharge, ESD) and the results of military research into electronic counter-measures (ECM) and the electromagnetic effects of nuclear explosions have led to a rather sudden increase in the significance of, and attention being paid to, the subject of electromagnetic compatibility. This is particularly true in the USA and in the EC but, whereas in the USA the main effect has been on the design of information technology equipment, the EC has adopted a more unified approach, attempting to introduce EMC requirements for all electrical and electronic equipment within a very short timescale: the requirements, many of which were still undefined at the beginning of 1991, were to come into force at the beginning of 1992. As a result of representations from member states, a transition period of four years, until the beginning of 1996, has been introduced. In this period, manufacturers have the option of complying with the relevant European Standard or with the national standard of the state in which the equipment is offered for sale. These national standards vary widely in their severity, and in the degree to which compliance is enforced.

The standardization of methods of measurement is all-important in the context of EMC, because a small change in procedure, or even the physical disposition of the equipment and its connecting cables, can considerably affect the measured values, particularly of radiation. The relevant international standards are produced partly by the IEC and partly by a subsidiary body of the IEC, the Comité International Spécial des Perturbations Radioélectriques, CISPR (International Special Committee on Radio Interference). A definitive list of standards relevant to microphones and microphone systems cannot be given, because a major programme of revision and extension of standardization in this field is under way at present, but these standards include:

● CISPR13 (1975) + Amendment 1 (1983) Limits and methods of measurement of radio interference characteristics of sound and television receivers. This deals with the emission characteristics, and applies to equipment and systems for household use. Corresponding European standard EN55013 (not identical) and British Standard BS905 Part 1 1991.
● CISPR16 (1987) CISPR specification for radio interference measuring apparatus and measurement method (*sic*). This standard is under complete revision in Technical Committee CISPR/A. Corresponding British Standard BS727 1983 (not identical, and obsolescent).
● CISPR20 (1985) Measurement of the immunity of sound and television broadcast receivers and associated equipment in the frequency range 1.5 to 30 MHz by the current-injection method. Guidance on immunity requirements

for the reduction of interference caused by radio transmitters in the frequency range 26 to 30 MHz. The concentration of interest on that frequency range is a legacy of the illegal use of AM Citizens' Band transceivers in Europe in the early 1980s. This standard applies to equipment and systems for household use. Corresponding European standard EN55020 (not identical) and British Standard BS905 Part 2 1991.

- IEC555 Disturbances in supply systems caused by household appliances and similar electrical equipment (3 Parts). This standard is under extensive revision in IEC TC77, and its scope is to be extended to cover all electrical and electronic equipment for connection to public mains electricity supplies. Corresponding British Standard BS5406. In view of the status of this standard, no issue dates can usefully be given, and no information on compatibility of texts. A revision of the corresponding European Standard EN60 555 is expected.

- IEC801 Electromagnetic compatibility of industrial-process measurement and control equipment (at least 6 Parts). The provisions of this standard are being applied to all kinds of equipment outside its formal scope. The Parts are subject to revision, and new Parts are in preparation, by IEC TC65. Corresponding British Standard BS6667. In view of the status of this standard, no issue dates can usefully be given, and no information on compatibility of texts. A European Standard EN60 801 is expected.

- IEC1000 Electromagnetic compatibility (many Parts). This is a new standard which is still being added to at the time of writing. Much of it is derived from IEC801. The Technical Committee concerned, IEC TC77, was originally concerned with electromagnetic compatibility between electrical equipment including networks, but extended its studies to the whole field of EMC. How this potential or actual overlap with the work of CISPR and other committees will be resolved is uncertain at present.

In the current absence of relevant IEC texts, some new draft standards refer to CCITT (Comité Consultatif International Télégraphique et Téléphonique) (International Telegraph and Telephone Consultative Committee) Recommendations, particularly 'Protection against Interference' Series K Recommendations, 'Blue Book', Volume IX.

At present, there are no EMC performance standards specifically for microphones for professional use. Within the EC, the 'generic' standards EN50081–1 and EN50082–1 would be expected to apply, as they are intended to apply to all products for which there is no specific standard. However, examination of the drafts of these standards, issued in 1990, caused such disquiet among UK manufacturers of audio and video equipment for professional use, that the feasibility of producing suitable standards at short notice is currently being explored. The situation is developing so rapidly at present that any printed text is bound to be greatly out of date: this is regrettable but not remediable.

10.4.2 Electromagnetic compatibility of microphones and microphone systems

Even passive microphones can emit electromagnetic interference (EMI), and suffer from inadequate immunity. The following list is not exhaustive, but should prove usefully comprehensive.

Passive microphones
Electrodynamic:

- may emit permanent magnetic field
- require immunity from external alternating magnetic fields.

Other types of passive microphone are not now of great importance. A very large number of carbon microphones remains in service in telephones; these generally do not emit (out-of-band) nor suffer from inadequate immunity, except in intense RF fields.

Active microphones
Externally and internally (electret) polarized capacitor microphones:

- do not normally emit from the microphone itself
- may emit supply frequency harmonics from an analogue power supply (APS), and/or switching frequency products from a switched-mode power supply (SMPS)
- require immunity from electrostatic discharge (ESD) at the signal output and at the terminals for the power supply.

Infra-red systems:

- considerations relevant to the type of microphone used still apply
- may emit supply frequency harmonics from an APS and/or switching frequency products from an SMPS
- may emit carrier-frequency products via the mains lead and/or by direct radiation
- require immunity from other infra-red systems, from ESD at all connectors, and from external magnetic and electromagnetic fields.

Radio microphone systems and RF (FM) capacitor microphones:

- considerations relevant to the type of microphone used still apply
- may emit supply frequency harmonics from an APS and/or switching frequency products from an SMPS
- may emit crystal frequency and/or multiplier chain and/or carrier-frequency products via the transmitter or receiver antenna and/or the receiver mains lead or by direct radiation from either unit
- require immunity from other radio systems, from ESD at all connectors, and from external magnetic and electromagnetic fields.

10.5 Requirements for radio microphones

There are at present no international standards for radio microphones, and these are being developed by ETSI. Performance requirements are set by national standards developed by the bodies (usually the Post, Telegraph and Telephone administrations) responsible for regulating spectrum usage. As a result, there are very great differences between the requirements for these systems, even in adjacent countries within the EC, and this has led to widespread use of unauthorized equipment in many countries. It is also a serious barrier to international trade.

In the UK, radio microphone systems must be type-approved. The specifications are issued by the Radiocommunications Agency, an organ of the Department of Trade and Industry. Specification MPT 1345 applies to radio microphone and radio hearing-aid systems which may be used without a licence, while specification MPT 1350 applies to radio microphone systems which require an individual licence.

10.6 Sources of published standards

Sufficient bibliographic information is given in the text of this chapter to identify each of the standards mentioned. Copies of national and some international standards are usually available for purchase from national standards bodies.

IEC and CISPR publications may be obtained from

International Electrotechnical Commission, Central Office Sales Department, P.O. Box No. 131, 3 rue de Varembé, CH-1211 Geneva 20, Switzerland

CCITT publications may be obtained from

International Telecommunication Union, General Secretariat – Sales Section, Place des Nations, CH-1211 Geneva 20, Switzerland

International, British and European Community Standards may be obtained from

British Standards Institution, Sales Department, Linford Wood, Milton Keynes MK14 6LE

and may be borrowed from most public libraries in the UK.

Radiocommunications Agency publications may be obtained from

Radiocommunications Agency, Information and Library Service, Waterloo Bridge House, Waterloo Road, London SE1 8UA.

11 Glossary

John Woodgate

11.1 Introduction

Standard terminology in general acoustics and electroacoustics appears in the International Electrotechnical Vocabulary, IEC50(801).[1] The corresponding British Standard is BS4727: Part 3: Group 08 (1985).[2] Microphone development has given rise to only a limited amount of specialized terminology, most of which is concerned with the content of specifications.

11.2 General terms in acoustics and electroacoustics

Acoustic (adj): Pertaining to the sense of hearing or the theory of sounds: operated by sound: (of musical instruments) producing sound directly, without ancillary equipment.
Note. Some authors draw a distinction in meaning between acoustic and acoustical. Such a distinction is not supported by most lexicographers and appears to be of little value.
Acoustic oscillation: Movement of particles in an elastic medium about an equilibrium position.
Acoustics (noun): (1) (plural in form but treated as singular) The study of sound and similar vibrations. (2) (treated as plural) Acoustic properties or characteristics (for example, of an enclosure).
Adiabatic: A thermodynamic process (such as the compression and rarefaction which constitute a sound wave) which occurs under conditions such that the medium can neither gain nor lose heat.
Audibility: The property of a sound (related to spectral content and relative loudness) which allows it to be identified among other sounds.
Clarity: The property of a sound (related to its freedom from distortion of all kinds) which allows its information-bearing components to be distinguished by a listener.
Diffuse (sound) field: Sound field in which the sound energy density is the same at all points and the mean acoustic power per unit area is the same in all directions.
Free (sound) field: Sound field in which the effect of the boundary is negligible.
Intelligibility: A measure of the proportion of the information content of a speech sound that can be understood correctly.
Note. Satisfactory intelligibility requires both adequate audibility and adequate clarity.
Isothermal: A thermodynamic process that occurs under conditions such that the temperature of the medium does not change.
Level: The logarithm of the ratio of a value of a quantity analogous to power to a stated or standard reference value.

Notes. (a) Quantities analogous to power include the squares of voltages, currents and pressures. (b) The definition applies to levels expressed in bel (base of logarithm = 10) or neper (base of logarithm = e). Levels expressed in decibel (base of logarithm = $10^{0.1}$) are naturally ten times larger numerically than the same levels expressed in bel. (c) The reference value for sound pressure in air is 20 μPa, or − 94 dB(Pa). The reference value in water is usually 1 Pa.

Noise: (1) Sound having no clearly-defined frequency components. (2) Electrical signal producing such a sound when applied to an electroacoustic transducer. (3) Any unwanted sound, or the electrical signal producing it.

Pink noise: Stationary, random noise having equal energy in each unit fractional bandwidth.

Note. Pink noise is used as a test signal in electroacoustics because its spectrum more closely resembles that of real audio signals (averaged over a long period) than does that of white noise, which contains more high-frequency energy. Pink noise also, by definition, has a level spectrum when measured with octave or third-octave filters.

Sound: Acoustic oscillation capable of exciting the sensation of hearing.

Sound field: Region of space containing sound energy.

Volume velocity: The product of the area of a surface element and the component of the particle velocity normal (perpendicular) to it.

Note. The surface may be notional. For instance, a surface may be imagined within a homogeneous body of fluid.

White noise: Stationary, random noise having a Gaussian probability distribution of amplitude.

Notes. (a) 'Stationary' means that the average properties of the noise do not change with time. A continuous sinusoidal signal is also a stationary signal. (b) 'Random' means that the amplitude of the signal at any instant cannot be predicted from a knowledge of previous values. A continuous sinusoidal signal clearly does not satisfy this requirement, because it is completely predictable. (c) 'Gaussian' means that the amplitude probability distribution (*P*) obeys the equation

$$P = \{\exp(-z^2/2)\}/\sqrt{(2\pi)} \tag{11.1}$$

where

$$z = X - \mu/\sigma \tag{11.2}$$

X is the instantaneous amplitude, μ is the mean value and σ is the standard deviation (usually normalized to unity). (d) White noise has the property of containing equal energy in each unit bandwidth. The definition does not preclude the existence of a DC component ($|\mu| > 0$), but in practice there must be an upper limit to the spectrum, because otherwise the energy would be infinite. So actually we are invariably dealing with band-limited white noise, and the bandwidth may be limited at the lower end as well as at the upper.

11.3 Electroacoustic systems and their elements

Acoustic admittance: Reciprocal of the acoustic impedance.

Acoustic compliance: (1) Synonym of 'acoustic stiffness (1)'. (2) Reciprocal of 'acoustic stiffness (2)' (q.v.).

Acoustic impedance: Quotient of the sound pressure at a surface (which may be notional) by the volume velocity through the surface.

Note. (a) In the frequency domain the impedance is in general a complex quantity (one having real and imaginary parts). (b) Acoustic systems which exhibit gyroscopic behaviour are rare, but some electroacoustic transducers behave in an analogous manner. (See 'Mechanical impedance'.)

Acoustic mass: (1) An element in which acoustic energy may be stored in kinetic form. (2) Quotient of the magnitude of the impedance of such an element at a given frequency by the angular frequency.

Note. This implies that for sinusoidal excitation the volume velocity through the element is in lagging quadrature with the sound pressure across it.

Acoustic reactance: Component of the acoustic impedance in which energy can be stored in kinetic or potential form.

Note. With sinusoidal excitation the pressure applied to an acoustic reactance is in quadrature with the resulting volume velocity, thus reactance is represented by an imaginary quantity in the frequency domain.

Acoustic resistance: Component of the acoustic impedance in which energy can be dissipated.

Note. With sinusoidal excitation the pressure applied to an acoustic resistance is in phase with the resulting volume velocity, thus resistance is represented by a real quantity in the frequency domain.

Acoustic stiffness: (1) An element in which acoustic energy may be stored in potential form. (2) Product of the magnitude of the impedance of such an element at a given frequency and the angular frequency.

Note. This implies that for sinusoidal excitation the volume velocity through the element is in leading quadrature with the sound pressure across it.

Acoustic system: System designed to generate, transmit, process or receive acoustic energy.

Active transducer: A transducer in which the emitted energy is derived in part from sources other than the received energy.

Blocked impedance: The driving-point impedance of a system when its output is presented with a load of infinite impedance.

Note. A load of infinite impedance is one which permits no flow; thus current, velocity or volume velocity is zero. Great care is necessary with this and allied concepts if mobility analogies are used.

Characteristic impedance of a medium: Quotient of the instantaneous sound pressure by the instantaneous particle velocity in a free plane-progressive wave.

Note. The characteristic impedance is equal to the product of the density of the medium and the speed of sound in it.

Compliance (mechanical): The reciprocal of stiffness.

Note. Energy is stored in a moving mass in the form of kinetic energy. A mass therefore possesses mechanical reactance. With sinusoidal excitation the force and velocity are in quadrature, with the phase of the velocity lagging.

Driving-point impedance: The impedance at the point of application of an excitation.

Electroacoustic force factor: Quotient of the output sound pressure of the blocked acoustic system by the input current of the electrical system.

Note. For a passive reciprocal transducer, the quotient of the open-circuit output voltage of the electrical system by the input volume velocity of the acoustic system is numerically equal to the electroacoustic force factor: if the coupling is gyroscopic the sign is reversed. Note that the electroacoustic coupling in an electrodynamic transducer is gyroscopic.

Electroacoustic transducer: Transducer designed to receive electrical excitation and provide an acoustic response.

Note. The term may also be used for a transducer which receives an acoustic excitation and provides an electrical response. However, some common transducers which provide an acoustic response to an electrical excitation are not reciprocal, and it is therefore preferable to use the term 'acoustoelectrical' for transducers having acoustic input and electrical output. Normally, of course, the term 'microphone' or 'hydrophone' would be used.

Electromechanical force factor: Quotient of the output force of the blocked mechanical system by the input current of the electrical system.

Note. For a passive reciprocal transducer, the quotient of the open-circuit output voltage of the electrical system by the input velocity of the mechanical system is numerically equal to the electromechanical force factor: if the coupling is gyroscopic (as in an electrodynamic transducer) the sign is reversed.

Electromechanical transducer: Transducer designed to receive electrical excitation and provide a mechanical response.

Note. The term may also be used for a transducer which receives a mechanical excitation and provides an electrical response. However, some transducers which provide a mechanical response to an electrical excitation are not reciprocal, and it is therefore preferable to use the term 'mechanoelectrical' for transducers having mechanical input and electrical output.

Excitation, syn. **Stimulus:** Energy presented to an input port of a system.

Free impedance: The driving-point impedance of a system when its output is presented with a load of zero impedance.

Note. A load of zero impedance is one which supports no tension; thus potential difference, force or sound pressure is zero. Great care is necessary with this and allied concepts if mobility analogies are used.

Linear transducer: A transducer in which the emitted energy is strictly (linearly) proportional to the received energy.

Load: A port of a system or an energy sink intended to receive energy from another system.

Note. The concept includes loads whose impedances are zero or infinite, which can therefore receive only infinitesimal energy from a finite system.

Loaded impedance: The driving-point impedance of a system when its output is presented with a load of specified impedance.

Mass (mechanical): The quantity of matter present in a body.

Mechanical admittance: Reciprocal of the mechanical impedance.

Mechanical immittance: Pantechnicon expression for 'impedance or admittance'.

Mechanical impedance: Quotient of the force at a point in a mechanical system by the velocity in the direction of the force at its point of application.

Notes. (a) In the frequency domain the impedance is in general a complex quantity (one having real and imaginary parts). (b) Mechanical systems exist in which the application of a force results in motion normal to the applied force. Such mechanical systems do not often occur in electroacoustics, but electromagnetic transducers exhibit analogous behaviour, which is termed 'gyroscopic'.

Mechanical reactance: Component of the mechanical impedance in which energy can be stored in kinetic or potential form.

Note. With sinusoidal excitation the force applied to a mechanical reactance is in quadrature with the resulting velocity, thus reactance is represented by an imaginary quantity in the frequency domain.

Mechanical resistance: Component of the mechanical impedance in which energy can be dissipated.

Note. With sinusoidal excitation the force applied to a mechanical resistance is in phase with the resulting velocity, thus resistance is represented by a real quantity in the frequency domain.

Motional impedance: The difference between the loaded impedance and the blocked impedance of a transducer.

Passive transducer: A transducer in which the emitted energy is derived entirely from the received energy.

Reciprocal transducer: A linear, reversible transducer.

Note. This is the IEV definition. Some authors use this term instead of 'reversible transducer' (q.v.), without implying strict linearity.

Response: Energy emitted by a system due to an excitation.

Reversible transducer: A transducer which will function with net energy-flow in either direction through it.

Notes. (a) Most transducers are reversible but many appear to be very inefficient when used in the opposite direction to that for which they are designed. In fact, the efficiencies in each direction are not independent. (b) Net energy flow is the long-term average of the difference between the energy entering a specified port and the energy emitted by it.

Sensitivity: Quotient of a specified quantity describing the response of a system by another specified quantity describing the excitation.

Note. For a microphone, the excitation is usually expressed in terms of sound pressure, and the response is then preferably expressed in terms of the (loaded) output voltage, since this is a quantity of the same kind as sound pressure and there is a quasilinear relationship between them. The sensitivity expressed in this way is substantially constant with respect to amplitude, and it is usually a design goal to make it substantially (but not necessarily completely) independent of signal frequency as well. Other logically satisfactory pairs of quantities may also be used to express sensitivity, such as sound power and the corresponding electrical power in the rated load impedance:

$$P = V^2/R \tag{11.3}$$

where V is the loaded output voltage and R is the rated load impedance (q.v.).

Specific acoustic impedance: Quotient of the sound pressure at a point by the particle velocity at that point.

Note. In the frequency domain the impedance is in general a complex quantity (one having real and imaginary parts).

Specific acoustic reactance: Component of the specific acoustic impedance in which energy can be stored.

Specific acoustic resistance: Component of the specific acoustic impedance in which energy can be dissipated.

Stiffness (mechanical): Quotient of change of force by change of displacement of an elastic element.

Note. Energy is stored in an elastic element in the form of potential energy. Such an element therefore possesses mechanical reactance. With sinusoidal excitation the force and velocity are in quadrature, with the phase of the velocity leading.

Transducer: Device designed to receive one form of energy and emit a different form, in such a manner that desired characteristics of the received energy (such as those conveying information) are transmitted.

Transfer function: Relationship between response and excitation in which excitation is the independent variable.

11.4 Transduction principles of microphones

Capacitor microphone, syn. **Electrostatic microphone**, syn. **Condenser microphone**: Microphone in which sound energy is converted to electrical energy through the variation of capacitance between a membrane and a fixed plate, or between two membranes.

Carbon microphone: Microphone in which sound energy is converted to electrical energy through the piezoresistive property of carbon.

Condenser microphone: See Capacitor microphone.

dynamic microphone: See Electrodynamic microphone.

Electret microphone: Capacitor microphone in which the polarizing voltage is produced by a permanently charged element (permanent electret).

Electrodynamic microphone, syn. **Dynamic microphone**: Microphone in which sound energy is converted to electrical energy through magnetic induction in a moving coil.

Electromagnetic microphone: Microphone in which sound energy is converted to electrical energy through magnetic induction in a fixed conductor (normally a coil or coils).

Electrostatic microphone: See capacitor microphone.

Piezoelectric microphone: Microphone in which sound energy is converted to electrical energy through the piezoelectric property of an element.

Ribbon microphone: Microphone in which sound energy is converted to electrical energy through magnetic induction in a thin, flexible, moving conductor.

Strain-gauge microphone: Microphone in which sound energy is converted to electrical energy through the piezoresistive property of a thin, flexible, conductor under stress.

Transistor microphone: Microphone in which sound energy is converted to electrical energy through the mechanoelectric properties of a semiconductor junction or element under stress.

11.5 Acoustic principles of microphones

Note. For a perfectly linear transducer, the definitions given below are sufficient. For a real transducer, it is necessary to specify the electrical quantity to be taken as the measure of response. Normally, this is the (loaded) output voltage, but for certain purposes the output current or available power may be specified.

Boundary-layer microphone: See Pressure-zone microphone.

(First-order) pressure-gradient microphone: Microphone that responds substantially proportionally to the sound pressure gradient to which it is exposed.

Gun microphone: See Interference-tube microphone

Higher-order (nth order) pressure-gradient microphone: Microphone system comprising n first-order pressure gradient microphones whose outputs are relatively delayed and combined in an appropriate way, so that the response is proportional to the nth spatial differential of the sound pressure.

Interference-tube microphone, syn. **Rifle microphone,** syn. **Line microphone,** syn. **(Shot)gun microphone**: Microphone system having an acoustic wave-interference structure as the principal acoustic entry, intended to produce an extremely restricted directional response.

Line microphone: See Interference-tube microphone.

Pressure microphone: Microphone that responds substantially proportionally to the sound pressure to which it is exposed.

Pressure-zone microphone, syn. **Boundary-layer microphone**: Microphone in which the transducer is mounted, within a small fraction of the shortest wavelength of relevant sound signal, near a reflecting surface.

Note. Trade-mark registration is claimed for one or both of these terms.

Pressure/pressure-gradient microphone: Microphone that responds substantially proportionally to a weighted sum of the sound pressure and sound pressure gradient to which it is exposed.

Probe microphone; Microphone in which the principal acoustic entry is in the form of an orifice in the end of a very small tube.

Rifle microphone: See Interference-tube microphone.

Shotgun microphone: See Interference-tube microphone.

Sound-field microphone: Microphone system, comprising four transducers arranged on the faces of a tetrahedron, whose output contains full directional information concerning the source(s) of the sound field.

Note. This term is claimed to be a registered trade mark.

Velocity microphone: Microphone that responds substantially proportionally to the sound volume velocity to which it is exposed.

11.6 Classification by application

Boom microphone: Microphone intended to be used on a boom or pole.

Close-talking microphone: A microphone designed to be used very close to the mouth of the user.

Commentator's microphone: A lip-microphone designed to produce broadcast-quality sound.

Hand microphone: Microphone intended to be held in the hand.

Note. This application makes particular demands on freedom from sensitivity to handling and on resistance to mechanical shock.

Handset: Combination of microphone and earphone intended to be held in the operating position by hand.

Headset: Combination of microphone and one or more earphones, intended to be held in the operating position by a headband, chinband or equivalent.

Infra-red microphone: Microphone system comprising one or more (usually fixed) microphones, amplification and modulation equipment, an infra-red emitter panel or panels (or equivalent) and one or (usually) more receivers, often for personal wear.

Lavalier microphone: Microphone intended to be held in front of the user by means of a neck cord.

Lip microphone: A close-talking microphone provided with a guarding or spacing structure, designed to ensure a constant spacing between the microphone body and the lips of the user.

Noise-cancelling microphone: A microphone, normally a close-talking microphone, designed to be insensitive to distant sounds.

Radio microphone: See Wireless microphone.

Roving microphone: Hand microphone equipped with a long cable, so that it can be used in many places at will.

Stand microphone: Microphone intended for use on a floor, banquet or table stand.

Throat microphone: Microphone intended to be held in contact with the throat of the user.

Tie-clip microphone: Microphone intended to be clipped to the front of the user's upper clothing.

Wireless microphone, syn. **Radio microphone**: Microphone system comprising a microphone and radio transmitter unit, often for personal wear, and a radio receiver, normally of the fixed type.

11.7 Microphone accessories

Banquet stand: A structure intended to stand on a table and support a microphone at the correct height for a standing user. Height adjustment may be provided.

Boom: A movable cantilever arm, normally a long metal tube, with provision for mounting a microphone at one end, supported by a substantial base structure which may be movable. There may be mechanical or electrical provision for twisting and/or tilting the microphone.

Floor stand: A structure, usually of metal, intended to stand on the floor and support a microphone at the correct height for one or more standing users. Height adjustment may be provided.

Pneumatic stand: A floor stand whose height can be remotely controlled by means of air pressure.

Pole: A cantilever arm, normally a long metal tube, with provision for mounting a microphone at one end, the other end intended to be held by an operator who positions the microphone as required.

Table stand: A structure intended to stand on a table and support a microphone at the correct height for one or more seated users.

11.8 Electroacoustic characteristics

Bidirectional microphone, syn. **figure-of-eight microphone**: A microphone whose directional response resembles a lemniscate (figure-of eight) curve ($r = a \cos \theta$ in polar coordinates) over most of its frequency range.

Note. Plotted on a linear amplitude scale, the lemniscate curve consists of two equal circles touching at the origin.

Cardioid microphone: A microphone whose directional response resembles a cardioid (heart-shaped) curve ($r = a(1 + \cos \theta)$ in polar coordinates) over most of its frequency range.

Note. The term unidirectional applied to a cardioid microphone is deprecated, because the response over the whole of the front hemisphere varies by only 6 dB: the response is zero, if at all, only at 180° incidence.

Figure-of-eight microphone: See Bidirectional microphone.

Hypercardioid microphone: A microphone whose directional response resembles the limaçon curve $r = a(1 + 3 \cos \theta)$ over most of its frequency range.

Note. Such a microphone has the least random energy sensitivity of all first-order directional microphones.

Limaçon curve: A curve described by the equation $r = a(1 + b \cos \theta)$ in polar coordinates.

Note. In pure mathematics, the two branches of the curve lie to the right (positive side) of the origin. In plotting microphone directional responses, the polarity inversion of the minor branch is normally not taken into account, so that this branch lies to the left (negative side) of the origin.

Omnidirectional microphone, syn. **Non-directional microphone**: A microphone whose directional response resembles a circle ($r = a$ in polar coordinates), or a sphere, over most of its frequency range.

Note. A directional microphone which has an axis of symmetry is likely to be non-directional in a plane perpendicular to that axis.

Reference axis: A line passing through the reference point, indicating a recommended direction of sound incidence, specified by the manufacturer.

Note. The reference axis should preferably be perpendicular to the plane of the principal sound entry of the microphone and pass through the centre of the entry.

Reference point: A point on the microphone specified by the manufacturer.

Note. In order to allow unambiguous specification of the reference point, reference axis and polarity, the manufacturer should designate a principal sound entry even for a bidirectional microphone.

Supercardioid microphone: A microphone whose directional response resembles the limaçon curve $r = a(1 + 2 \cos \theta)$ over most of its frequency range.
Note. Such a microphone has the highest front-to-random sensitivity index of all first-order directional microphones.

References

1. IEC Publication 50(801) *International Electrotechnical Vocabulary*: Chapter 801, Acoustics and Electroacoustics, International Electrotechnical Commission, Geneva (1984).
2. BS4727: Part 3: Group 08, *Glossary of Electrotechnical, Power, Telecommunications, Electronics, Lighting and Colour Terms*. Acoustics and electroacoustics terminology, British Standards Institution, London (1985).

Index

AKG C522 microphone, 391, 294
AMS ST250 MS microphone, 394, 395
Acoustic boundary microphones, 52
Acoustic cavities, 3
Acoustic impedance, 34, 431
Acoustic measurements, 1
Acoustic parameters, measurement of, 236
Acoustic transfer function, 236
Acoustics, terminology, 430
Acousto-optic modulators, 153, 154
Aerials, 190, 192
  of radio microphone transmitters, 217
  receiving, 191
Ambisonics, 367, 406–418, 419
  encoding and decoding, 409
  figure-eight components, 414
  pan rotate unit, 411
  principles of, 406
  signal formats, 407
    A, 407, 414
    B, 407, 408, 413, 415
    C, 408, 410
    D, 409, 411
  Soundfield microphone, 411, 412
    capsules, 412, 413
    control, 416
    elevation, 416, 417
    polar patterns, 415
  source of signals, 407
  transcoders, 411
Ambisonic microphone, 407
Ambisonic systems, 350, 367
Ambisonic signals;
  UHJ decoding, 375
  W,X,Y,Z signals in, 372
Amplifiers, 277–312
  balanced, 335
  BBC, 277, 286
    type A & B, 286
    type C & D, 287, 288
  circuit diagrams, 287
  design, 281
  electronically balanced input circuits,
    305
  equivalent input noise concept, 304, 306
  flicker noise and, 291, 299, 300
  for RF condenser microphone, 161
    phase modulation design, 172
    push-pull bridge design, 182

Amplifiers – *continued*
  for ribbon microphones, 42
  fundamentals, 277
  history of, 277
  Johnson noise, 291, 293, 306, 308
  low noise transistors in, 298
  noise in, 291, 303, 296, 305, 306, 308
    causes, 301
    representation of, 292
  noise figure, 294
  noise performance requirements, 303
  OBA/8, 283
    output, 284
  power supply, 283, 286
  random noise, 290
  shot noise, 291, 293
  signal-to-noise ratio, 286
  total noise, 292
  transistors in, 286, 298
    parameters, 299
  transistor noise in, 296
  voltage noise, 303
Amplifier reference voltage, 124
Analogues:
  circuits, 28
  direct and inverse, 25
  impedance and admittance, 29
  use of, 25
Analogue quantities, 7
Anti-shock mounting, 138
Anti-turbulence screens, 119
Anti-vibration mountings, 22
Anti-vibration protection, 23
Atmospheric pressure, effect on
  microphones, 97

Background noise, ribbon microphones
  for, 50
Baffles, 16
  microphones inserted into, 52
Barkhausen noise, in transformers, 327
Beam splitters, 141
Bidirectional microphones, 437
  definition, 437
  ribbon type, 8, 16, 42
Bifilar windings for transformers, 333, 340,
  346
Bi-Morph piezoelectric microphone, 69

Bi-Morph piezoelectric sandwich elements, 66
Binaural delay, 356
Binaural microphone techniques, 403
Binaural mixers, 371
Binaural signals:
　processing of, 371
　turned into stereo signals, 370
Binaural techniques, 350
　in holophonics, 369
Blind people:
　aural visualization, 359
　infra-red systems for, 230
Blumlein/Clark theories, 361, 396
Blumlein shuffler network, 370
Blumlein's stereophonic system, 362, 364
Boom microphones, 26, 436
Boom mountings, 23, 26
Boundary-layer microphones, 52
Brain, assessing direction of sound, 255, 258, 351, 352
British Broadcasting Corporation (BBC), 277
　amplifiers, 277
　　type A and B, 286
　　type C, 287
　　type D, 287
　microphone amplifiers, 282
　outside broadcast equipment, 283, 284
　ribbon microphones, 283
　stereo and, 288
Broadcast commentaries, ribbon microphone for, 50
Broadcasting, 188
　digital, 223
　frequency bands, 188, 193
　microphones for, 13
　outside broadcasts, 283, 285
　precision condenser microphones for, 130
　radio microphones for, 222
　radius of, 193
　regulations, 188
　stereo, 288
　television bands, 210
　transmitters for outside use, 221

Cables, 18
　for condenser microphone preamplifiers, 82
　interference in, 134
　requirements, 21
　resistance of, 334
　size and flexibility, 21
Calibration:
　checking, 121
　comparisons, 108
　equipment and accessories, 100
　in the field, 100

Cables – *continued*
　hydrophone, 126
　insert voltage, 109, 116
　laboratory, 102
　multipiston pistonphone, 126
　outdoor assembly, 75, 102
　pistonphones, 100, 101, 109, 119
　reciprocity, 108, 126
　windshield corrections, 117
Calrec Soundfield microphone, 406, 407, 411, 412
　azimuth control, 416, 417
　control of, 416
　dominance control, 416, 418
　elevation control, 416, 417
　inversion of, 416, 417
　polar patterns, 415
Cantilever bimorph piezoelectric microphone, 66, 69
Capacitor microphone *see* Condenser microphone
Carbon microphones, 14, 278, 435
Cardioid condenser microphone, 68
　characteristics, 38
　pressure differences, 302
　responses, 67
　specification, 131
Cardioid microphones, 8, 437
　directional characteristics, 17
　pressure gradient forces, 10
　specifications, 131
Cardioid moving coil microphones:
　impedance characteristics, 301
　noise criteria, 300
　response characteristics, 31
Cardioid ribbon microphones, two-phase shift, 18
Children, speech training, 221, 230
Cinema surround systems, 367
　signals in, 373
Clipping diodes, 216
Close talking, 436
　ribbon microphones for, 50
Compact discs, 223, 231
　white noise on, 290
Concert hall broadcasts, 63
Condenser microphones, 4, 56, 216, 273, 434
　accessories, 117
　analogue circuit, 28, 35, 37, 70
　calibration chart, 76, 77
　cardioid, 68
　　characteristics, 38
　　pressure differences, 302
　　responses, 67
　specification, 131
　diaphragm, 59
　directional, 63
　directional characteristics, 67, 132
　dual unit, 38

Condenser microphones – *continued*
  equivalent circuit, 79
  figure-of-eight, 302
  force/deflection characteristic, 37
  for concert hall broadcasts, 63
  for sound measurement, 62, 159
    calibration, 76, 77
    construction, 137
    dimensions, 78
    theory of, 69
  free-field and pressure responses, 74
  half-wave-length distance, 17
  high quality RF, 158–186
    see also Condenser microphones, RF
  high-stability electret polarization, 133
  impedance, 301
  internal interference, 21
  let into wooden squares, 52
  line matching unit, 134
  low frequency response, 71, 72
  maximum sound pressure level, 72
  mechanical layout, 70
  noise elimination, 20
  noise criteria, 300
  noise performance, 302
  omnidirectional, 34–36, 63, 68
    internal construction, 34
    responses, 132
  output characteristics, 133
  pm-type MKH, 173
  phase responses, 133
  phase shift, 36
  polar curves, 272
  polarizing voltage, 32
  power supplies, 18, 134
    specifications, 135
  preamplifiers, 20
  precision:
    directional characteristics, 132, 136
    for broadcasting, 130
    output charateristics, 133
    phase responses, 133
    specifications, 130, 131
  pressure differences in, 302
  pulse responses, 93, 274
  push-pull MKH, 183
  RF:
    AM push-pull bridge design, 174
    acoustic impedance design, 162
    amplifiers for, 161, 172
    benefits of, 158
    bridge design, 159, 174, 175
    circuit design, 170, 171
    clip-on types, 174
    design aspects, 160
    diaphragm, 180
    difference frequency test, 164
    directional characteristics, 162
    discriminators, 170, 180
    feeding circuits, 174, 182

Condenser microphones – *continued*
  RF – *continued*
    frequency determining circuit, 172
    frequency response, 163, 174
    high source impedance, 158
    history of, 159
    improving noise performance, 161
    linearity, 163
    MKH series, 161, 173, 174, 175, 178,
      184, 185
    noise levels, 160
    omnidirectional types, 174
    oscillators, 174
    pm-type, 173
    phase modulation circuit, 170
    phase modulation design, 165
    phase modulated types, 174
    principles of, 158
    push-pull circuit design, 178, 181
    push-pull design, 164, 183
    push-pull types, 184
    sensitivity, 161, 185
    technical data, 173, 183
    transformerless output, 165
    transistors for, 172
    transmission factors, 161
    types of, 174
  sectional views of, 65
  specifications, 130
  standards for, 428
  theory of, 69
  transformers in, 332
  transient response, 137
  transistorized, 159, 160
  unidirectional, 35, 36–38
    construction, 36
    polar-response curves, 35
    response of, 35
  variable polar response, 32
Condenser microphone cartridge, 73
  calibration chart, 73, 76, 77
  distortion limits for output capicitances,
    83
  factors affecting response, 77
  for sound measurement, 80
  free-field and pressure response, 74
  free-field correction curves, 104, 105, 106
  inherent noise specifications, 82
  nose cones, 79, 80
  preamplifier, 79, 80
    cables, 82
    capsules, 82, 83, 84
    combinations, 81
    dynamic range, 88
    response, 80
    specifications, 84, 86
  random incidence diffuse field
    corrections, 77, 79
  sizes and applications, 73
  stability, 73

Condenser microphone cartridge –
  continued
  vent construction, 73, 74
  weatherproof kits, 75
  with preamplifiers,
    effect of humidity, 92
    phase response, 91, 92
    responses, 89
Conference systems, 224
  infra-red, 229
Connections, 18

Daylight, infra-red, 227
Decca tree, 400
Dehumidifiers, 102
Design of microphones, 25
  see also specific types
Design techniques, 13
Diaphragm, 12
  behaviour of, 140
  combined with transducers, 13
  deflections, 142
  direct analogue-to-digital conversion,
    149
  force feedback to, 156
  measurement of movement, 143
  mounting tension, 59
Difference frequency test set-up, 164
Digital audio broadcasting, 223
Digital signal processing, 405
Directional characteristics, 13, 110, 111, 112,
    113
  limitations, 16
Directional line microphones, 16
Directional microphones, 54–56
  response of, 162
  tubular line, 56
  types of, 17–18
Directional perception, 349
  by blind people, 359
  free field condition, 351
  front-back or median plane, 358
  lateral plane, 351
  level difference, 351
  phase difference, 353
  role of pinna, 357, 358, 359, 403
  shadowing of head, 352, 353
  time-of-arrival difference, 355
  in vertical plane, 359
Discriminator circuits, for RF condenser
    microphones, 170, 180
Distortion characteristics, 115
Dolby Prologic decoder, 375, 376
Dolby system, 367, 374
Dummy head techniques, 403

Earthing, 20
Earth leakage contact breakers, 20
Echoic chambers, 244

Eddy current loss in transformers, 324, 331
Electret condenser microphone, 56
  power supplies, 18
Electrical interference, 57
Electroacoustics, terminology, 430, 431
Electroacoustic characteristics, terminology,
    437
Electromagnetic compatibility, 426
Electromagnetic fields, interference from,
    57
Electromagnetic waves, propagation
    problems, 189
Electrostatic actuator, 109
  design of, 109, 115
Electrostatic fields, interference from, 57
Equalization, 287
Equivalent input noise concept, 304, 306
Europe, standardization in, 423
European Union, 423
Evolution of microphones, 25

Fabry-Perot interferometer, 150
  modified, 151
  photodiode current, 152
  problems, 155
Faulkner pair of stereo microphones, 397,
    398
Ferromagnetism, 327
Figure-of-eight capacitor microphones, 302
Flicker noise, 291, 299, 300
Force feedback, 140, 155–157
  application of, 156
  maximum available, 156

Gradient microphones:
  frequency response, 163
  properties of, 50
  wind noise and, 23
Gradient-operated microphones, 7, 8
Grain orientated silicon sheet for
    transformer cores, 342
Grid stoppers, 281

Haas effect curve, 357
Half-wave plates, 141
Handicapped:
  infra-red systems for, 230
  radio aids for, 221
Head, shadowing effect of, 353
Headphones:
  binaural system for, 403
  for stereophonic sound, 419
  reproduction, 369
  two channels on, 369
Hearing, 1
  binaural delay, 356
  directional, 350
  directional perception, 349
  from two loudspeakers, 363
  Haas effect curve, 357

Hearing – *continued*
  role of pinna, 357, 358, 359, 403
  shadowing effect of head, 352, 353
  time-of-arrival difference, 355
    in stereophonic sound, 361
Hearing aids, 57, 187
Hearing impairment:
  infra-red systems for, 223
  radiomicrophones for, 187, 222
Highly directional microphones, 54, 56
Holophonics, 369
Hot wire microphone, 12
Humbucking, 45, 47
Humidity:
  effects of, 2, 97, 102
  microphone testing and, 237
Hydrophones, 1, 125
  calibrations, 126
  piezoelectric microphones for, 63
  receiving characteristics, 127
  specifications, 128
  spherical, 127
  structure of, 126
Hypercardioid microphone, 437
Hysteresis loss in transformers, 325

Impedance relationships, 7
Input circuits, 18, 19
Inductance, 315
Information systems, infra-red for, 229
Infra-red emitting semiconductor diodes,
    188, 223
  optical bandwidth, 226
Infra-red light:
  applications of, 188
  audio signal properties, 228
  from daylight, 227
  general aspects of, 224
  generation and propagation, 224
  limitations of power, 225
  modulation and channel bandwidth, 226
  spurious emission, 226
Infra-red microphones, 226, 436
Infra-red systems:
  applications, 228–231
  electromagnetic compatibility, 428
  for blind people, 230
  for deaf people, 223
  for handicapped, 230
  for TV sound, 224
  future aspects of, 231
  headphones, 228
  immunity against interference, 227
  information and conference, 229
  interpreter, 226
  loudspeakers, 228
  in medical education, 230
  multi-channel, 229
  pulse modulation, 226

Infra-red systems – *continued*
  stereo, 223
  vehicle applications, 230
Infra-red wireless links, 223–231
Inherent noise level, 94
Insert voltage calibration, 109, 116
Integrated circuit microphones, 56
Interference, within microphones, 21
Interference microphones, 163
Interferometers and interferometry,
  dual-wavelength, 153
  dynamic ranges, 149
  Fabry-Perot, 150, 152
    photodiodes, 152
  fundamentals of, 144
  heterodyne type, 141
  light sources, 140
  Michelson, 144, 145
    beam splitters, 146, 147, 148
    dynamic type, 149
    photodiode current, 146
    problems, 155
    using half-mirror, 147
  Michelson and Fabry compared, 152
  multiple beam, 150
  single wavelength, 144

Johnson noise, 291, 293, 306, 308

K factor, 124
Kundt's tube, 241

Lagrange's equation, 40
Lasers, 140
  diodes, 231
  output power, 143
  microphones, 12
  scan systems, 6
  transverse Zeeman, 153, 154
Lavalier microphones, 53–54, 436
Linearity, 59
Line microphones, 54
  basic action of, 55
Lip microphones, 43, 436
Loudspeakers:
  in binaural systems, 369
  for mcirophone testing, 238–240
  infra-red, 228
  operating as microphones, 346
  placement for stereo, 372
  in stereophonic sound, 362
    more than two channels, 367
    two channels, 362
  surround format, 368
  two channels on, 362
  wide-ranging system, 362

Magnetic fields, response to, 94
Magnetophone, 278, 279
Magnets, 5
Marconi-Reisz transverse current carbon
    microphone, 280
Marconi-Sykes magnetophone, 278, 279
Maximal-length sequences, 255, 257
Measurement microphones, RF, 159
Mechanical vibrations, sensitivity of
    microphones to, 256
Michelson interferometer, 144, 145, 152
    beam splitters, 146, 147, 148
    dynamic range, 149
    photodiode current, 146
    problems, 155
    using half-mirror, 147
Microphones:
    see also under types etc.
    acoustic principles of, 435
    appearance and finishes, 59
    applications, terminology, 436
    calibration *see under* Calibration
    design of, 13
    development of, 12
    electromagnetic compatibility, 427
    evolution sand design, 25
    quality of, 236
        see also Testing of microphones
    specifications, 64
    techniques, 1–61
    terminology, 436
    testing *see* Testing of microphones
    transduction principles, 434
    types of, 14, 436
        see also specific types
    uses of, 1
Microphone accessories, terminology, 436
Microphone transducer relationships, 30,
    434
Microphony, 279, 281
Mirrors, for optical microphones, 141
Mirror galvanometer, 142
Mode hopping, 140
Mountings, 27
    anti-vibration, 22
    noise prevention, 26
    noise protection, 23, 26
Mouth, artificial, 244, 247
Moving coil microphones, 38–41
    analogous electrical circuit, 33, 39
    analysis of, 39
    cardioid, 31, 38, 300, 301
    construction of, 29, 40
    diagram, 28
    directional characteristics, 17
    flux density, 58
    impedance, 301, 319
    membranes of, 271
    noise criteria, 300
    polar curves, 272

Moving coil microphones – *continued*
    pulse response, 274
    omnidirectional, 38
        construction of, 29, 39
    spherical case, 53
    unidirectional, 33
Moving coil units, 15
Moving iron microphones, 15
Multifilar windings for transformers, 341
Mumetal:
    as transformer core material, 324, 342
    cans, 332
    saturation flux density, 330, 332
    screening, 332
Music, ribbon microphones for, 49, 267

Negative feedback:
    circuit types, 338
    transformers and, 335
Neumann boundary-layer microphones, 53
Neumann KU6li dummy head, 403
Neumann RSM 19li microphone, 386, 388,
    392, 394
    polar patterns, 389
Neumann USM 69i microphone, 392, 393
NOS pair of stereo microphones, 397, 398
Noise, 431
    Barkhausen, 327
    causes of, 20, 301
    electronically balanced input circuits
        and, 305
    elimination:
        methods of, 20
        shielding and protection, 22
    flicker, 291, 299, 300
    Johnson, 291, 293, 306, 308
    pink, 431
    popcorn, 292
    precautions against, 22
    prediction in transistors, 296
    random, 290
    representation of, 292
    RF condenser microphones and, 160
    shot, 291, 293
    terminology, 431
    valve, 281
    waveforms, 290
    white, 431
    wind-induced, 120
Noise cancelling microphone, 436
Noise figure, 294
Nose cones, 80, 100

Oesophageal stethoscopes, 230
Omni microphones, 437
    in spaced pairs, 401
    outriggers, 402
ORTF pair of stereo microphones, 397, 398

Optical heterodyning, 141, 144, 155
Optical isolators, 141
Optical measuring systems, 140
  mirror galvanometer, 142
  noise, 143
  techniques, 142
Optical microphones, 140–157
  see also under Interferometers
  acousto-optic modulators, 153, 154
  components, 140
  demodulated, 155
  force feedback, 155–157
    applications, 156
    dynamic range of, 157
  frequency response, 155
  heterodyning, 144
  interferometry, 144
  mirrors, 141
  mode hopping, 140
  noise, 148, 155
  optical isolators, 141
  photodiodes, 143
  quarter and half-wave plates, 141
  sensors, 142
  sensor noise, 148
  source noise, 148
  sources of light, 140
Oscillators, for RF condenser microphones, 172, 182
Otological hearing measurement, 56

Pan potting, 364, 373
Pan rotate units, for ambisonics, 411
Particle movement detectors, 62
Phase-locked-loop technology, 217
Phase locking, 353
Phase shift microphones, 30
Piezoelectric disc, 14
Piezoelectric earphone, 14
Piezoelectric foils, 13
Piezoelectric microphones, 14, 125–129, 435
  for sound measurement, 63
Piezoelectric sandwich elements, 66
Piezo junction microphones, 5
Pinna, role in hearing, 357, 403
Pipe resonances, 3
Pistonphones, 100
  design of, 119
  double piston, 101
  electrodynamic form of, 109
  multi-piston, 126
Plane-wave concepts, 3
Polarization effects, 95, 96
Ponging, 279
Popcorn noise, 292
Pop effects, 8, 20, 23
  ribbon microphones and, 267
  testing for, 258–262, 263
    calibration, 263

Pop shields, 21, 22
Power supplies, 18
Preamplifiers,
  circuits, 95
  for condenser microphones:
    cables, 82
    capsules, 82, 83
    dynamic range, 88
    feedback circuits, 92
    phase response, 91, 92
    specifications, 84, 86
  for condenser microphone cartridges, 80
    frequency response, 82
    inherent noise, 88
  with switched heated circuit, 92
Precision microphones, 1
  for measurement and reproduction, 62
Pressure gradient forces, 17
Pressure gradient operated microphones, 435
  directional 29, 30
Pressure microphones, 3, 7, 245, 435
Pressure phase loss effects, 12
Probe microphone, 124
  kit, 123
Pulse response, measurement, 273, 274

Quadraphonic sound, 406
Quarter wave plates, 141

Radio aids for the handicapped, 221
Radio frequencies, wireless links on, 189–224
Radio interference, 57
Radio microphones, 187–224, 436
  applications, 207
  audio performance characteristics, 212
  data on, 234
  dynamic range, 212
  electrical performance stability, 208
  electromagnetic compatibility, 428
  for deaf, 221
  for tour guide systems, 221
  future aspects, 222
  general aspects, 200
  high-power links, 219
  medium or narrow band, 219
  multi-channel operation, 209
  multiple path reception, 210, 211, 212
  noise contributions, 213
  noise from processing signal, 214
  power supply, 208
  range of, 193
  S/N ratio, 213, 214
  simultaneous use of, 210, 211
  single-channel applications, 208
  synopsis of technology, 235
  technical possibilities, 207

Radiomicrophones – *continued*
 uses of, 221
 wide band, 219
 wired, 207
Radio microphone receivers, 205, 217, 218
 block diagram, 219
 characteristics, 213, 217, 233
 for pocket, 220
 for tour guide systems, 221
Radio microphone transmitters:
 aerials for, 217
 block diagram, 216
 clipping diodes in, 216
 for outside broadcasts, 220
 for tour guide systems, 221
 phase-locked-loop technology and, 217
 voltage controlled oscillators in, 217
Radio multi-channel conference systems, 224
Radio receivers, 205–207
 EM series, 205, 206
 portable, 207
Radio transmission:
 capture effect, 200
 heterodyne, 197
 high-fidelity, 195
 interference in, 198
 intermodulation, 198, 199, 200
 power of, 193
 RF bandwidth, 193
 RF non-linearity, 195
 S/N ratio, 195
 spurious emissions, 197
Radio transmitters:
 hand-held, 203, 205
 miniature, 200–205
  circuit, 201, 202
 pocket, 200, 201, 203, 205
 SK series, 200, 201, 202, 203, 204
 semi-conductors in, 202
Radio waves:
 aerial systems, 290, 292
 applications of, 187
 free field propagation, 189
 modulation and RF bandwidth, 193
 propagation problems, 189
 radiated power, 193
Rain covers, 75, 101, 120
Random incidence corrector, 122
Rayleigh disc, 62
Recalibration, 121
Reciprocity calibration, 108, 126
 apparatus for, 106
Reflectors, paraboloid, 16
Reisz microphone, 280
 response, 283
Reisz microphone amplifier, 281
Residual loss, in transformers, 328
Ribbon microphones, 42–52, 267–276, 435
 acoustic impedance, 43

Ribbon microphones – *continued*
 advantages of, 269, 271
 amplification, 42
 basic theory, 43
 bidirectional, 8, 16, 42
  pressure gradient forces, 10
 cardioid, 18
 close talking noise reducing, 50, 51
 construction, 45, 47
 corrugations, 59
 design aspects, 42, 46
 diaphragm, 271
 directional characteristics, 17
 directivity pattern, 270
  at high frequencies, 271
 dual unit, 48
 flux distribution, 59
 force acting on, 269
 function of, 268
 general layout, 44
 high-quality noise-reducing, 50
 history of, 267
 humbucking, 45, 47
 impedance, 43, 275, 319
  characteristics, 49, 301
 induced voltage, 268
 musicians and, 49, 267
 noise criteria, 300
 of BBC, 283
 polar curves, 272
 polar response, 46
 pop effect, 267
 pulse characteristics, 271
 pulse response, 274
 resonance, 46
 responses, 43, 46, 49
  assessment of, 47
 size and weight, 49, 267
 stereo recording with, 270, 276
 structure, 268, 269
 transformers, 275–276
 two phase shift, 18

Safety standards, 425
Sanken CMS-2 microphone, 392, 394
Sanken MS microphone, 383, 384
Saturation flux density values, 330, 332
Semi-conductors, in radio transmitters, 202
Sensitivity, 434
Short range wireless transmission, 187
'Shot gun' microphones, 16, 435, 436
 wind shield, 24
Shot noise, 291, 293
Signal-to-noise ratio, 302
Silicon diodes, 189
Sound, 431
 directional perception *see* Directional perception
 directional reproduction, 359

Sound – *continued*
  free-field condition, 351
  stereophonic *see* Stereophonic sound
Sound fields, 430, 431
  artificial diffuse, 244
  diffuse, 243
  for microphone testing, 237, 238
Sound field microphone, 367, 436
Sound measurement:
  amplifier reference voltage, 124
  calibration equipment and accessories,
      100
  condenser microphones for, 62
    advantages of, 62
    calibration, 76, 77
    construction, 137
    lower limit, 80
    theory, 69
  developments in, 62
  lower limit of, 80
  microphones for, 62
  piezoelectric microphones for, 63
  RF condenser microphones for, 159
Sound pressure gradient relationships, 9
Sound pressure measurements, 125
  microphones for, 56
Sound reproduction, precision
    microphones for, 66
Sound waves:
  deflected, 10
  displacement, pressure and
      pressure-gradient, 2
  effect of diffraction on, 11
  effects on microphones, 2
  microphones and, 1
  obstacles to, 19
Soundfield microphone *see* Calrec
    Soundfiled microphone
Spatial perception, 350
Specifications, 64
Speech:
  close-range, 50
  ribbon microphones for, 50
  training, 221
Spherical case microphones, 52
Spherical wave proximity effects, 10
Stabilized transverse Zeeman laser, 153,
    154
Standards and standardization, 421–429
  industrial, 421
  international, 423
  national, 422
  origins, 421
  regional, 422
  safety, 425
  sources of publication, 429
Stereo microphones, 377–406
  AB (XY), 391
    pairs, 378, 381, 383
    relating to MS pairs, 385

Stereo microphones *continued*
  M, 383
  MS pairs:
    applications, 390
    relating to AB pairs, 385
  back-to-back cardioids, 381, 382
  binaural pairs for headphones, 403
  binaural recording, 404
  cardioid crossed pairs, 380, 381, 382
  coincidental pair principles, 378
  commercial implementations, 392
  crossed pairs, 380
  definition, 377
  end-fire configurations, 390, 391
  Faulkner pair, 397, 398
  figure-of-eight pair, 379, 380, 385
    advantages of, 380
  hypercardioid pattern, 382, 387
  multi-microphone arrangements, 405
  near-coincident configurations, 396
  NOS pair, 397, 398
  operational considerations, 395
  orientation, 390
  ORTF pair, 397, 398
  pan pot, 404, 405
  polar patterns, 379, 380
  side fire, 390
  'side fire' AB, 384
  spaced arrays, 397–405
    analysis of, 399
    binaural pairs, 403
    Decca tree, 400
    dimensions, 401
    dummy head techniques, 403
    omni microphones, 401
    omni outriggers, 402
    positioning, 399
    principles of, 397
  susceptibility to noise, 395
  wind affecting, 395
  with LED, 384
Stereophonic sound, 359
  see also Ambisonics
  3/2 surround formats, 368, 419
  binaural mixers, 371
  binaural signals, 371
  broadcasting and, 288
  cinema surround systems, 367, 373
    l, C.R and S signals, 373
  digital signal processing, 405
  Dolby system, 367, 374
  full-surround, 371
  historical development, 360
  loudspeaker placement, 372
  multi-channel, 367
    encoded to two channels, 374
  multiple-point source, 360
  multi-source system, 361
  'pan-potting', 364, 373
  room acoustics, 372

Stereophonic sound – *continued*
  three-channel system, 362
  three-microphone system, 362
  transaural, 371
  turning binaural signals into, 370
  two channel, 362
    on headphones, 369
    on loudspeakers, 362
  two loudspeaker system, 362, 418
  wide band speech signals, 364
  wide-imaging speaker system, 362
  wide spaced microphones, 418
  with ribbon microphones, 276
Stereophony, 349–420
  see also under Stereo
  directional perception in, 349
  M-S, 276
Stereo signals, 372–377
  A, S, M and S, 372
  L, C, R and S, in cinema surround
    systems, 373
  W, X, Y and Z, in ambisonic systems,
    372
  conversion between formats, 374
  cross-talk, 377
  derivation of, 373
  Dolby system, 374
  effects of misalignment, 377
  multichannel into two channels, 374
  terminology, 372
  UHJ decoding, 375
Supercardioid microphone, 438
Switches, 57
  requirements for, 21
Symbols, 7

Talking, close, 50
Telephone operator's headset, 19
Television, 210
  infra-red transmission, 224
Television cameras, zoom microphones
  with, 18
Temperature:
  effects of, 115
  effect on microphones, 97
  microphone testing and, 237
Testing of microphones, 236–266
  acoustic parameters, 236
    artifical diffuse fields, 244
    diffuse field testing, 243
    directional characteristics, 246
    echoic chambers, 244
    environment for, 238
    free-field testing, 238
    frequency response, 247, 248, 252, 253,
      254
    impulse measurements, 250
    loud speaker for, 238
    maximal length sequences, 255, 257
    measurement in duct, 241

Testing of microphones – *continued*
  nearfield measurements, 244
  phase measurements, 247, 249
  polar patterns, 247, 248
  random noise, 253
  with artificial mouth, 244
  comparisons, 236
  effectiveness of wind screens, 265
  non-acoustic parameters, 256–265
    pop effect, 258–262, 263
    vibration, 256
    wind sensitivity, 262
  reference microphones, 236
  simultaneous comparison methods, 237
  sound fields, 237, 238
    diffuse, 243
  substitution methods, 237
  vibration, 256
  wind sensitivity, 262
Time-of-arrival difference, 355
  from multi-point source, 361
Tour guide systems, 221
Transaural sound, 371
Transducers, 4, 5, 6, 434
  combined with diaphragm, 13
  electro dynamic, 6
  magneto-strictive, 6
  piezoelectric, 5
  principles, 4, 5, 6
Transformers, 277, 312–347
  1000:1 test, 345
  Barkhausen noise, 327
  B-H loops, 325
  bifilar windings, 333, 340, 346
  common-mode interference, 334
  core, 315
  core materials, 342
  core size and performance, 314
  distortion in, 328
  eddy current loss, 324, 331
  effect of altering input size, 331
  electrostatic screens, 333
  for higher-impedance microphones, 243
  for loudspeaker talk-back, 346
  frequency response, 316
  grain-orientated silicon steel, 342
  hum pick-up, 332
  hysteresis loss, 325
  inductance, 315
  interference voltages, 335
  leakage inductance, 322
  low frequency equalization, 324
  microphone equalization, 322
  mounting, 332
  multifiler windings, 341
  mumetal as core material, 342
  mumetal cans, 332
  negative feedback and, 335
  noise levels, 337
  OBA/8, 343

Transformers – *continued*
  primary impedance, 321
  residual loss in, 328
  resistance of wirings, 316
  response curves, 317, 318, 319
  saturation flux density values, 330, 332
  shunt winding capacitance, 322
  simple design, 312–316
    probing tests, 317
    tests of, 316
  toroidal, 332
  transistors, 337
Transformerless amplifiers, 307
Transistors:
  advent of, 286
  in RF condensor microphones, 159, 160, 172
  for radio receivers, 205
  low noise, 298
  microphones, 435
  parameters, 299
  prediction of noise, 296
  with transformers, 337
Tubular line directional microphone, 56
Turbulence, 119

Under water microphones, 57
Units, 7

Valve noise, 281
Variable distance principle, 17
Variable phase shift microphone, 17
Velcoity microphones, 3, 436
Vent constructions, 73, 74
Vibration, prevention, 22, 23, 26
Vibration excitor, 256
Vocalists, microphones for, 22
Voltage controlled oscillators, 217

Wave forms, 2
Wente's capacitor microphone, 277
Wind induced noise levels, 120, 121
Windscreens, 22, 24, 100, 101
  corrections for calibration, 117
  effectiveness of, 263
  materials for, 23
  mechanism of, 23
Wind sensitivity, 262
Wind tunnels, 253, 264
Wireless microphone *see* Radio microphone
Wireless transmission, short range, 187

Zoom microphone, 18